Cellular and Molecular Biology of Nitric Oxide

Cellular and Molecular Biology of Nitric Oxide

Jeffrey D. Laskin

University of Medicine and Dentistry of New Jersey–
Robert Wood Johnson Medical School
Piscataway, New Jersey

Debra L. Laskin

Environmental and Occupational Health Sciences Institute
Rutgers University and
University of Medicine and Dentistry of New Jersey–
Robert Wood Johnson Medical School
Piscataway, New Jersey

MARCEL DEKKER, INC. NEW YORK • BASEL

ISBN: 0-8247-1965-4

This book is printed on acid-free paper.

Headquarters
Marcel Dekker, Inc.
270 Madison Avenue, New York, NY 10016
tel: 212-696-9000; fax: 212-685-4540

Eastern Hemisphere Distribution
Marcel Dekker AG
Hutgasse 4, Postfach 812, CH-4001 Basel, Switzerland
tel: 44-61-261-8482; fax: 44-61-261-8896

World Wide Web
http://www.dekker.com

The publisher offers discounts on this book when ordered in bulk quantities. For more information, write to Special Sales/Professional Marketing at the headquarters address above.

Current printing (last digit):
10 9 8 7 6 5 4 3 2 1

PRINTED IN THE UNITED STATES OF AMERICA

Preface

For most of the twentieth century, nitric oxide gas has been well known as an intermediate in the manufacture of many chemicals, as well as an urban air pollutant. In fact, little more than 30 years ago the idea that nitric oxide could serve in any physiological process or be useful in human medicine was simply dismissed. Prior to the 1970s, researchers capitalized on the free radical nature of nitric oxide in protein crystallographic studies. By virtue of its ability to bind iron, nitric oxide could be used as an aid in pinpointing precise structural features of iron-containing proteins. In protein characterization studies in the early 1970s, Ferid Murad discovered that nitric oxide could bind to and activate one of these proteins, the heme-containing guanylate cyclase (an enzyme responsible for producing the important signaling molecule cGMP), a critical molecule regulating smooth muscle relaxation. Murad also showed that nitrovasodilating drugs release nitric oxide. In the early 1980s, Robert Furchgott first described an endothelial-derived relaxing factor that could mediate the action of acetylcholine in regulating smooth muscle tone in blood vessels. Furchgott and Louis Ignarro hypothesized that this material was nitric oxide and indeed, in 1987, Ignarro and Salvador Moncada demonstrated that this was the case. For their pioneering work in the nitric oxide field, Murad, Furchgott, and Ignarro were awarded the 1998 Nobel prize in Physiology and Medicine. To say the least, this field has been explosive since their discoveries. Over the past decade, thousands of papers have appeared on nitric oxide in almost every discipline in the medical sciences. Nitric oxide is now known to control such diverse physiological processes as blood pressure, neurotransmission, renal function, gastric motility, respiration, nonspecific host defense, and wound healing.

This book presents a summary of some of the more recent discoveries in the field of nitric oxide biology, biochemistry, and toxicology. It is hoped that the book will further fuel the interest in the rapidly expanding field of nitric oxide biology and encourage additional research into understanding the role of nitric oxide in the pathogenesis of disease and the development of therapeutic agents that will be useful in human medicine. Nitric oxide is now a broad discipline covering almost every field of physiology, pharmacology, biochemistry, and medicine. The reader will immediately recognize that the chapters in this volume have been written by world experts in the field. A special effort has been made by these authors to concentrate on new findings in the nitric oxide field. It is through their vision that new leads will develop to more completely understand the cell and molecular biology of nitric oxide,.

Finally, we express our gratitude to all the contributors to this volume. Special thanks go to our many graduate students, postdoctoral fellows, research associates, and colleagues who have spurred our interest in the nitric oxide field and made our research efforts so successful. We would also like to acknowledge the support of the National Institutes of Health, particularly the NIEHS and NIGMS, the Burroughs Wellcome Fund, and the Environmental and Occupational Health Sciences Institute of Rutgers University and UMDNJ–Robert Wood Johnson Medical School.

Jeffrey D. Laskin
Debra L. Laskin

Contents

Contributors

Ruth E. Billings, Ph.D. Department of Environmental Health, Colorado State University, Fort Collins, Colorado

Joan P. Blonder, Ph.D. RxKinetix, Inc., Louisville, Colorado

Marion Man-Ying Chan, Ph.D. Department of Microbiology and Immunology, Temple University School of Medicine, Philadelphia, Pennsylvania

David T. Denhardt, Ph.D. Department of Cell Biology and Neuroscience, Rutgers University, Piscataway, New Jersey

Jonathan R. Diamond, M.D. Department of Medicine, Pennsylvania State University College of Medicine, Hershey, Pennsylvania

Roger A. Egolf, Ph.D. Department of Chemistry, Pennsylvania State University, Fogelsville, Pennsylvania

Dunne Fong, Ph.D. Department of Cell Biology and Neuroscience, Rutgers University, Piscataway, New Jersey

Carol R. Gardner, Ph.D. Environmental and Occupational Health Sciences Institute, Rutgers University, and University of Medicine and Dentistry of New Jersey–Robert Wood Johnson Medical School, Piscataway, New Jersey

Diane E. Heck, Ph.D. Department of Pharmacology and Toxicology, Rutgers University, Piscataway, New Jersey

Ned D. Heindel, Ph.D. Department of Chemistry, Lehigh University, Bethlehem, Pennsylvania

Lori A. Hoerrner Department of Cell and Developmental Biology, Rutgers University, Piscataway, New Jersey

Harry Ischiropoulos, Ph.D. Department of Pediatrics, Biochemistry and Biophysics, Stokes Research Institute, Philadelphia, Pennsylvania

Aaron J. Kowalski Department of Microbiology and Molecular Genetics, Rutgers University, Piscataway, New Jersey

Debra L. Laskin, Ph.D. Department of Pharmacology and Toxicology, Environmental and Occupational Health Sciences Institute, Rutgers University, and University of Medicine and Dentistry of New Jersey–Robert Wood Johnson Medical School, Piscataway, New Jersey

Jeffrey D. Laskin, Ph.D. Department of Environmental and Community Medicine, University of Medicine and Dentistry of New Jersey–Robert Wood Johnson Medical School, Piscataway, New Jersey

Andrew Lubeskie Department of Pharmacology, University of Medicine and Dentistry of New Jersey–Robert Wood Johnson Medical School, Piscataway, New Jersey

Elizabeth E. Mannick, M.D. Department of Pediatrics, University of Rochester, Rochester, New York

Thomas M. Mariano, Ph.D. Department of Environmental and Community Medicine, University of Medicine and Dentistry of New Jersey–Robert Wood Johnson Medical School, Piscataway, New Jersey

Danette R. Miller, Ph.D. Department of Environmental Health, Colorado State University, Fort Collins, Colorado

Mark J. S. Miller, Ph.D. Department of Pediatrics, Albany Medical College, Albany, New York

Sidney M. Morris, Jr., Ph.D. Department of Molecular Genetics and Biochemistry, University of Pittsburgh School of Medicine, Pittsburgh, Pennsylvania

William J. Murphy, Ph.D. Department of Pathology and Laboratory Medicine and Wilkinson Laboratory of the Kansas Cancer Institute, University of Kansas Medical Center, Kansas City, Kansas

Robert A. Naddelman Department of Pharmacology, University of Medicine and Dentistry of New Jersey–Robert Wood Johnson Medical School, Piscataway, New Jersey

Michael R. Quinn, Ph.D. Department of Developmental Biochemistry, New York State Institute for Basic Research in Developmental Disabilities, Staten Island, New York

Karen A. Ribbons, Ph.D. Progen Industries, Ltd., Darra, Queensland, Australia

Susan A. Rotenberg, Ph.D. Queens College, City University of New York, Flushing, New York

Darren A. Saks, M.D. Department of Pharmacology, University of Medicine and Dentistry of New Jersey–Robert Wood Johnson Medical School, Piscataway, New Jersey

Manuel Sandoval, Ph.D. Department of Pediatrics, Physiology and Cell Biology, Albany Medical College, Albany, New York

Georgia B. Schuller-Levis, Ph.D. Department of Immunology, New York State Institute for Basic Research in Developmental Disabilities, Staten Island, New York

David J. Sieg, Ph.D. Department of Immunology, The Scripps Research Institute, La Jolla, California

Stephen R. Thom, Ph.D., M.D. Department of Emergency Medicine, University of Pennsylvania, Philadelphia, Pennsylvania

J. Brice Weinberg, M.D. Veterans Affairs Medical Center and Duke University Medical Center, Durham, North Carolina

Patricia D. Wilson, Ph.D. Department of Medicine, Mount Sinai School of Medicine, New York, New York

Donald J. Wolff, Ph.D. Department of Pharmacology, University of Medicine and Dentistry of New Jersey–Robert Wood Johnson Medical School, Piscataway, New Jersey

1

Transcriptional Regulation of the Genes Encoding Nitric Oxide Synthase

William J. Murphy
University of Kansas Medical Center, Kansas City, Kansas

The discipline of nitric oxide biology is arguably among the most intensively studied, and intensively reviewed, in the biological sciences. Many recent reviews have covered the area of nitric oxide synthase (NOS) gene expression (1–5). In the interest of being timely, this chapter will focus on new developments in the area of transcriptional control over expression of the three NOS isoforms. It is recognized that post-transcriptional processes—e.g., alternate transcriptional starting points, alternative splicing, and mRNA stability—can contribute greatly to the control of NO production. However, transcriptional regulation of gene expression is a novel and exciting aspect of control of the so-called "constitutively" expressed NOS isoforms and is the primary mode of regulation of the inducible isoform which is transcriptionally silent in most unstimulated tissues.

I. NITRIC OXIDE SYNTHASE GENES

Complementary cDNAs for each of the three NOS isoforms have been cloned and sequenced. A nomenclature has been set based on the order of cloning of these three genes; hence, the neuronal NOS is referred to as NOS1, the inducible isoform as NOS2, and the endothelial form as NOS3. The genomic copies of these genes have also been sequenced extensively, including the 5′-flanking regions. A comparative analysis of the relative sizes, intron/exon organization and chromosomal locations of the three human NOS genes has been reviewed recently, and the reader is directed to these fine reviews for detailed information (4,5). To capsulize these analyses the intron/exon organization of these three genes is strikingly similar, suggesting an evolutionary relatedness among them. However, sequence analysis of the promoter regions of these three genes reveals features that are unique to each, indicating differing mechanisms of transcriptional control.

To date the 5′-flanking region sequence of only one NOS1 gene has been reported—that from human beings (6). Both the human (7) and bovine (8) NOS3 gene 5′-flanking regions have been cloned and sequenced, and a comparative analysis reveals a strong sequence conservation between the two (Fig. 1). Despite this sequence identity, the transcriptional start points identified in human endothelial cells (ECs; position –22 from the ATG start codon) versus bovine ECs (positions –170 and –240 from the ATG start codon) differ dramatically. The human and bovine NOS3 genes both lack an identifiable TATA box, however, which directs precise initiation of transcription from a single nucleotide position. Multiple transcription start points is a characteristic of TATA-less promoters (9).

The 5′-flanking region of the NOS2 gene has been cloned and sequenced from the genomes of the mouse (10), human being (11), rat (12), and chicken (13). Figure 2 shows a comparative sequence analysis of the mouse, rat, and human promoter regions. The enhancer and basal promoter—including the transcriptional start points and TATA boxes—are remarkably conserved in these three genes (notable exceptions are discussed in detail below). The chicken NOS2 5′-

---

Figure 1 Sequence comparison of the NOS3 genes from human and bovine sources. The human and bovine NOS3 genes (GenBank/EMBL Data Bank accession numbers are L10693-L10709 and L27056, respectively) were cloned and sequenced by Marsden et al. (7) and Venema et al. (8), respectively. The sequences were aligned using the GAP and BESTFIT programs from Genetics Computer Group (Madison, WI) on a Silicon Graphics supercomputer, model IRIS 4D/340VGX. Vertical lines represent conserved sequence between the two species. Candidate transcription factor binding sites (7,8) are shown in boldface type and are underlined. Those binding sites known to be active are shown in boldface type and are double-underlined. The arrowheads represent the known transcriptional start points for the two genes. Since the two genes are not transcribed from the same transcriptional start point, the sequences are numbered such that the ATG start codon of each gene is nucleotide position +1. Upstream sequences are numbered negatively. Abbreviations: AP1, binding consensus for transcription factor AP1; Sterol, sterol regulatory element; SP1, binding consensus for transcription factor SP1; CACCC, sequence element necessary for proper GATA transcription factor function; NF-1, binding consensus for transcription factor NF-1; AP2, binding consensus for transcription factor AP2; GATA, binding consensus for GATA family of transcription factors; APE, acute-phase element; SSRE, shear stress-responsive element; MRE, heavy metal-responsive element; CRE, cAMP-responsive element; CAAT, binding consensus for CAAT family of transcription factors; p53, binding consensus for transcription factor p53.

```
Hu  ......TCTGATGCTGCCTGTCACCTTGACCCTGAGGATGCCAGTCACAGCTCCATTAAC  -1,568
Bo  ......tctgcggttcgttgccgctttgtccctggggaggccagtcacagctg..ttagc  -1,631

Hu  TGGGACCTAGGAAAA.TGAGTCATCCTTGGTCATGCACATTTCAAATGGTGGCTTA.ATA  -1,510
Bo  tggcatct.ggaaaaaggagtcattttcagtcatgtgcattgta...ggtggtggctaca  -1,575
                    AP1

Hu  TGGAAGCCAGACTTGGGATCTGT.TGTCTCCTCCAGCATGGTAGAAGATGCCTGAAAAGT  -1,451
Bo  tgggagccagaaatggggtccatctgtttc.tgccgtggtggggagaatgtctgagaggt  -1,516

Hu  AGGGG.CTGGATCCCATCCCCTGCCTCACTGGAAGGCGAGGTGGTGGGGTGGGGTGGGG  -1,392
Bo  aggggcctggaccccgtctcctgcctccggtggca.............agtggggtgggg  -1,469
                                                  STEROL

Hu  CCTCAGGCTT..................GGGGTCATGGGACAAAGCCCAGGCTGAATGC  -1,351
Bo  cctcgggtttcagggccagcggaggggtggcagtgggggggaacaggcccagttgaacac  -1,409

Hu  CGCCCT........TCCATCT......CCCTCCTCCTGAGACAGGGGCAGCAGGGCACAC  -1,305
Bo  aggtctcatacttctgaatctcacctgccccctggctgagggaggggcagcaaagtttct  -1,349
     SP1

Hu  TAGTGTCCAGGAG...CAGCTTAT...........GAGGCCCCTTCACCCTCCATCCTCC  -1,259
Bo  gggagcccaggaggtcaggcttgtccaggcatacaagggccccctcaccctcttccctgc  -1,289
                                                CACCC

Hu  AAAACTGGCAGACCCCACCTTCTTGGTGTGACCCCAGAGCTCTGAGCACAGCCCGTTCCT  -1,199
Bo  .agaccggcagaccccaccttcttggtgtgaccccagaactcgcagcacggcccgttcct  -1,230
                              NF1

Hu  TCC.........GCCTGCCGGCCCCCCACCCAGGCCCACCCCAACCTTATCCTCCACTGC  -1,148
Bo  tccggttggtctgcctgctcgcccccccacccaggcccgcacccatcatatcttccactgc  -1,170
                          AP2          CACCC      GATA

Hu  TTTTCAGAGGAGTCTGGCCAACACAAATCCTCTTGTTTGTTTGTCTGTCTGTCTGCTGCT  -1,088
Bo  ttttcgaaggagctcagccaacacaaatctacatgtttgtttgtctgcctgtctgcctct  -1,110
```

(a)

```
Hu  CCTAGTCTCTGCCTCTCCCAGTCTCTCAGCTTCCGTTTCTTTCTTAAACTTTCTCTCAGT  -1,028
Bo  cccggtctctgcccatcccagtcgctcaacttctgtttctttctcgcagtttctcttggt  -1,050
                   APE

Hu  CTCTGAGGTCTCGAAATCACGAGGCTTCGACCCCTGTGGA..CCAGATGCCCAGCTAGTG   -970
Bo  ctgcgtggtctcagatccatgaggctggaacccctgtagatgctgaaggcccagcttacg   -990
        SSRE

Hu  GCCTTTCTCCAGCCCCTCAGATGGCACAGAACTACAAACCCCAGCATGCACTCTGGCCTG   -910
Bo  gcccttctccagcccct.agatggcttagaactacaaaccccagcatgcactatggctca   -931
      NF1                                          MRE

Hu  AAGTGCCTGGAGAGTGCTGGTGTACCCCACCTGCATTCTGGGAACTGTAGTTTCCCTAGT   -850
Bo  tggtgcctgaggagtgtctgagtacaccacctgcattctgggaatcgtagtttccctcac   -871
                                          APE

Hu  CCCCCATGCTCCCACCAGGGCATCAAGCTCTTCCCTGGCCGGCTGACCCTGCCTCAGCCC   -790
Bo  cccacacgct..taccagggatccatggtctccccag..tggcagtccctgcctcagccc   -815

Hu  T.......AGTCTCTCTGCTGACCTGCGGCCCC...........GGGAAGCGTGCGTC   -750
Bo  tgaccgctgggtctatctgctgacctacccccacccccacccctggggaagcatgcgtc   -755
                                                           CRE

Hu  ACTGAATGACAGGGTGGGGG.TGGAGGCACTGGAAGGCAGCTTCCTGCTCTTTTGTGTCC   -691
Bo  actggatgacagggtggggggtggaggccctgaatcgcagcttcctgccctttgtgtcc   -695
               CACCC

Hu  CCC.ACTTGAGTCA............TGGGGGTGTGGGGGTTCCAGGAAATTGGGGCTGG   -644
Bo  cccaacttgagtcacaggggtggcggggtggggcggggttccaggaaataggggccgg   -635
           A P1                                   CAAT    APRE

Hu  GAGGGGAAGGGATACCCTAATGTCAGACTCAAGGACAAAAAGTCACTACATCCTTGCTGG   -584
Bo  gagggaatggaataccctaatgccagacccaaggacaaagggtcagagcgcccttgctgg   -575
              GATA

Hu  GCCTCTATCCCCAAGAACCCAAAAGGACTCAAGGGTGGGGATCCAGGAGTTCTTGTATGT   -524
Bo  gcctgtatccccaggaacccgaaaggactcaacagtaggagtccaggagtgcttagatgt   -515
         GATA
```

(b)

```
Hu  ATGGGGGGAGGTGAAGGAGAGAACCTGCATGACCCTAGAGGTCCCTGTGGTCACTGAGAG   -464
Bo  cagcaaggtggtcaagcagagccctggggtaaccctagcagtccctgtggtcactgagag   -455

Hu  TGTGGG..CTGCCATCCCCTGCTACAGAAACGGTGCTCACCTTCTGCCCAACCCTCCAGG   -406
Bo  caagggacttgttgtcacctgctgcggccacagtgcccaccttctgcttgacgcctcagg   -395

Hu  GAAAGGCACACAGGGGTGAGGCCGAA..CCTTCCGTCTGGTGCCACATCACAG.AAGGAC   -349
Bo  gtgaggccagcaagggcgaggtggaaggcctcccctctggtcagacacccccgcagggtt   -335

Hu  CTTTATGACCCCCTGGTGGCTCTACCCTGCCACTCCCCAATGCCCCAGCCCCCATGCTGC   -289
Bo  ctttcgggcactctggctagcccaccccgccatcccacttcaccccagtccctgcgctct   -275

Hu  AGCCCCAGGGCTCTGCTGGACACCTGGGCTCCCACTTATCAGCCTCAGTCCTCACAGCGG   -229
Bo  ggccccagggttctgccagacacctgggttccccCttatcaccctc.cttctcagagcgg   -216
                                      ➤GATA

Hu  AACCCAGGCGTCCGGCCCCCCACCCTTCAGGCCAG.CGGGCGTGGAGCTGAGGCTTTAGA   -170
Bo  aacccaggcgtctggcgctccaccctctgggccagccgggcatggAgccgaggctctgga   -156
                     AP2                         ➤

Hu  GCCTCCCAGCCGGGCTTGTTCCTGTCCCATTGTGT..ATG..GGATAGGGGCGGGGCGAG   -114
Bo  gcctcccggccgggctagttcctgtcccattgtgtgtatgagggac.ggggcgg..cgag    -99
       APE       p53                              GATA    SP1

Hu  GGCCAGCA.CTGGAGAGCCCCCTCCCACTGCCCCCTCCTCTCGGTCCCCTCCCTCTTCCT    -55
Bo  ggcccccctcct..accgccccctccc.cttcc.taaggaaaagg.cccgaagctctg.ct    -45

                                       ➤                  +1
Hu  AAGGAAAAGGCCAGGGCTCTGCTGGA.GCAGGCAGCAGAGTG.GACGCA.CAGTAACATG     +3
Bo  .ggga....gccacaga...gca.gacggaggccgccc.gtccgccccagcg...acATG     +3
```

(c)

flanking sequence (13) differs considerably from the other three and, hence, is not included in this analysis. In spite of this, certain structural features of the basal promoter, including the κB and octamer elements, are conserved in the chicken promoter (see below), although a recognizable enhancer appears to be missing. The following sections discuss in more detail the transcriptional activation of

```
Hu  tattatgctgaatttaacttgggaaagacaagaaggaaatgagtggacagtggtagcaaa -1,178
Mo  CACGAGGCTGAGCTGACTTTGGG...GACCATGCGAAGATGAGTGGACCCTGGCAGGA..   -981
Ra  cacgaggctgagctgaatttggg...aaccatgggatgatgagtggaccctggcggga..   -958

Enhancer

Hu  gtgttgggacggtgagatcaagggtgactttttttccctttgcttctcaacttctcccta -1,118
Mo  ..............TGTGCTAGGGGGATTTTCCCTCTCTCTGTTTGTTCCTTTTCCCCTA   -935
Ra  ..............tatgccagggggattttccctctctctgtttgttcctttttcccccta   -912
                           κB                             GAS

Hu  atgtagttgttattttactttatgatggaaaaatgcat.atgtatgggaatactgtatt  1,059
Mo  ACACTGTCAATATTTCACTTTCATAATGGAAAATTCCATGCCATGTGTGAATGCTTTATT   -875
Ra  atactgtcaatatttcactttcataatggaaaatcccatgccatgtatgaatcgttgtag   -852
               ISREa       ISREb
    <---------------------------------------------------------->

Hu  gtgcctctatctta..........gcagccaccctgctgatgaactgccaccttggactt   -254
Mo  CCTAGTGAGTCCCAGTTTTGAAGTGACTACGTGCTGCCTAGGGGCCACTGCCTTGGACGG   -194
Ra  cctagtgagtcccagttttcaagaggccactcgctgccaagggaccattgccttggactg   -198
                               HRE

Hu  gggaccagaaagaggtgggttgggtgaagaggcaccacacagagtgatgtaacagcaaga   -158
Mo  GCGACCAGGAAGAGATGGCCTTGCATGAG.GATACACCACAGAGTGATGTAAT.......   -142
Ra  gggaccaggaagaggtggcct.caccgaa.gatacaccacagagtgacgtaat.......   -147

Basal promoter

Hu  tcaggtcacccacaggccctggcagtccacagtcataaattagctaact..gtacacaag   -100
Mo  ......CAAGCACACAGACTAGGAGTGTCCA.TCATGAATGAGCTAACTTGCACACCCAA    -89
Ra  ......aatgcatacagactaggagtgtcca.tcgcgaatgagctaacttgcacacccta    -94

Hu  ctggggacactccctttggaaacc▾gagaccttt.....tatgcaaaaacaactctctgga    -45
                            (A)21
Mo  CTGGGGACTCTCCCTTTGGGAAC.......AGT.....TATGCAAAA...TAGCTCTGCA    -44
Ra  ctggggactctccctttgggaac.......agtgactttatgcaaaa...cagctctgca    -44
       κB                                 Octamer
                                            ➤
Hu  tggcatggggtgagtataaatacttcttggctgccagtgtgttcataactttgtagcgag    +16
Mo  GAGCCTGGAG.GGGTATAAATACCTGATGGCTGCTGCCAGGGTCACAACTTTACAGGGAG    +16
Ra  gagcgtggat.gggtataaatacctgatggctactgtcagggccacagctttacagggag    +16
               TATA
```

these three genes and the cis-acting DNA elements that are currently known to be necessary for their expression.

II. REGULATED EXPRESSION OF THE NOS1 AND NOS3 ISOFORMS

Constitutively expressed NOS isoforms were originally isolated from central neurons (15) and vascular endothelium (16). More recently, however, NOS1 and NOS3 have been demonstrated in a variety of other tissue and cell types. A partial listing of these is shown in Table 1.

Although NOS1 and 3 are constitutively found in the tissues described above, it is increasingly clear that expression of these isoforms is regulated and that both increases and decreases in expression can be observed, depending upon the stimulus and cell type. The NOS1 and 3 genes can be regulated at the level of transcription initiation as well as by posttranscriptional mechanisms such as mRNA stability changes. A large variety of forces have been demonstrated, to date, to alter expression of these NOS isoforms and these will be considered under two general headings—external stimuli and physiologic/pathologic states.

A. External Stimuli

Bacterial lipopolysaccharide (LPS) and the cytokines IFN-γ, IL-1, and TNF-α have each been shown to decrease expression of NOS3. Systemic LPS (15 mg/

Figure 2 Sequence comparison of the NOS2 genes from human, mouse, and rat sources. The human, mouse, and rat NOS2 genes (GenBank/EMBL Data Bank accession numbers are L36031, L23806, and X95629, respectively) were cloned and sequenced by Lowenstein et al. (10), Zhang et al. (14), and Eberhardt et al. (12), respectively. The sequences were aligned using the GAP and BESTFIT programs from Genetics Computer Group (Madison, WI) on a Silicon Graphics supercomputer, model IRIS 4D/340VGX. Vertical lines represent conserved sequence between the two species. Only those transcription factor-binding sites known to be active are shown in boldface type and are double-underlined. The sequences are numbered such that the transcription start point of each gene, denoted by the arrowhead, is nucleotide position +1. Upstream sequences are numbered negatively. The gene's enhancer and basal promoter are indicated in the left margin. Abbreviations: κB, binding consensus for the transcription factor NF-κB; GAS, IFN-γ activated site; ISRE, interferon-stimulated response element; HRE, hypoxia-responsive element; $(A)_{21}$, a 21-nucleotide adenosine tract in the human NOS2 gene which has been abbreviated as a matter of convenience for comparing the three sequences; Octamer, binding consensus for the octamer family of transcription factors; TATA, the core promoter of the NOS2 genes.

Table 1 Tissue Distribution of the "Constitutive" Isoforms of NOS

NOS1	NOS3
Central neurons (15)[a]	Endothelial cells (16)
Peripheral neurons (17)	Osteoclasts (27)
Skeletal muscle (18)	Various epithelial cells (28–31)
Pancreatic islet cells (19)	Certain neurons (32)
Kidney macula densa cells (20)	Platelets (33)
Certain epithelial cells (20)	Hippocampal pyramidal cells (34)
Astrocytes (21)	Cardiomyocytes (35)
Adrenal gland (22)	Adipocytes (36)
Glial cells (20)	Syncytiotrophoblasts (37)
Mast cells (23)	Skeletal muscle (38)
Osteoblast (24)	Fibroblasts (37,39)
Chondrocytes (25)	Leydig and Sertoli cells of the testis (31)
Retina (26)	

[a]Reference numbers in parentheses.

kg) decreases NOS3 mRNA abundance in rat heart, lung, and aorta, and also decreases NOS1 mRNA fourfold in the brain (40). LPS also decreases NOS3 expression, ex vivo, in bovine coronary venular ECs (41). In contrast to these reports, the study of Lee et al. (42) reveals an opposite effect on NOS1 gene expression in rats challenged with lower concentrations of systemic LPS (100 μg/kg), i.e., that which results in no endotoxic shock and minimal disruption of the blood-brain barrier. Under these conditions hypothalamic NOS1 mRNA increases by approximately 50%. The reason for this apparent biphasic response to LPS is unclear but may reflect differing cellular responses to high versus low concentrations of LPS. IFN-γ, IL-1, TNF-α, and LPS have each been shown to decrease NOS3 mRNA abundance in human aortic ECs treated ex vivo (43). The downregulatory effect of TNF-α has also been demonstrated in bovine aorta ECs (BAECs) (44) and human umbilical vein ECs (HUVECs) (45).

Growth factors and hormones have the opposite effect, to that cited above, on NOS3 expression. TGF-β increases both NOS3 mRNA and protein in BAECs (46) while bFGF increase NOS3 protein in bovine microvascular ECs (47). Estradiol increases NOS3 protein in the myometrium of nonpregnant sheep (48) and in human aortic ECs (49). NOS3 mRNA is also increased in the aorta of pregnant, as compared with nonpregnant rats (50). In this latter study, estradiol replacement in gonadectomized rats also causes an increased accumulation of NOS3 mRNA. In addition, the administration of gonadotropin increases NOS3 protein in ovary granulosa cells of the rat (51). In spite of the upregulatory effects of the steroid hormone estradiol on NOS3 expression, the coadministration of systemic LPS and the corticosteroid dexamethasone (40) does not alter the downregulatory effect of LPS observed in heart, lung, or aorta (see above).

The expression of NOS1 is also hormonally controlled. In contrast to the study of NOS3 gene expression by Goetz et al. (50), ovarectomy and castration have the effect of increasing NOS1 mRNA accumulation in the rat anterior pituitary gland (52). Estrogen or testosterone replacement therapy returns NOS1 mRNA to its lower, basal level. Conversely, administration of the thyroid hormone T3 causes an increase in NOS1 mRNA accumulation in both the periventricular and supraoptic nuclei of the hypothalamus, while 6n-propyl-2-thiouracil-induced hypothyroidism resulted in decreases mRNA synthesis (53). As observed for NOS3 expression, dexamethasone does not affect the LPS-induced downregulation of NOS1 in rat brain (40).

The oxidative status of cells and tissues can also affect the level to which NOS1 and 3 are expressed. Oxidative damage due to exposure of rat skin to UV irradiation causes an increase in NOS1 protein in cutaneous nerve fibers (54). An increase of pO_2 from normoxic increases NOS3 protein and mRNA in ex vivo cultures of ovine fetal pulmonary artery ECs (55) and bovine pulmonary artery cells or BAECs (56). Conversely, hypoxia reduces NOS3 mRNA levels in HUVECs (57,58) and bovine pulmonary artery cells or BAECs (56). In sharp contrast to these studies, exposure of BAECs to hypoxic conditions (1% oxygen) doubled NOS3 mRNA levels within 6 h in the experiments of Arnet et al. (59). Similarly, exposure of rats to chronic hypoxia (3 weeks) led to an increase in NOS3 mRNA accumulation in lung tissue, when compared with normoxic rats (60). Ischemia induced by clamping shut one of the renal arteries also causes an increased synthesis of NOS1 mRNA in the kidney macula densa (61). The reason for the difference between the observations of different laboratories is not clear.

Conflicting reports have been published regarding the effect of oxidized low-density lipoprotein (oxLDL) and the important diffusable component of oxLDL, lysophosphatidylcholine (lysoPC) (62). Liao et al. (63) have demonstrated that treatment of human saphenous vein ECs with oxLDL reduces NOS3 mRNA accumulation. Furthermore, a higher degree of LDL oxidation correlates with a stronger decrease in NOS3 mRNA. Surprisingly, however, treatment of HUVECs with lysoPC causes a rapid (within 3 h) and dramatic (approximately 11-fold) induction of NOS3 mRNA (64). Induction occurs in a time- and dose-dependent fashion.

Two second messengers of extracellular signals, calcium and cAMP, vary in their abilities to affect transcription of the two NOS genes. Depolarization of cultured mouse cerebellar neurons with extracellular potassium—which causes an increase in intracellular calcium—dramatically reduces NOS1 protein expression whereas the blockade of voltage-dependent calcium channels causes an increase of NOS1 expression (65). Thus, calcium is a potentially important negative regulator of the NOS1 gene. Calcium ionophore (A23187), however, does not affect the expression of NOS3 in HUVECs (64). Furthermore, calcium chelation does not affect the ability of forskolin to decrease expression of NOS3 in rat ventricular myocytes (66), indicating that forskolin functions through the induction of intracellular cAMP. This is confirmed by the fact that 8-bromo-cAMP mimics the effect of forskolin and that inhibition of type III phosphodiesterase activity in vivo, with the pharmacological agent milranone, also results in decreased myocyte NOS3 mRNA and protein accumulation (66). These effects of forskolin on NOS3 gene expression are not observed in cardiac microvascular ECs, however (66), nor do cAMP elevating agents affect the expression of NOS1 in primary cerebellar granule cells (65). Thus, calcium and cAMP may have cell type-specific effects and may affect the regulation of NOS1 versus NOS3 differently.

The role played by protein kinase C (PKC) also appears controversial. Treatment of HUVECs with PMA, a PKC activator, does not affect NOS3 mRNA accumulation whether used alone or in combination with A23187 (64). In BAECs, however, PMA-induced activation of PKC increases NOS3 mRNA accumulation in a dose-dependent manner (67). Surprisingly, though, inhibition of PKC in BAECs—with either staurosporine, chelerythrine (67), or calphostin (68)—has the same effect of increasing NOS3 mRNA accumulation. In the latter study, calphostin inhibition of PKC did not affect shear stress-induced synthesis of NOS3 mRNA (see below). Thus, while homeostatic levels of PKC would appear to affect basal expression of the NOS3 gene, the overall molecular role of PKC in NOS3 gene regulation remains uncertain.

The external application of stress or strain has been found to influence NOS gene expression. Shear stress—the application of laminar fluid force across the surface of a cell culture monolayer—and cyclic strain—the cyclical stretching and relaxing of cell monolayers attached to a flexible membrane—both cause increased expression of NOS3 mRNA in BAECs (44,69) and HUVECs (70,71). A similar increase in NOS3 mRNA accumulation has been observed in an aortacaval fistula model of chronic high blood flow (72). Interestingly, emotional stress induced by exposure of rats to constant light results in decreased pineal gland NOS1 protein expression to virtually undetectable levels (73). However, immobilization stress, produced by wrapping rats in flexible wire gauze, induces NOS1 mRNA accumulation in the hypothalamic-pituitary-adrenal axis (74).

B. Physiologic/Pathologic States

The growth, death, and repair of neurons is associated with many changes in NOS1 expression. In the developing rat brain NOS1 immunoreactive neurons are detectable at 15 days of gestation in a few cells of the hypothalamus and pons (75). Four days later these immunoreactive cells increase in size and number and immunoreactive neurons can additionally be found in the striatum and medulla. Different isoforms of NOS1 are present in the rodent brain during embryogenesis and after birth (76). A truncated 144-kDa protein is present in synaptosomes during embryogenesis but disappears postnatally. After birth the usual 155-kDa form appears, which increases until day 7 and then slowly decreases to a lower level thereafter. However, coincident with this day 7 peak is a 155-kDa isoform that is more highly sensitive to inhibition by the arginine analogue N^{ω}-nitro-L-arginine than is the dominant 155 kDa NOS1 isoform. The appearance of the 155 kDa isoforms is coincident with the onset of synaptogenesis in specific brain regions. Thus, expression of NOS1 is under developmental control in the brain.

Both NOS1 and NOS3 isoforms are developmentally regulated in the rat lung (77). NOS1 and NOS3 protein levels peak at day 20 of gestation and then fall to lower levels after birth. NOS3 mRNA accumulation parallels the protein abundance profile. However, NOS1 mRNA levels peak at day 16 of gestation and fall to a nadir at day 20, when protein levels are at their highest. NOS1 mRNA accumulation subsequently increases after birth, while protein levels decline. Thus, NOS1 mRNA and protein levels may reflect regulation at the levels of transcription initiation and mRNA stability as well as at the levels of protein translation efficiency and stability.

Damage to neurons—root avulsion or axotomy—results in increased expression of NOS1 protein (78–83) and mRNA (84). Neurotoxins, such as ibogaine (85), capsaicin (86), or tetrodotoxin (65), also increase NOS1 protein expression. In addition, NOS1 can be differentially regulated in the hypothalamus and caudal ventrolateral medulla of the rat brain during hypertension induced by clamping shut one of the two renal blood vessels (87). In this model, hypothalamic NOS1 mRNA levels decrease 3 weeks after clamping but increase 6 weeks after clamping. Conversely, caudal ventrolateral medulla NOS1 mRNA levels increase 3 weeks after clamping but decrease 6 weeks after clamping. This reversal of expression levels may be related to the temporal phases of different physiologic states induced by this hypertension model. The opposing effect of this hypertension model on NOS1 gene expression in the hypothalamus versus the caudal ventrolateral medulla indicates cell type-specific regulation of this gene. This hypertension model also causes increased NOS1 mRNA accumulation in the macula densa of the affected kidney, when compared with the normal kidney (61).

Chronic salt loading in rats—ad libitum drinking of water containing approximately 2% NaCl—results in increased expression of hypothalamic NOS1 mRNA (88,89) and protein (89). However, a similar treatment of rats results in decreased accumulation of NOS1 mRNA in the macula densa of the kidney (61,90). The reason for this difference is unclear but may reflect cell type-specific responses to hypernatremia. Consistent with the studies of Kadowaki et al. (88) and Villar et al. (89) are those that show that water deprivation of rats, which also causes hypernatremia, leads to increased NOS1 mRNA accumulation in the hypothalamus, in comparison with control rats (91,92). Food deprivation, on the other hand, has the opposite effect—NOS1 mRNA is decreased in the rat hypothalamus in comparison with controls (92,93). Interestingly, combined food and water deprivation leaves NOS1 mRNA levels unchanged, compared with control rats (92). Whether or not the opposing effects of water versus food deprivation simply cancel each another out when combined, is unknown.

Normal physiologic states such as pregnancy (50) and lactation (94) can increase NOS3 (in rat aorta) and NOS1 (in rat hypothalamus) mRNA synthesis, respectively. Similarly, chronic exercise increases NOS3 mRNA in canine thoracic aorta ECs (95). Pathologic states such as cirrhosis (96), portal hypertension (97), and atherosclerosis (98) increase NOS3 mRNA or protein levels whereas diaphragmatic hernia decreases mRNA and protein levels (29). NOS1 protein also increases in chondrocytes of patients with osteoarthritis (25) and decreases in diabetic rats (99), compared with healthy counterparts. Finally, NOS1 mRNA levels also decrease in the hypothalamus and brainstems of rats in heart failure (100).

C. Transcriptional Control of NOS1 and NOS3 Expression

In spite of numerous studies, many of which are cited above, that document a change in NOS1 or NOS3 expression in response to a given external stimulus or physiologic/pathologic change, relatively few studies have been extended to include a more fundamental examination of the molecular bases for these changes. To date, no studies have demonstrated control of NOS1 gene expression at the level of transcription initiation (i.e., by nuclear run-on analysis). However, oxygen tension has been shown to affect NOS3 gene transcription rates in BAECs and HUVECs. Increased oxygen (95%) causes a twofold increase in NOS3 gene transcription (in BAECs) within 12 h (56), whereas hypoxia causes a threefold decrease in transcription (in HUVECs) within 48 h (57) and a 10-fold decrease in transcription (in BAECs) within 12 h (56), when compared with transcription rates under normoxic conditions. Again, the transfection studies of Arnet et al. (59) contrast with these findings. A reporter-gene construct, containing the NOS3 promoter linked to the luciferase gene, was induced twofold after transfection into BAECs and subsequent incubation of transfectants under hypoxic conditions (1% oxygen). The increase in reporter-gene expression reflects a linear rise in luciferase activity during the first 24 h after change in oxygen tension.

Interestingly, oxLDL induces a biphasic transcriptional effect on the NOS3 gene in saphenous vein ECs which may, in part, explain the conflicting reports of the effects of oxLDL and lysoPC on NOS3 gene expression. The NOS3 gene transcription rate decreases to 75% of basal level within the first 6 h of treatment and subsequently increases to twice the basal level after 24 h of treatment (63). LysoPC treatment also increases NOS3 gene transcription twofold in HUVECs treated for 3 h (64). Lastly, application of cyclic strain to BAECs increases NOS3 gene transcription 2.6-fold within 4 h (69).

Increases in NOS3 transcription rates, after proper cellular stimulation, appear to occur without a need for prior protein synthesis. Simultaneous treatment of HUVECs with lysoPC and cycloheximide (CHX; a protein synthesis inhibitor) induces NOS3 mRNA accumulation equivalent to that of lysoPC alone (64). Similarly, CHX has no negative effect on hyperoxia-induced expression of NOS3 mRNA in BAECs (55). Rather, in these latter experiments CHX augments the level of mRNA induced by hyperoxia. Under conditions that downregulate NOS3 mRNA accumulation—TNF-α treatment of HUVECs (45) or exposure of BAECs to hypoxia (56)—CHX cotreatment reverses the mRNA downregulatory effect. These results suggest that mRNA stability is a key factor in lowering accumulated NOS3 mRNA levels and that a ribonuclease may be involved, the induction of which is inhibited by CHX. Induction of NOS3 mRNA, however, is independent of protein synthesis, suggesting an underlying transcriptional cause mediated by latent transcription factors.

A few laboratories have begun to dissect the NOS3 promoter for constitutive and regulated transcriptional control regions and functional elements. Deletion of sequences surrounding the candidate Sp1 site at position –117 to –126 of the human NOS3 gene (see Fig. 1) dramatically reduces the constitutive level of expression of a linked reporter-gene in HUVECs (101) and the human EC line tHUE-2 (102). Mutation of this Sp1 sequence causes a marked reduction of human promoter activity in transfected BAECs (103) and HUVECs (101). Gel shift and anti-Sp1 antibody supershift experiments confirm that Sp1 can bind to the –117 to –126 sequence in nuclear extracts prepared from BAECs (103), HUVECs (101), and the tHUE-2 cell line (102). Additional, unidentified Sp1 element-binding proteins are also detected in the nuclear extracts employed by each of these authors. As shown in Figure 1, this Sp1 site is also reasonably well conserved in the bovine NOS3 gene. The transcription factor Sp1 is important for the constitutive regulation of many viral and cellular genes. The latter include the so-called housekeeping genes which, like the NOS3 gene, lack a recognizable TATA box (9). Sp1 can activate these "TATA-less" promoters more than 100-fold in in vitro transcription assays (104).

A second cis-acting promoter element, that for the GATA transcription factor, has also been identified in the human NOS3 promoter (103). The bottom-strand sequence of this element is found at positions –249 to –252 of the human gene (Fig. 1) and is conserved in the bovine promoter sequence. Deletion of DNA sequence containing this element, or specific mutation of this sequence, results in decreased constitutive expression of a NOS3 promoter-linked luciferase reporter gene in BAECs. However, the contribution to constitutive expression of this gene by the GATA element was considerably less than that of the aforementioned Sp1 site. Gel shift and antibody supershift experiments confirmed the presence of the GATA-2 transcription factor in nuclear extracts of these cells. The GATA-2 transcription factor is found in erythroid cells and a limited number of other cell types, including ECs (105,106). Other GATA family proteins either are absent from ECs (GATA-1) or present in small quantities (GATA-3) (106). GATA cis-acting elements are associated with, and necessary for the transcription of, several genes that are selectively expressed in ECs, including: endothelin-1 (105,106), P-selectin (107), and von Willebrand factor (108). In addition, another member of the GATA family, GATA-1, is known to activate gene transcription in a cooperative fashion with the transcription factor Sp1 (109–112). This suggests a possible cooperative interaction of the GATA and Sp1 transcription factors in the constitutive expression of the NOS3 gene.

One report has addressed the regulated expression of the bovine NOS3 gene in BAECs (46). These authors noted an increase in expression of a luciferase reporter gene, linked to the bovine NOS3 promoter when transfected BAECs where treated with TGF-β, as compared with untreated control transfectants. Deletion or specific mutation of the candidate NF-1 element located between nucleotide positions –1254 and –1266 of the bovine gene (see Fig. 1) resulted in a decrease in TGF-β inducibility. This sequence is absolutely conserved in the human gene, as well. Gel shift and antibody supershift analyses confirmed that a transcription factor immunologically related, but likely nonidentical, to NF-1 could interact specifically with this NOS3 element in nuclear extracts prepared from untreated BAECs. TGF-β treatment of BAECs prior to nuclear extract preparation did not affect the binding of this NF-1-like protein to the NOS3 cognate binding sequence. A similar observation has been made for the involvement of a NF-1-like factor that regulates the expression of the type 1 plasminogen activator inhibitor gene (113). Thus, TGF-β treatment may result in the modification of NF-1-like proteins which increases their ability to trans-activate a given promoter without altering their DNA binding properties. Alternatively, TGF-β may alter the ability of a given NF-1-like factor to recruit other trans-activating factors to the promoter.

Although transcriptional regulation of the NOS1 promoter at the level of cis-acting DNA elements has not been rigorously explored, transfection of NOS1 promoter, β-galactosidase reporter gene constructs into HeLa cells have demonstrated constitutive expression of this promoter (114). Two promoters, separated by approximately 300 bp, were demonstrated functionally by transfection. The first, more proximal promoter (5′1) initiated transcription in a GC-rich region that lacked an identifiable TATA box. Multiple transcriptional start points, characteristic of TATA-less promoters, were detected in this promoter region. The second promoter (5′2), which transcribed a different first exon, initiates from a single start point 25 bp downstream of a potential TATA box (TATTTA). Although the TATA-less, GC-rich nature of the promoter for exon 5′1 is suggestive of those controlling constitutively expressed, "housekeeping" genes, these two promoters can be physically separated from one another and each is constitutively expressed in HeLa cells.

Deans et al. (115) have demonstrated that the 5′1 promoter, but not the 5′2 promoter, is activated by transcription factor Oct-2. This is in spite of the fact that no canonical octamer sequence can be found in the vicinity of the 5'1 promoter. Although Oct-2 is expressed only in a limited number of cell types (see below), it has been demonstrated in neuronal cells of the brain (116,117). Transactivation of thc 5′1 promoter by Oct-2 is dependent upon a 20 amino acid C-terminal region which is part of the activation domain of the Oct-2 transcription factor.

A third possible promoter has also been characterized upstream of 5′2 (6) and encodes a first exon that begins upstream of 5′2 but ends at the same 5′ splice donor site. This promoter initiates from a single start site 26 bp downstream of a candidate TATA box (TATAAA). In total, eight different first exons of the NOS1 gene have been demonstrated in different tissues, some of which are tissue-specific (4). These first exons are dispersed over a region of approximately 200 kb upstream of the second exon. The tissue-specific expression of some of these first exon variants suggests that tissue-specific, cis-acting promoter elements may regulate the expression of these variants. However, a detailed characterization of any of these potential NOS1 promoters remains to be performed.

III. INDUCIBLE NOS OR NOS2

A. Agents that Induce NOS2

The inducible form of NOS, NOS2, was originally identified in mouse macrophages (118). It is now recognized, however, that virtually all nucleated cells can be induced to express NOS2 (see Table 2). Inducers of NOS2 typically include one or more cytokines (and/or LPS) plus interferon (type II, usually, but type I

Table 2 Tissue Distribution of the Inducible Isoform of NOS-NOS2

Macrophages (118,119)[a]	Lipocytes (128)
Neurons (21)	Osteoblasts (24)
Neutrophils (120)	Cardiomyocytes (129)
Mast cells (121)	Platelets (33)
Endothelial cells (121-123)	Keratinocytes (130)
Hepatocytes (124)	Pancreatic islet cells (131)
Chondrocytes (125)	Kidney epithelial cells (132)
Smooth muscle cells (126)	Colon and squamous cell carcinomas (133,134)
Kupffer cells (122,123)	Fibroblasts (135,136)
Astrocytes/Microglial cells (127)	Sertoli cells (137)
Osteoclasts (27)	Nonmetastatic melanoma cells (138)

[a]Reference numbers in parentheses.

can substitute in some cases). In some cell or tissue types complex combinations of cytokines, LPS and interferon are required for maximal expression of NOS2. The combination IFN-γ+IL-β+TNF-α+LPS induces high levels of NO production or NOS2 mRNA synthesis in rat (126) and human (43) vascular smooth muscle (VSM) cells, human hepatocytes (124), the A549 human alveolar epitheliumlike cell line (139), nonmetastatic mouse melanoma cells (138), rat neurons (140), rat Sertoli cells (137) (IL-1α was used rather than IL-1β), and rat cardiomyocytes (141). In hepatocytes, a nearly equivalent effect on NO production and NOS2 mRNA synthesis is observed in the absence of LPS (ie., IFN-γ+IL-β+TNF-α) (124) whereas this latter condition induces no NOS2 mRNA in human aortic VSM cells (43). IFN-γ+IL-β+TNF-α is also an effective inducer in rat ECs (142), mouse lung fibroblasts (143) (IL-1α was used rather than IL-1β), human islets (144), mouse (145) and human (146) epithelial cells, rat primary osteoblasts (147), and human neutrophils (148). The combination of IFN-γ+IL-β+TNF-α+IL-6 will induce NO production and/or NOS2 expression in the colorectal carcinoma cell line DLD-1 (149), the A549 human alveolar epithelial-like cell line and primary cultures of human bronchial epithelial cells (150), and in human pancreatic islet cells (144). In the latter cells, however, IL-6 is somewhat inhibitory compared with cellular stimulation with IFN-γ+IL-β+TNF-α.

Most cell types can be induced, at least to some degree, by simpler combinations of IFN-γ plus either, LPS, TNF-α, or IL-1. Mouse macrophages produce high amounts of NO and dramatically increase intracellular NOS2 mRNA in response to IFN-γ+LPS or IFN-γ+TNF-α. These cells do not respond to IL-1(α or β), however, most likely because this cell type lacks the appropriate IL-1 receptor (151). Mouse macrophages respond well to LPS alone but the autocrine/paracrine effect of IFN-β is obligatory (152,153). In this regard, many cell types can deliver strong NO or NOS2 mRNA responses as a consequence of stimulation with single mediators. IL-1 seems to be the most effective stimulus, inducing NOS2 mRNA in pancreatic islet cells of the hamster (131), mouse (154), rat, and human being (155), and in human chondrocytes (125), rat VSM cells (156), and rat ECs (157). IFN-γ can, by itself, stimulate the production of NO in peritoneal macrophages from some strains of mice but not others (158,159) and can induce NOS2 mRNA in rat VSM cells (160,161), rat type II pneumocytes (162), and human keratinocytes (130). TNF-α can also effectively induce NOS2 mRNA or NO production in rat VSM cells (161), mouse ECs (163), and human chondrocytes (125).

A large variety of other agents induce the production of cellular NO or NOS2 mRNA, either by themselves or in combination with cytokines, LPS, or IFN-γ. Ouabain, an inhibitor of the Na^+/K^+-ATPase, augments the NOS2 mRNA levels induced by either TNF-α or IFN-γ in murine ECs (163). CHX, which inhibits NOS2 mRNA accumulation in response to cytokine/interferon signals in many cell types, induces NOS2 mRNA in rat ECs (157) and rat primary neonatal hepatocytes (164), and augments mRNA production induced by IL-1β and phorbol ester, respectively. Rat VSM cells can be induced to accumulated NOS2 mRNA by either neopterin or H_2O_2 (165). Similarly, the O_2- generating compound pyrogallol induces the rapid accumulation of NOS2 mRNA in the A549 epithelial cell line (166). Platelet-activating factor, either alone or to a greater extent in combination with LPS, can induce NOS2 mRNA in rat Kupffer cells (167). Silica can strongly induce NOS2 mRNA in RAW 264.7 macrophages (168). Finally, picolinic acid (PA), a catabolite of L-tryptophan, which alone induces little or no NO production by mouse ANA-1 macrophages (169), strongly synergizes with IFN-γ for the induction of NO synthesis.

The effects of cAMP-elevating agents on NOS2 gene expression are complex. Rat VSM cells can be induced to accumulated NOS2 mRNA by forskolin or 8-bromo-cAMP (156). In this work, forskolin and IFN-γ acted synergistically in the induction of NOS2 mRNA. Forskolin is also able to induce NOS2 mRNA in the Swiss 3T3 fibroblast cell line (135). Dibutyryl-cAMP (dbcAMP) can induce NOS2 mRNA accumulation in rat cardiac myocytes where it also augments IL-1β-induced mRNA production (170). In rat Kupffer cells, cAMP-elevating agents (PGE_2 or dbcAMP) can augment the ability of LPS to induce NO production, and synergize with IL-1β or TNF-α in this capacity even though IL-1β, TNF-α, PGE_2, and dbcAMP have no effects by themselves (171). Interestingly, dbcAMP augments the LPS-induced production of NO in rat peritoneal macrophages (172), but 8-bromo-cAMP inhibits IFN-γ+LPS-induced NO production in the mouse J774 macrophagelike cell line (173). Similarly, PGE_2 and LTB_4 inhibit NO production and NOS2 mRNA accumulation induced by IL-1β or IFN-γ+IL-1β+TNF-α+LPS in rat hepatocytes (174).

The effects of phorbol esters on NOS2 gene expression are also complex. Phorbol esters, by themselves, can induce NOS2 mRNA in Swiss 3T3 fibroblasts (135), the human FLG29.1 pro-osteoclast cell line (27), rat cardiomyocytes (175), and rat neonatal hepatocytes (164). In addition, while phorbol esters have no effect by themselves in the following cell types, they can affect the induction of NOS2 gene expression by other stimuli. Phorbol ester can augment NOS2 expression induced by LPS in rat Kupffer cells (171), and by IFN-γ (176) or thapsigargin (177) in mouse peritoneal macrophages. Conversely, phorbol ester can inhibit NOS2 gene expression induced by low (but not high) concentration of LPS (10 ng/mL) in the mouse macrophage cell line J774 (178), by IL-1β (179) or IFN-γ +TNF-α (180) in rat VSM cells, and by IL-1β in rat renal mesangial cells (181).

Finally, inhibitors of NOS2 induction in certain cell types can function as inducers in others. TGF-β (182) and IL-10 (183) both inhibit the induction of NO in mouse macrophages. Likewise, IL-8 inhibits the induction of NOS2 activity in rat peritoneal neutrophils (184). However, TGF-β can induce NOS2 mRNA in human keratinocytes as long as differentiation promoting factors—

Ca^{2+} and fetal bovine serum—are present (130). Furthermore, either IL-8 or IL-10 can induce NOS2 mRNA in avian osteoclastlike cells (185).

The plethora of NOS2 activating agents and their various effects, or lack thereof, on cells of differing lineage, are difficult to comprehend on a purely cellular basis. Thus, a number of laboratories have embarked upon an analysis of the molecular mechanisms by which transcription of the NOS2 gene is controlled. These analyses have uncovered a limited number of transcription factor binding sites upstream of the NOS2 transcription unit that bind trans-activating factors which are activated, or induced, by the above-mentioned stimuli. These transcription factors, and their cognate binding proteins, are discussed in the following section.

B. NOS2 Gene Promoter Structure

Initial analyses of the sequence and functional organization of the mouse NOS2 promoter identified two upstream regions that were necessary for full IFN-γ+LPS-stimulated induction of a linked reporter gene (10,186). The sequences of these two regions are shown in Figure 2. The promoter-proximal region is necessary for induction by LPS—IFN-γ has no apparent effect on promoter constructs that contain only this region of the promoter. The more distal region confers additional responsiveness to LPS and is also responsive to IFN-γ. The distal control region functions independent of orientation or distance from the transcription start point (187). Thus, this control region constitutes the NOS2 gene's enhancer. Conversely, the proximal control region must be closely linked with the TATA box and transcription start point (the so-called core promoter) and is functional only in its prototypic orientation (187). This region is, thus, the basal promoter of the NOS2 gene.

The mouse and rat NOS2 genes are virtually identical in the immediate enhancer and basal-promoter regions (see Fig. 2). Outside these regions nucleotide changes are more numerous. The human gene's enhancer and basal promoter are similar to that of the two rodent counterparts, and proximal transcription factor binding elements are highly conserved. However, the enhancer-linked transcription factor-binding elements have numerous changes. The structure and function of the human NOS2 gene will be discussed in more detail below.

The bipartite nature of the mouse NOS2 promoter has also been demonstrated in the rat A7r5, thoracic aorta VSM cell line (188). In this cell line the wild-type promoter can be induced eightfold by the combination of IL-1 (the type is not specified), TNF-α, and IFN-γ. Deletion of the enhancer, however, results in induction of only approximately 2.5-fold. Interestingly, the mouse promoter can be induced 10-fold by IL-1β in normal rat aortic VSM cells (189), yet in these cells deletion of the enhancer has no negative effect on promoter activity. The

mouse promoter is also inducible in the murine proximal convoluted tubule cell line (MCT) by IFN-γ+LPS (190), the human glioblastoma cell line, A-172, by IFN-γ+IL-1β+TNF-α+LPS (191), normal mouse mesangial cells by IFN-γ+LPS (192), rat cardiac myocytes by IL-1β (170), and the human epithelial cell line A549/8 by IFN-γ+IL-1β+TNF-α (193). The rat promoter is inducible in the Swiss 3T3 cell line after stimulation with either IL-1β or dbcAMP (12). Lastly, the chicken NOS2 promoter is induced, after transfection into the chicken macrophage cell line HD11, after stimulation with LPS (13).

C. Transcriptional Control of the NOS2 Gene

Numerous candidate transcriptional control elements have been identified, by computer-assisted analysis, in the NOS2 promoter sequence, a few of which are located within the boundaries of the enhancer and basal promoter, as defined by deletional analysis (10,186). To date, however, only six of these have been demonstrated to be functional and necessary for full gene induction, in mouse macrophages, by IFN-γ+LPS. A seventh element has been identified which is responsive to exogenously added picolinic acid as well as to hypoxic cell culture conditions. These elements, and their cognate binding proteins, will be discussed in more detail in the following sections. Many of the studies done to identify functional transcription factor binding elements have been carried out initially in mouse macrophages, but corroborating evidence of functionality in other cell types will be given where available.

1. κB:NF-κB/c-Rel

The κB-binding proteins, which are among the best-studied of latent transcription factors, are a family of proteins designated p50, p52, p65, c-Rel, and RelB. Homo- and heterodimeric combinations of these proteins (e.g., p50/p50, [$(p50)_2$], or p50/p65) are present in the cytoplasm of nonactivated cells complexed with an inhibitor protein, IκB (194). This association prevents the nuclear transport of NF-κB factors and, hence, their ability to trans-activate gene transcription. When certain stimuli, for example, LPS, are added to a given cell culture, IκB is rapidly phosphorylated and then degraded (195,196), thereby releasing NF-κB. Degradation of IκB is *via* the ubiquitin-proteasome pathway (197). The NF-κB transcription factor then translocates to the nucleus, where it can initiate transcription of any gene with one or more κB binding sites. The concentration of proteins that bind the κB element are increased substantially in the nuclei of cells treated with appropriate stimuli (198). Interestingly, treatment of many cell types with protein synthesis inhibitors, such as CHX, leads to the activation of NF-κB and accumulation of nuclear NF-κB at levels in excess of those typically seen by other activating agents, so-called superinduction.

The NOS2 promoter contains two candidate κB sites. One is in the gene's basal promoter, referred to here as κBI, and the other, κBII, is enhancer-linked. Evidence that κBI is functional was first provided by Xie et al. (199). A construct containing only the basal promoter of the mouse NOS2 gene (starting at position –85 in Fig. 2) could be activated by LPS in a dose-dependent fashion in transfected RAW 264.7 macrophages. However, removal of the κBI site, by truncation of the basal-promoter construct to position –75 (see Fig. 2), resulted in a virtual loss of basal-promoter activity. The dependence on κBI for full promoter activity was confirmed by Spink et al. (188) and Murphy et al. (200)—the former by transfecting full-length NOS2 promoter constructs into the VSM cell line A7r5 and the latter by transfecting full-length NOS2 promoter constructs into the mouse macrophage cell line RAW 264.7γNO(–). In both cell lines, mutation of the κBI site led to a 35% to 40% reduction in promoter activity.

LPS induces the binding of NF-κB to the κBI sequence in nuclear extracts prepared from RAW 264.7 macrophages (199,201) as well as RAW 264.7γNO (–) macrophages (200). IFN-γ, by itself, was unable to induce NF-κB binding to κBI in RAW 264.7γNO(–) cells and affected neither positively nor negatively the ability of LPS to induce binding (200). IL-1β also induces the binding of NF-κB proteins to κBI in rat aortic VSM cells (189) and rat renal mesangial cells (202). Antibody supershift analysis confirmed that the NF-κB/Rel family proteins p50, p65 (199–202), and c-Rel (199,200) are able to interact with the κBI sequence after appropriate cellular stimulation.

In addition to κBI, the κBII sequence also participates in the regulation of NOS2 expression. Mutagenic inactivation of κBII causes a 65% reduction in NOS2 promoter activity in A7r5 cells (188) and an 80% reduction in RAW 264.7γNO(–) macrophages (200). Thus, κBII is as functionally important, or more so, than is κBI, as one would expect of an enhancer-linked, cis-acting element. It is noteworthy that in transfected RAW 264.7γNO(–) macrophages, mutation of κBII renders the NOS2 promoter unresponsive to IFN-γ, when added in combination with LPS, compared with LPS treatment alone (200). Thus, while IFN-γ does not by itself activate transcription factor NF-κB, LPS-induced NF-κB is essential for full promoter responsiveness to IFN-γ+LPS. A likely explanation for this observation is that a cooperative interaction of NF-κB, IFN-γ-induced transcription factors (see below), and the NOS2 enhancer DNA is a prerequisite for maximal stimulation of the NOS2 promoter. Lastly, NF-κB can be fully activated in RAW 264.7γNO(–) macrophages at concentrations of LPS (<1 ng/mL) that induce no NOS2 gene induction. Thus, while NF-κB is necessary for the induction of the NOS2 gene in mouse macrophages, it alone is not sufficient.

NF-κB complexes can also be formed with the κBII sequence, in vitro. Goldring et al. (201) demonstrated the binding of both $(p50)_2$ and p50/p65 to κBII using nuclear extracts from RAW 264.7 macrophages. In A7r5 cells treated with

IFN-γ+IL-1+TNF-α, only p65 can be demonstrated, immunologically, to bind to κBII although UV-crosslinking reveals a faint-50 kDa protein, which is not a NF-κB/Rel family member, bound to this sequence (188). Thus, either $(p65)_2$ or a heterodimer containing p65 and the 50-kDa protein binds to the κBII site in stimulated A7r5 cells. In RAW 264.7γNO(–) macrophages, however, the binding pattern is more complex (200). In addition to three complexes that bind to κBI ($[p50]_2$, p50/p52, and p50/p65), a fourth, p65/c-Rel, binds only to the κBII sequence. The binding of p65/c-Rel to κBII is not efficiently inhibited by an excess of unlabeled κBI oligonucleotide, further suggesting specific binding of p65/c-Rel to κBII. Insight into the significance of the p65/c-Rel complex found in macrophages and the putative $(p65)_2$ complex found in VSM may be provided by the fact that a κB sequence identical to that of κBII is found upstream of the IL-6 gene. This sequence is bound by $(c\text{-Rel})_2$ in human lymphoid cells (203) and by $(p65)_2$ in human monocytelike cell lines (204). Thus, both p65 and c-Rel can bind to either half-site of the κBII sequence and, thus, p65/c-Rel and $(p65)_2$ could be the trans-activating species in macrophages and VSM, respectively. Furthermore, resident macrophages from c-Rel knockout mice are attenuated in their ability to accumulate NOS2 mRNA in response to LPS (205), further suggesting a critical role for c-Rel in macrophage NOS2 gene induction.

In addition to the above cited analyses, in vivo footprinting has also revealed, directly in living cells, the involvement of both NOS2 κB elements in gene induction (200,206). Guanosine residues in both elements are either protected from in vivo methylation by dimethyl sulfate (due to direct NF-κB-DNA contacts), or augmented in their methylation (due to NF-κB binding-induced unwinding of DNA) after cellular stimulation with LPS. In footprinting experiments employing RAW 264.7γNO(–) macrophages (200), the κBI site is footprinted (i.e., bound by protein) even in unstimulated cells. This observation advances the possibility that NOS2 gene expression is repressed, through the action of a κB-binding protein complex, in unstimulated mouse macrophages.

It is widely recognized that a vast array of inducing agents can activate NF-κB in a wide variety of cell types (198). Agents that, as reviewed above, induce NOS2 gene expression are also known to activate NF-κB include LPS, CHX, phorbol ester, TNF-α, IL-1, cAMP, and oxygen radicals (H_2O_2, and O_2^-). In addition to these, ouabain (163), neopterin (165), and silica (168) are also able to activate NF-κB in the specific cell types discussed above. In certain cell types the activation of NF-κB appears to be sufficient for partial induction of the NOS2 gene whereas in others NF-κB activation is necessary but not sufficient. Thus in BALB 3T3 fibroblasts TNF-α will activate NF-κB and, in turn, induce the NOS2 gene (207). Conversely, in cell types such as the mouse macrophage, TNF-α is a strong inducer of NF-κB (208,209) but does not by itself induce NOS2 gene expression (210,211).

There are at least three explanations for the difference in NOS2 gene responsiveness in different cell types: (a) certain NOS2 promoter elements may be requisite in some cell types but dispensable/nonfunctional in others; (b) some cell types may contain constitutively active transcription factors necessary for NOS2 gene induction, which would lessen or eliminate the need for stimulation by multiple cytokines/interferons; or (c) certain transcription factors necessary for NOS2 gene induction might be induced by a given stimulating agent in some cell types but not in others. Evidence exists to support each of these possibilities. Evidence for possibility c is provided by the fact that IFN-γ fails to activate NF-κB in most cell types including some macrophages (200). However, IFN-γ strongly activates NF-κB in the murine renal epithelial cell line, MCT (190), and in rat hepatocytes (212). IFN-γ also activates NOS2 gene expression in these latter cell types whereas it fails to do so in macrophages from many sources (158,159). Evidence for possibility b cited above stems from observations made in VSM cells. Both human and bovine VSM express a constitutively active and transactivating NF-κB-like activity (213,214). Stimulation of VSM with IFN-γ induces IRF-1 (215), a transcription factor that is also necessary for NOS2 gene induction (see below). Thus, IFN-γ alone is presumably able to induce the expression of NOS2 in VSM by virtue of inducing IRF-1, which, in turn, would act in concert with the constitutively present NF-κB. Evidence for the remaining possibility (a) is cited below in the discussion of the NOS2 octamer sequence.

2. ISRE:IRF-1

The NOS2 enhancer contains two immediately juxtaposed interferon stimulated response elements (ISREs) (216). Binding sites comparable to these ISREs have been found in the 5′-flanking regions of many IFN-inducible genes (217). Several proteins can interact with the ISRE after cells are stimulated with either IFN-γ or -α (218). The factor relevant to the discussion here is the positively acting interferon regulatory factor-1 (IRF-1; a.k.a. interferon stimulated gene factor-2, or ISGF-2), which is found constitutively in most cell types in small amounts (219) and is inducible by interferon (220) and to a lesser degree by TNF-α (221–223), IL-1 (221), IL-6 and LIF (224), and LPS (225).

Initial evidence that the two NOS2 ISREs had functional activity came from a report wherein peritoneal macrophages from IRF-1 knockout mice failed to produce NO when stimulated with LPS/IFN-γ (216). Peritoneal macrophages from these knockout mice failed to accumulate NOS2 mRNA in response to IFN-γ+LPS. However, in other studies with these mice, a low level of IFN-γ+LPS-inducible accumulation of NOS2 mRNA could be detected (226). In the studies of Kamijo et al. (216), an oligonucleotide containing the more distal NOS2 ISRE sequence (ISREa in Fig. 2) competed, in vitro, with a previously characterized ISRE sequence for the binding of IRF-1. The more proximal ISRE (ISREb in

Figure 2), however, did not effectively compete for IRF-1 binding. The two NOS2 ISREs also caused trans-activation of a heterologous promoter (that of the IFN-β gene) when this reporter gene plasmid was cointroduced into P19 embryonal carcinoma cells along with a vector that constitutively expressed the IRF-1 protein. Mutation of either of the two ISRE sequences resulted in a dramatic decrease in promoter activity. Thus, the distal ISRE represents a strong transcriptional activator sequence, due to its strong binding interaction with IRF-1, whereas the proximal ISRE is a weak activator.

Further evidence for the function of the strong, distal NOS2 ISRE came from mutational analysis of the NOS2 gene promoter (227,228). Mutation of this ISRE dramatically decreased NOS2 promoter activity in response to added IFN-γ. The direct binding of IRF-1 to this ISRE was also demonstrated by gel shift and antibody inhibition studies as well as by UV crosslinking studies (227).

The constitutive levels of IRF-1 are balanced, under normal proliferative conditions, by equivalent or slightly higher constitutive levels of the negative regulatory protein IRF-2 (219). Although a direct negative role of IRF-2 in NOS2 gene transcription has not been demonstrated, IRF-2 does play a positive role in NOS2 mRNA accumulation (226). Macrophages from IRF-2 knockout mice accumulate less NOS2 mRNA, in response to IFN-γ+LPS, than do their wild-type counterparts. Thus, IRF-2 appears to play a posttranscriptional role in NOS2 gene expression.

IFN-γ induces the synthesis of an excess of IRF-1 transcripts, compared with that of IRF-2, and the synthesis of new IRF-1 protein. Shifting the balance of IRF-1/IRF-2 to a higher relative abundance of IRF-1 allows competitive binding of the positive transcriptional regulator to genes containing one or more ISREs and, hence, gene induction. The effectiveness of CHX at blocking NOS2 gene induction in many cell types investigated is thought, at least in part, to be causally related to its inhibition of new IRF-1 synthesis. However, CHX does not inhibit NOS2 mRNA accumulation in all cell types or under all conditions of activation. In the Swiss 3T3 fibroblast cell line, phorbol ester induces rapid accumulation of NOS2 mRNA, which is not inhibited by CHX (135). However, this may be explained by the fact that quiescent cell cultures were used. Horiuchi et al. (229) demonstrated that IRF-1 is more highly expressed in quiescent versus proliferating R3T3 cell cultures allowing for expression of the angiotensin II type 2 receptor gene—which is dependent upon trans-activation by IRF-1—in quiescent cells. Thus, as noted above, the constitutive presence of IRF-1 in quiescent fibroblasts may eliminate the need for an interferonlike signal in NOS2 gene induction.

Whether or not a similar mechanism explains the ability of CHX to augment or induce NOS2 gene expression in other cell types is not known. CHX has no effect on NOS2 gene induction in mouse (154) or rat (230) islet cells induced with IFN-γ+TNF-α or IL-1β, respectively, or in rat VSM stimulated with either IFN-γ (231) or IL-1β (232), or rat mesangial cells induced with LPS (233). Furthermore, CHX can by itself induce NOS2 gene expression in the following cell types and augment NOS2 gene expression induced by the following agents: 8-bromo-cAMP (234) or LPS (235) in rat VSM; phorbol ester in rat hepatocytes (164); and IL-1β in rat aortic ECs (157). The induction of NOS2 mRNA accumulation by CHX alone and its ability to augment induction by other stimuli is likely due, at least in part, to the ability of CHX to superinduce NF-κB binding activity (see above).

Although it has not been demonstrated that IRF-1 is induced by the above-cited agents, or that IRF-1 is already present at levels in excess of IRF-2 in the cell types investigated, it is unlikely that new IRF-1 protein is produced in the presence of CHX. Thus, it appears that de novo synthesized IRF-1, and its interaction with the ISRE, are important for NOS2 gene expression in cells like the mouse macrophage, but may be less critical for expression or even dispensable in other cell types. A compelling example of the apparent paradoxical effect of CHX on NOS2 gene expression is observed in rat VSM cells. In this one cell type from this single species, CHX: (a) is without effect on IFN-γ induced increases in NOS2 mRNA (231); (b) has no effect on (231) or partially inhibits (231) IL-1β-induced mRNA expression; (c) markedly inhibits TNF-α-induced mRNA accumulation (231); (d) augments NOS2 gene expression induced by 8-bromo-cAMP (234) or LPS (235); and (e) induces NOS2 gene expression on its own (234,235). Findings such as these underscore the complexity of the "circuitry" regulating NOS2 gene expression.

That certain inhibitors of NOS2 gene induction may function through the inhibition of IRF-1 induction has been suggested. IL-1β strongly induces IRF-1 mRNA accumulation in rat islet cells (236). This induction is completely prevented by treatment of cell cultures with high doses (20 mM) of nicotinamide. This inhibition of IRF-1 induction by nicotinamide is accompanied by a nearly complete inhibition of NOS2 gene expression and a complete abrogation of NO production. However, as noted above, the experiments of Kwon et al. (230) indicate that CHX has no effect on IL-1β-induced NOS2 mRNA accumulation in rat islet cells, suggesting that de novo synthesized IRF-1 plays no role in NOS2 gene induction in this cell type. Different rat strains were used in these two studies, which may account for the conflicting results. These types of controversies will need to be resolved before a thorough understanding of NOS2 gene regulation can be achieved. IRF-1 induction is also inhibited by the antioxidant chrysin (5,7-dihydroxyflavone) in RAW 264.7 macrophages activated with IFN-γ+LPS (237). Chrysin concurrently inhibits NOS2 mRNA accumulation (by approximately 97%) but fails to inhibit NF-κB activation (in fact, in some experiments NF-κB activation was augmented by chrysin treatment).

3. GAS:Stat1α

A second type of DNA binding site that is responsive to IFN is the IFN gamma-activated site (GAS) (238). It is bound and trans-activated by a 91-kDa protein of the STAT (signal transducers and activators of transcription) family of transcriptional activators, called Stat1α (a.k.a. p91 or IFN-gamma activation factor [GAF]). Stat1α exists as a latent cytoplasmic transcription factor in unstimulated cells (239). Interaction of IFN-γ with its receptor causes Stat1α to become tyrosine-phosphorylated, allowing it to dimerize and then translocate to the nucleus (240). Once there, it activates transcription of genes containing a GAS in their promoters.

STAT factors are also activated by type I interferons, IFN-α and β. IFN-α/β treatment of most cell types causes the tyrosine phosphorylation of Stat1α and another latent factor, Stat2 (a.k.a. p113). Stat1α and Stat2 then heterodimerize with each other and associate with a third factor, p48 (ISGF-3γ), to form the ISGF-3 transcription factor (241). ISGF-3 binds to a class of ISRE that is distinct from the NOS2 enhancer-linked ISRE in that nucleotide positions that flank the IRF-1-binding "core" sequence are required for ISGF-3 binding (242). Excess phosphorylated Stat1α, accumulated after IFN-α/β treatment (i.e., that not associated with ISGF-3) can homodimerize forming GAS-binding activity (243–246). The result is that fewer Stat1α homodimers are available to bind GAS elements after IFN-α/β stimulation than after IFN-γ stimulation, and, thus, GAS-driven gene expression is lower after cellular treatment with IFN-α/β than IFN-γ (246). This phenomenon is also observed for NOS2 gene induction in mouse macrophages—IFN-γ+LPS is a stronger inducer of NOS2 gene expression than is IFN-β+LPS (159). The molecular explanation for this observation may involve differing levels of Stat1α homodimers induced by these two sets of mediators (see below).

An initial report on the NOS2 gene's candidate GAS element concluded that it was not functional (227). However, subsequent analysis of the GAS demonstrated a functional role for this sequence in regulation of the mouse NOS2 gene (228). Mutation of this sequence resulted in a two- to threefold reduction in NOS2 promoter activity achieved by stimulation of transfected RAW 264.7 cell cultures with IFN-γ, LPS, or IFN-γ+LPS. In addition, four tandemly linked copies of the NOS2 GAS could direct expression of a downstream thymidine kinase promoter. Gel shift and antibody supershift analyses demonstrated the inducible binding of Stat1α to a radiolabeled oligonucleotide containing the NOS2 GAS sequence. Other STAT family members—Stat3, 4, 5, or 6—were not detected in the inducible binding complex observed. Neither were the transcription factors Stat2 or p48 (unpublished observations). This binding was induced not just by IFN-γ, but by LPS as well. The difference between IFN-γ- or IFN-γ+LPS-induced activation of Stat1α and that induced by LPS was temporal and quantitative (228). IFN-γ or IFN-γ+LPS induced maximal Stat1α activation within 30 min of exposure whereas LPS treatment required 2 to 4 h. LPS also induced much less binding activity than did IFN-γ.

The activation of Stat1α caused by LPS is due to the autocrine/paracrine production of IFN-β (279). Supernatants from mouse macrophage cell cultures treated with LPS for 4 h can be transferred to a naive cell culture and cause rapid—within 30 min—Stat1α activation. The activity in these transferred culture supernatants can be completely abolished by the addition of neutralizing anti-IFN-β antibodies. This observation is consistent with the proposed role of Stat1α in induction of the NOS2 gene (see above). Autocrine IFN-β (induced by LPS) causes the activation and accumulation of less Stat1α than does IFN-γ+LPS such that NOS2 gene expression is lower when induced by LPS than with IFN-γ+LPS.

Stat1α is most likely an indirect activator of the NOS2 gene as well. The IRF-1 gene has a single GAS in its promoter that binds activated Stat1α after cell induction with IFN-γ (246). Trans-activation by Stat1α causes strong induction of the IRF-1 gene. Although the IRF-1 promoter also has a κB-like site, and several other candidate transcription factor-binding elements, none of these have been shown to be active to date. Interestingly, the κB-like sequence can bind NF-κB proteins in vitro and is active when linked to a heterologous (c-fos) promoter (247). However, this κB sequence confers no transcriptional induction on the wild-type IRF-1 promoter and does not, by in vivo footprinting analysis, bind protein after induction of the IRF-1 gene.

4. Octamer: Oct-1 and Oct-2

The octamer family of transcription factors includes Oct-1, a factor which is expressed in near ubiquitous fashion—it is not expressed in nondividing, mature neurons (248)—and other octamer-binding proteins that either have a more limited tissue distribution or are developmentally regulated (249). Oct-2 is a tissue-specific factor that is expressed only in lymphoid cells (249), neuronal cells of the brain (116,117), and macrophages (250). In tissues where Oct-1 or Oct-2 is expressed, they can be found constitutively in the nuclear compartment. The activity of these two transcription factors can be altered through either alternative splicing events that produce Oct-1 or Oct-2 isoforms (251,252), through phosphorylation (253), or by interaction with a number of accessory proteins that are often required for functional activity of Oct-1 or Oct-2 (254).

A candidate octamer sequence is absolutely conserved in sequence, and similarly located, in each of the mouse (position -61), rat (position -61), and human (position -65) promoters (see Fig. 2). Remarkably, this sequence is also conserved in sequence and relative position in the chicken NOS2 gene (position -54) (13), albeit in the opposite orientation. Circumstantial evidence that the

NOS2 octamer sequence might be functional came initially from studies that tried to assess the spatial relationship between the basal-promoter regulatory elements and the core promoter (187). The octamer motif, like the majority of cis-acting DNA elements, is functional in either orientation (255). However, the octamer motif is somewhat unique in that it can function at a great distance from the core promoter in lymphoid cells but must be located very close to this regulatory element to be functional in nonlymphoid cells (256). Consistent with these previous observations, the octamer binding site in the mouse NOS2 promoter was functional in its native position, but when the entire basal promoter was reversed, promoter activity decreased considerably, presumably because the location of octamer was moved 134, rather than 31, bp upstream of the TATA box.

Other proof that the NOS2 gene octamer is functional comes from in vivo footprint analysis. In RAW 264.7 (206) and RAW 264.7γNO(–) (unpublished observations) macrophage cultures the octamer sequence is bound by protein after stimulation with LPS or IFN-γ+LPS. This binding blocks the methylation of a guanosine residue found on the bottom strand of the octamer sequence. Oligonucleotides containing the NOS2 octamer sequence also bind three protein complexes in unstimulated and stimulated macrophage cultures (206; unpublished observations). The proteins interact specifically with antibodies raised against Oct-1 and Oct-2 (unpublished observations). In RAW 264.7γNO(–) macrophages, these octamer-binding proteins are found in extracts prepared from unstimulated cell cultures and are not further induced by cell treatment with either IFN-γ, LPS, or IFN-γ+LPS—in spite of the fact that in vivo footprinting reveals protein binding to the NOS2 octamer sequence *only* after cellular stimulation with LPS or IFN-γ+LPS. Thus, the constitutively present octamer binding activity gains access to the NOS2 octamer sequence only after macrophage activation by LPS±IFN-γ. One possible explanation for this observation is that activation of other transcription factors necessary for NOS2 gene induction must be achieved before the octamer-binding protein can access the gene—perhaps through protein-protein interactions with these other factors.

Definitive proof that the NOS2 octamer sequence is function comes from mutational analysis (257; unpublished observation). Mutation of two base pairs within the octamer sequence lowers inducible promoter activity (in response to IFN-γ+LPS) by threefold, after transfection of reporter gene constructs into RAW 264.7γNO(–) macrophages, when compared with a wild-type construct (unpublished observation). Interestingly, transfection of the wild-type and mutant constructs into the fibroblast cell line L929 reveals no difference in activity. One notable difference between macrophages and fibroblasts is that the former express Oct-2 whereas the latter do not (258–260). However, our preliminary experiments indicate that over xpression of Oct-2 isoforms—Oct-2.2 and Oct-2.5 (252)—in L929 fibroblasts *does not* increase NOS2 promoter activity, while overexpression in RAW 264.7γNO(–) macrophages partially inhibits NOS2 promoter inducibility. These results suggest that the NOS2 octamer sequence is not utilized in fibroblasts, and suggests that the octamer motif is a tissue-specific promoter element functional in mouse macrophages.

The octamer sequence also appears to be important in the differentiation of mouse macrophages from immature precursors (257). IL-6 can induce differentiation in the M1, myeloid precursor cell line. Differentiation is accompanied by the expression, 48 to 72 h poststimulation, of the NOS2 gene. The expression of the NOS2 gene coincides temporally with the appearance of transcription factor Oct-2 and with a significant decrease in Oct-1, as measured by gel shift and antibody inhibition-of-binding analyses performed using an oligonucleotide containing the NOS2 octamer sequence. Additionally, at least three other octamer-binding proteins, not related immunologically to Oct-1 or Oct-2, interact with the NOS2 octamer sequences in nuclear extracts prepared from this cell line. Mutation of the eight nucleotide octamer sequence specifically inhibits interaction with these octamer binding proteins and abrogates induction, by LPS, of a basal promoter, reporter gene construct transfected into RAW 264.7 macrophages. The coincidental appearance of Oct-2 and NOS2 gene products in differentiating macrophages suggests, but does not prove, that this transcription factor is necessary for NOS2 gene expression in this maturing cell population.

5. HRE:HIF-1

The hypoxia responsive element (HRE) is active in a variety of cell types in response to low oxygen tension (261). The HRE is bound by a transcription factor called hypoxia-inducible factor-1, or HIF-1 (262,263). HIF-1 is composed of two different subunits, HIF-1α and HIF-1β (a.k.a., aryl hydrocarbon receptor nuclear translocator or ARNT) (264). Expression of each of these subunits is increased at the transcriptional level by hypoxia (265). The DNA-binding activity of HIF-1 also requires protein phosphorylation (263) and reducing conditions (266).

Demonstration that the NOS2 gene promoter contains a functional HRE came from experiments that examined the mechanism of induction of this gene by picolinic acid (PA) (267). As discussed above, PA induces no expression of NO in mouse ANA-1 macrophages but strongly synergizes with IFN-γ in this capacity (169). Deletional analysis of the NOS2 promoter identified a 127-bp region just upstream of the basal promoter (nucleotide positions –201 to –328 in Fig. 2) that is responsive to PA in transfected ANA-1 macrophages (IFN-γ responsiveness mapped to the NOS2 enhancer, and presumably involves the ISRE, GAS, or both). Computer-assisted analysis of this region of DNA revealed a candidate HRE at positions -219 to -226 (see Figure 2), the deletion or mutation of which led to a drastic reduction in promoter activity induced by PA or PA+IFN-γ in transfected ANA-1 macrophages (267). Three tandemly linked copies of the NOS2 HRE were also able to direct the transcriptional initiation of a heterologous thymidine kinase promoter in response to PA. This latter construct was also induced by hypoxic conditions (1% oxygen). While the wild-type NOS2 promoter construct was induced some twofold by either IFN-γ or hypoxia, the combination of these two stimuli dramatically increased promoter activity in transfected ANA-1 cell cultures. Thus, in a manner similar to macrophage induction by IFN-γ+LPS, two independent signals are required for NOS2 gene induction under hypoxic conditions.

Treatment of ANA-1 macrophages with PA also induced the expression of a nuclear factor which bound to a ^{32}P-labeled oligonucleotide containing the NOS2 HRE sequence. Binding of this protein complex to the NOS2 HRE was inhibited by addition of an excess of unlabeled NOS2 HRE oligonucleotide as well as by an oligonucleotide representing a previously characterized HRE—that from the human erythropoietin gene. An oligonucleotide containing a NOS2 HRE mutated at three nucleotide positions highly conserved in HRE-containing genes failed to inhibit binding. Additional protein-binding activity, constitutively present in nuclear extracts but increased by PA treatment, was also detected by the NOS2 HRE oligonucleotide. Although antibody supershift or binding inhibition studies were not performed to confirm that the inducible protein complex was in fact HIF-1, the pattern of detectable protein bands and the aforementioned DNA binding characteristics were consistent with those observed by other laboratories.

It is interesting to note that the HRE sequence is not highly conserved in either the human or rat NOS2 genes. Even if one considers the more loosely representative consensus sequence—(C/G/T)ACGT(C/G)(C/G)(G/T) (268)—the human and rat genes fail to conform. Whether the mouse NOS2 gene is uniquely responsive to hypoxic conditions or an HRE or similar such sequence is located elsewhere in the human or rat promoter is unknown.

D. Model of NOS2 Gene Induction in Mouse Macrophages by IFN-γ+LPS

The work cited above has led us to propose a working model for the activation of NOS2 gene expression in mouse macrophages (Fig. 3). In unstimulated macrophages (Fig. 3a) the NOS2 gene is transcriptionally silent. Low, basal levels of IRF-1 gene expression occur. The IFN-β gene is transcriptionally active in unstimulated mouse macrophages, but the mRNA is unstable and rapidly degraded (269). The latent transcription factors Stat1α and NF-κB are located in the cytoplasm, the latter in a complex with IκB, and Oct-1/2 is located in the nucleus but is not associated with the NOS2 octamer sequence.

Upon stimulation with LPS (Fig. 3b), NF-κB is released from its inhibitor IκB. LPS treatment of mouse macrophages leads to a stabilization of IFN-β mRNA (269) (S.W. Russell, personal communication) and subsequent production of autocrine/paracrine IFN-β. IFN-β, in turn, binds the IFN-β receptor caus-

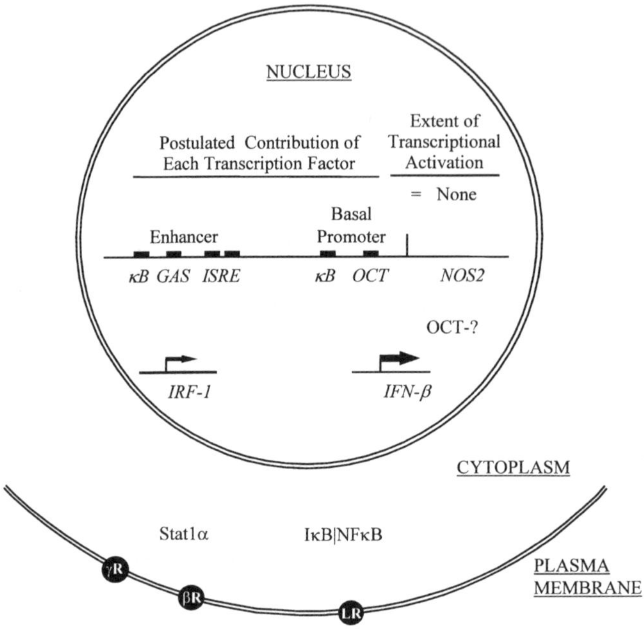

(a)

Figure 3 Model for the transcriptional activation of the NOS2 gene, by LPS or IFN-γ+LPS in mouse macrophages. The figure shows a representation of a macrophage with the nucleus, cytoplasm, and plasma membrane, labeled accordingly. γR, βR, and LR denote the IFN-γ, IFN-β, and LPS receptors, respectively. IκB/NF-κB (Fig. 3(a)) represents the NF-κB transcription factor bound to its inhibitor protein IκB. The "free" IκB (Fig. 3(b) and (c)) indicates protein that is in the process of degradation after release from NF-κB. The horizontal lines represent the NOS2, IRF-1, and IFN-β genes. The vertical line associated with each gene represents the transcription start point and affixed arrows depict ongoing transcription, the relative extent of which is represented by the size of the arrow. Other lines/arrows represent signal transduction pathways leading to the activation of transcription factors necessary for NOS2 gene expression. The dashed lines indicate less relative transcription factor accumulated compared with continuous lines. The postulated contribution of each transcription factor, in turn, is summarized by the +++ and + symbols at the tops of each figure. The extent of transcriptional activation is, thus, based on the abundance of each transcription factor associated with the NOS2 gene. All other descriptions of these figures are found in the text.

ing activation of the JAK family tyrosine kinases associated with this receptor, (270) and subsequent phosphorylation of Stat1α. IFN-β treatment also leads to the phosphorylation of Stat2 (not shown) whereupon Stat1α, Stat2, and p48 form the ISGF-3 transcription factor complex. Excess Stat1α forms homodimers which

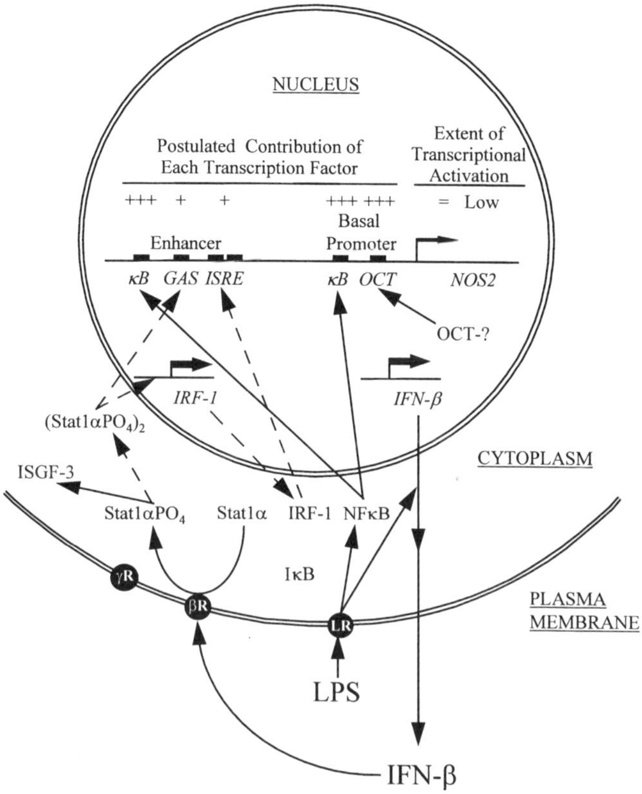

(b)

translocate to the nucleus, where they activate IRF-1 gene transcription. With the accumulation of excess IRF-1, all transcription factors (i.e., NF-κB, IRF-1, Stat1α, and Oct-1 or 2) presumed to be necessary for NOS2 gene transcription are present in the nucleus, assemble on the NOS2 basal promoter and enhancer, and effect transcription initiation. Because the levels of Stat1α homodimers are low (due to the codistribution of Stat1α between both ISGF-3 and Stat1α homodimers), IRF-1 and NOS2 gene transcription are also low, which in turn results in less than maximal expression of NOS2 mRNA, NOS2 protein, and NO.

After stimulation of mouse macrophages with IFN-γ+LPS (Fig. 3c), Stat1α is phosphorylated by JAK kinases associated with the IFN-γ receptor and subsequently forms homodimers. These homodimers then activate strong IRF-1 gene transcription (226). NF-κB is released from IκB, due to LPS signaling, and translocates to the nucleus. IFN-γ+LPS stabilize IFN-β mRNA to a degree greater than

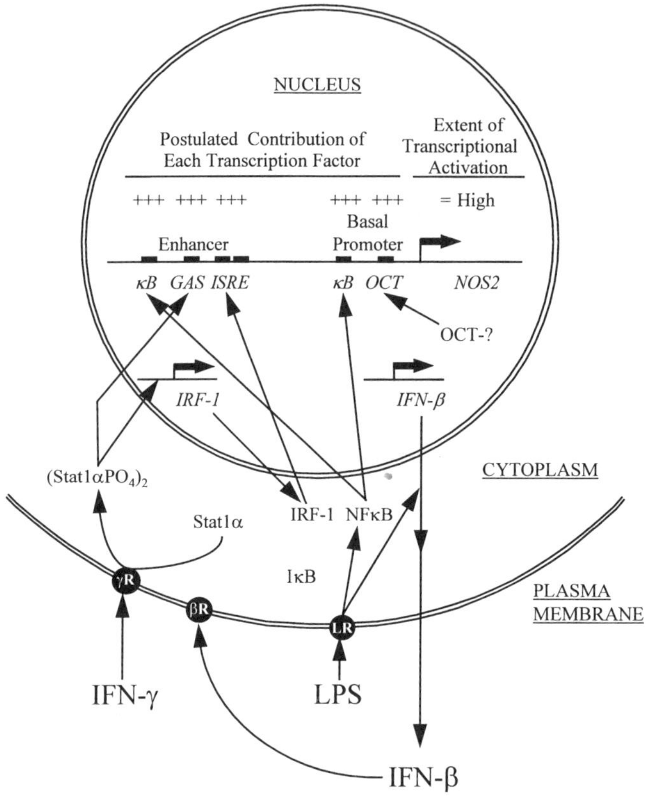

(c)

that of LPS alone (269) (S.W. Russell, personal communication), and, thus, IFN-β is produced in higher abundance. However, since the appearance of autocrine/paracrine IFN-β lags behind the initial cellular exposure to IFN-γ, the interaction of IFN-β with its receptor does not appear to affect the molecular distribution of Stat1α—homodimers predominate. Thus all transcription factors necessary for NOS2 gene induction are present at maximal amounts, and, hence, maximal NOS2 mRNA, protein, and NO are produced.

This is a reasonable working model for other mouse cell types as well. Our experience with mouse fibroblasts, where the octamer site fails to contribute to NOS2 gene activation, suggests that the octamer component of this model may indeed be specific to macrophages. As another example, the fact that IFN-γ alone is a strong activator of NOS2 gene expression in VSM cells might lead to the conclusion that LPS-like signaling, via the action of NF-κB, is not requisite. However, the constitutive presence of trans-activating NF-κB (213,214) may obviate LPS signaling in these cells, and provide gene expression effects equivalent to those induce by LPS in other cell types through interaction with the NOS2 gene κB sequences.

E. Expression of the Human NOS2 Gene

As shown in Figure 2, the mouse and human NOS2 genes share considerable sequence identity in both the basal-promoter and enhancer regions. However, whereas the κB and octamer elements present in the basal promoter of these two genes are conserved at 17 out of 18 positions (94%), the κB, GAS, and ISRE elements in the enhancer of these genes are conserved at only 31 out of 41 positions (76%). Based on mutational evidence in the literature, the nucleotide changes in the human enhancer sequence would likely render it nonfunctional (14, and references therein). Indeed, the κB-like site located in the human gene's enhancer fails to bind NF-κB protein, using nuclear extracts derived from either human or mouse macrophage sources. Furthermore, the human enhancer sequence fails to contribute transcriptional activity when linked either to the human or mouse basal promoter when such reporter gene constructs are introduced into the mouse macrophage cell line RAW 264.7γNO(–) followed by cellular stimulation with LPS±IFN-γ (14). Thus, the human NOS2 enhancer has been rendered nonfunctional, either through natural selection or genetic drift.

The κB site in the human NOS2 basal promoter (huκBI), on the other hand, is functional (14). A reporter gene construct containing only the human NOS2 basal promoter (upstream to nucleotide position –335 in Fig. 2) is induced twofold in mouse RAW 264.7γNO(–) macrophages treated with LPS, and this level of induction is increased more than threefold by linking the mouse enhancer to the human basal promoter. In addition, the huκBI sequence bound NF-κB proteins in nuclear extracts prepared from either mouse or human macrophage cell lines stimulated with IFN-γ+LPS. The binding affinity for different NF-κB dimers was different for the human and mouse κBI sites, however, which was probably a reflection of their slightly differing sequences (see Fig. 2).

The transfection results of Zhang et al. (14) are confirmed by Nunokawa et al. (271) who stimulated transfected DLD-1 cells with IL-1β. This cytokine induced a twofold increase in promoter activity of all constructs containing a functional human NOS2 basal promoter. Mutation of the huκBI site, however, eliminated promoter induction by IL-1β, providing definitive proof that this element contributes to human NOS2 promoter activity. Interestingly, the same level of basal-promoter inducibility was not observed in transfected DLD-1 cells stimulated with the combination of IFN-γ+IL-β+TNF-α+IL-6 (149) or in the human liver epithelial cell line AKN-1 stimulated with IFN-γ+IL-β+TNF-α (272). The human basal promoter is functional in the rat VSM cell line A7r5, where high levels of constitutive activity, not further inducible by LPS, were observed (273).

Deletion of the huκBI site greatly abated this constitutive promoter activity. This is again consistent with the expression of constitutive trans-activating NF-κB in this cell type (213,214). Interestingly, while the human promoter was not induced by LPS, IFN-γ, or IL-1β in transfected A7r5 cells or rat VSM cultured ex vivo, transfected human VSM (ex vivo cultures) showed a twofold stimulation of promoter activity in response to IL-1β (273).

In spite of a significant effect of IFN-γ on human NOS2 gene induction (124), the 3.7 kb 5′-flanking sequence of this gene failed to show responsiveness to added IFN-γ in DLD-1 (271) and RAW 264.7γNO(–) (14) cell lines as well as in rat and human VSM (273), indicating that IFN-γ -responsive transcriptional elements must lie much farther upstream. This was confirmed by de Vera et al. (272), who identified three DNA regions, each responsive to IFN-γ+IL-β+TNF-α, between nucleotide positions –3800 to –5800, –5,800 to –7,000 and –7000 to –16000. Stimulation of transfected AKN-1 cells with pairwise combinations of these cytokines showed that the 16 kb 5′-flanking region contained cis elements highly responsive to IFN-γ, IL-1β, and, to a lesser extent, TNF-α.

IV. FUTURE CHALLENGES

Although a considerable amount of information has been generated about the transcriptional regulation of these three genes, there is much yet to learn. The discovery of transcriptional regulation of the "constitutive" isoforms of NOS—both upregulation and downregulation—challenges our understanding of homeostasis and redefines our notions of constitutive, or "housekeeping," genes. The fact that the NOS1 and NOS3 genes are expressed in a variety of disparate tissues complicates the analysis of up- and downregulation of these two genes. Furthermore, conflicting reports of NOS1 and 3 gene up- or downregulation must be resolved.

It is conceivable that, like the NOS2 promoter, a common set of cis-acting regulatory elements, and cognate binding proteins, may govern the constitutive expression of these genes. If this is the case, then variation in the expression of transcription factors that interact with these DNA elements may modulate absolute levels of constitutive expression from one tissue to the next. These levels may be modulated further by transcription factors that are expressed exclusively in certain tissues—or during tissue-specific developmental programs—provided that a cognate DNA-binding site is present in the NOS1 or NOS3 promoter.

The ability to upregulate either of these genes will also depend on the ability of the cell type to respond to the stimulus in question and the availability or inducibility of appropriate transcription factors within the cell type in question. Thus, shear stress, for example, may be a strong inducer of NOS3 gene expression in vascular ECs but an ineffective inducer in another cell type. An inability on the part of a given cell type to respond to this kind of stress may relate to its ability to sense stress forces or to respond by induction or activation of the transcription factor that binds the candidate shear stress response element.

Similarly, the mechanisms of downregulation of the NOS1 and NOS3 genes should prove interesting. Is downregulation controlled solely at the level of transcript stability or does downregulation stem, at least in part, from the inactivation of transcription factors necessary for constitutive transcriptional control? Is an active repressor molecule induced which binds specifically to a sequence within the promoter in question? Furthermore, in cell types that do not constitutively express either the NOS1 or NOS3 genes, are the transcription factors necessary for constitutive expression missing or inactivated or, again, are these genes actively repressed in these cells via a trans-acting factor?

Several facets of regulation of the mouse NOS2 gene promoter also require further investigation. The discovery of the HRE and its synergistic interdependence on IFN-γ signaling raises the following questions: (a) does the IFN-γ signal depend upon Stat1α, IRF-1 or both; (b) can the hypoxia- and cytokine-induced pathways of NOS2 gene activation function additively or synergistically or are they mutually exclusive (i.e., inhibitory) of one another; (c) are there additional pathways of NOS2 gene activation akin to PA/hypoxia mediated induction; and (d) if so, are they also dependent upon two separate signaling pathways?

The generality of the model depicted in Figure 3 is also of interest. Will this model, or a close alternative of such, explain the complex regulation of NOS2 observed in cell types like VSM? As a working model it should prove useful in resolving the many controversies surrounding NOS2 gene expression in various tissues. It is a strong possibility that other cis-regulatory elements, and their cognate binding proteins, may play an active role in NOS2 gene expression. Indeed, other candidate transcription factor-binding sites are found in the mouse NOS2 5′-flanking sequence (10,186). Additional factors, such as architectural proteins necessary for DNA bending or for coalescing transcription factors bound to the widely separated enhancer and basal-promoter motifs (for the purpose of recruiting basal transcription factors and RNA polymerase), are also likely to be involved. As discussed above, the octamer sequence may be functional only in macrophages. Hence, tissue-specific cis-acting elements may take the place of octamer in other cell types.

The model in Figure 3 also provides a molecular framework for the synergy observed between IFN-γ and cytokines or LPS. Presumably, an increase in NOS2 transcription occurs only when cytokine(LPS)-stimulated transcription factors (e.g., NF-κB) and interferon-stimulated transcription factors (e.g., Stat1α and IRF-1) coordinately accumulate in the nucleus. Is cooperative assembly of some or all of these transcription factors and their cognate DNA-binding sites necessary for NOS2 gene induction? This would be analogous to assembly of the "enhansosome"—a multicomponent complex of transcription factors NF-κB,

ATF-2/c-Jun, IRF-1, and HMG I(Y)—required for transcriptional synergy displayed by the IFN-β gene enhancer (274). Does the temporal appearance of these factors provide a window of opportunity for NOS2 gene expression, and if so, what happens if the cytokine(LPS)- and IFN-γ-activated transcription factors accumulate at different times—does this attenuate NOS2 gene induction? Similarly, since many of these transcription factors dimerize or undergo other protein-protein interactions, can competition for a given protein among different transcription factor complexes affect NOS2 gene expression? The answer would appear to be yes if the model outlined in Figure 3b, for the induction of the NOS2 gene by LPS alone, is correct. This model could also be applied to the observation of an inverse, often antagonistic, relationship between growth factor-induced proliferation and cytokine(LPS)/IFN-γ-induced antiproliferative effects and nitric oxide production (275–277). The growth factors EGF (275) and PDGF (276) both inhibit NOS2 gene induction, and each of these growth factors induces the phosphorylation of Stat1α and another STAT protein, Stat3 (278). Stat1α and Stat3 not only homodimerize but heterodimerize with each other as well. These hetero- and homodimers can bind to the growth-related serum inducible element (SIE) found in genes like that encoding c-Fos. Thus, growth factors may reduce NOS2 gene expression by reducing the availability of Stat1α homodimers (via formation of Stat1α/Stat3 heterodimers) necessary for IRF-1 and NOS2 gene induction, in a manner similar to that proposed above for LPS and autocrine/paracrine IFN-β.

Lastly, the structural organization of the human NOS2 promoter poses strategic problems for the identification of functional cis-acting elements. These control sequences appear to be scattered over some 12,000 bp of genomic DNA, which may necessitate a molecular tour de force for complete identification. Nonetheless, elucidation of these controlling elements should provide mechanistic clues to the regulation of this gene in human beings, and may provide a molecular explanation for the difficulty encountered when trying to induce this gene in human macrophages.

ACKNOWLEDGMENTS

The writing of this chapter was supported, in part, by a program project grant from the National Cancer Institute (P01 CA 54474) and by the Wilkinson Endowment for Cancer Research.

REFERENCES

1. Xie Q-w, Nathan C. The high-output nitric oxide pathway: role and regulation. J Leukoc Biol 1994; 56:576–582.

2. Young AP, Murad F, Vaessin H, Xie J, Rife TK. Transcription of the human neuronal nitric oxide synthase gene in the central nervous system is mediated by multiple promoters. Adv Pharmacol 1995; 34:91–112.
3. Harrison DG, Venema RC, Arnal JF, Inoue N, Ohara Y, Sayegh H, Murphy TJ. The endothelial cell nitric oxide synthase: is it really constitutively expressed? Agents Actions Suppl 1995; 45:107–117.
4. Wang Y, Marsden PA. Nitric oxide synthases: gene structure and regulation. Adv Pharmacol 1995; 34:71–90.
5. Förstermann U, Kleinert H. Nitric oxide synthase: expression and expressional control of the three isoforms. Naunyn Schmiedebergs Arch Pharmacol 1995; 352:351–364.
6. Hall AV, Antoniou H, Wang Y, Cheung AH, Arbus AM, Olson SL, Lu WC, Kau C-L, Marsden PA. Structural organization of the human neuronal nitric oxide synthase gene (NOS1). J Biol Chem 1994; 269:33082–33090.
7. Marsden PA, Heng HHQ, Scherer SW, Stewart RJ, Hall AV, Shi X-M, Tsui L-C, Schappert KT. Structure and chromosomal localization of the human constitutive endothelial nitric oxide synthase gene. J Biol Chem 1993; 268:17478–17488.
8. Venema RC, Nishida K, Alexander RW, Harrison DG, Murphy TJ. Organization of the bovine gene encoding the endothelial nitric oxide synthase. Biochim Biophys Acta 1994; 1218:413–420.
9. Dynan WS. Promoters for housekeeping genes. Trends Genet 1986; 2:196–197.
10. Lowenstein CJ, Alley EW, Raval P, Snowman AM, Snyder SH, Russell SW, Murphy WJ. Macrophage nitric oxide synthase gene: two upstream regions mediate induction by interferon g and lipopolysaccharide. Proc Natl Acad Sci USA 1993; 90:9730–9734.
11. Chartrain NA, Geller DA, Koty PP, Sitrin NF, Nussler AK, Hoffman EP, Billiar TR, Hutchinson NI, Mudgett JS. Molecular cloning, structure, and chromosomal localization of the human inducible nitric oxide synthase gene. J Biol Chem 1994; 269:6765–6772.
12. Eberhardt W, Kunz D, Hummel R, Pfeilschifter J. Molecular cloning of the rat inducible nitric oxide synthase gene promoter. Biochem Biophys Res Commun 1996; 223:752–756.
13. Lin AW, Chang CC, McCormick CC. Molecular cloning and expression of an avian macrophage nitric–oxide synthase cDNA and the analysis of the genomic 5′-flanking region. J Biol Chem 1996; 271:11911–11919.
14. Zhang X, Laubach VE, Alley EW, Edwards KA, Sherman PA, Russell SW, Murphy WJ. Transcriptional basis for hyporesponsiveness of the human inducible nitric oxide synthase gene to lipopolysaccharide/interferon-γ. J Leukoc Biol 1996; 59:575–585.
15. Bredt DS, Glatt CE, Hwang PM, Fotuhi M, Dawson TM, Snyder SH. Nitric oxide synthase protein and mRNA are discretely localized in neuronal populations of the mammalian CNS together with NADPH diaphorase. Neuron 1991; 7:615–624.
16. Pollock JS, Förstermann U, Mitchell JA, Warner TD, Schmidt HHHW, Nakane M, Murad F. Purification and characterization of particulate endothelium-derived relaxing factor synthase from cultured and native bovine aortic endothelial cells. Proc Natl Acad Sci USA 1991; 88:10480–10484.

17. Aoki E, Takeuchi IK, Shoji R, Semba R. Localization of nitric oxide-related substances in the peripheral nervous tissues. Brain Res 1993; 620:142–145.
18. Nakane M, Schmidt HHHW, Pollock JS, Förstermann U, Murad F. Cloned human brain nitric oxide synthase is highly expressed in skeletal muscle. FEBS Lett 1993; 316:175–180.
19. Schmidt HHHW, Warner TD, Ishii K, Sheng H, Murad F. Insulin secretion from pancreatic B cells caused by L-arginine-derived nitrogen oxides. Science 1992; 255:721–723.
20. Schmidt HHHW, Gagne GD, Nakane M, Pollock JS, Miller MF, Murad F. Mapping of neural nitric oxide synthase in the rat suggests frequent co-localization with NADPH diaphorase but not with soluble guanylyl cyclase, and novel paraneural functions for nitrinergic signal transduction. J Histochem Cytochem 1992; 40:1439–1456.
21. Minc-Golomb D, Schwartz JP. Expression of both constitutive and inducible nitric oxide synthases in neuronal and astrocyte cultures. Ann NY Acad Sci 1994; 738:462–467.
22. Heym C, Colombo-Benckmann M, Mayer B. Immunohistochemical demonstration of the synthesis enzyme for nitric oxide and of comediators in neurons and chromaffin cells of the human adrenal medulla. Anat Anz 1994; 176:11–16.
23. Bacci S, Arbi-Riccardi R, Mayer B, Rumio C, Borghi-Cirri MB. Localization of nitric oxide synthase immunoreactivity in mast cells of human nasal mucosa. Histochemistry 1994; 102:89–92.
24. Riancho JA, Salas E, Zarrabeitia MT, Olmos JM, Amado JA, Fernández-Luna JL, González-Macías J. Expression and functional role of nitric oxide synthase in osteoblast-like cells. J Bone Miner Res 1995; 10:439–446.
25. Amin AR, Di Cesare PE, Vyas P, Attur M, Tzeng E, Billiar TR, Stuchin SA, Abramson SB. The expression and regulation of nitric oxide synthase in human osteoarthritis-affected chondrocytes: evidence for up-regulated neuronal nitric oxide synthase. J Exp Med 1995; 182:2097–2102.
26. Park CS, Gianotti C, Park R, Krishna G. Neuronal isoform of nitric oxide synthase is expressed at low levels in human retina. Cell Mol Neurobiol 1996; 16:499–515.
27. Brandi ML, Hukkanen M, Umeda T, Moradi-Bidhendi N, Bianchi S, Gross SS, Polak JM, MacIntyre I. Bidirectional regulation of osteoclast function by nitric oxide synthase isoforms. Proc Natl Acad Sci USA 1995; 92:2954–2958.
28. Tracey WR, Pollock JS, Murad F, Nakane M, Förstermann U. Identification of an endothelial-like type III NO synthase in LLC-PK_1 kidney epithelial cells. Am J Physiol 1994; 266:C22–C28.
29. North AJ, Moya FR, Mysore MR, Thomas VL, Wells LB, Wu LC, Shaul PW. Pulmonary endothelial nitric oxide synthase gene expression is decreased in a rat model of congenital diaphragmatic hernia. Am J Respir Cell Mol Biol 1995; 13:676–682.
30. Xue C, Botkin SJ, Johns RA. Localization of endothelial NOS at the basal microtubule membrane in ciliated epithelium of rat lung. J Histochem Cytochem 1996; 44:463–471.
31. Zini A, O'Bryan MK, Magid MS, Schlegel PN. Immunohistochemical localization of endothelial nitric oxide synthase in human testis, epididymis, and vas def-

erens suggests a possible role for nitric oxide in spermatogenesis, sperm maturation, and programmed cell death. Biol Reprod 1996; 55:935–941.
32. O'Dell TJ, Huang PL, Dawson TM, Dinerman JL, Snyder SH, Kandel ER, Fishman MC. Endothelial NOS and the blockade of LTP by NOS inhibitors in mice lacking neuronal NOS. Science 1994; 265:542–546.
33. Chen LY, Mehta JL. Further evidence of the presence of constitutive and inducible nitric oxide synthase isoforms in human platelets. J Cardiovasc Pharmacol 1996; 27:154–158.
34. Dinerman JL, Dawson TM, Schell MJ, Snowman A, Snyder SH. Endothelial nitric oxide synthase localized to hippocampal pyramidal cells: implications for synaptic plasticity. Proc Natl Acad Sci USA 1994; 91:4214–4218.
35. Balligand J-L, Kobzik L, Han X, Kaye DM, Belhassen L, O'Hara DS, Kelly RA, Smith TW, Michel T. Nitric oxide-dependent parasympathetic signaling is due to activation of constitutive endothelial (type III) nitric oxide synthase in cardiac myocytes. J Biol Chem 1995; 270:14582–14586.
36. Ribiere C, Jaubert AM, Gaudiot N, Sabourault D, Marcus ML, Boucher JL, Denis-Henriot D, Giudicelli Y. White adipose tissue nitric oxide synthase: a potential source for NO production. Biochem Biophys Res Commun 1996; 222:706–712.
37. Eis ALW, Brockman DE, Pollock JS, Myatt L. Immunohistochemical localization of endothelial nitric oxide synthase in human villous and extravillous trophoblast populations and expression during syncytiotrophoblast formation in vitro. Placenta 1995; 16:113–126.
38. Weiner CP, Lizasoain I, Baylis SA, Knowles RG, Charles IG, Moncada S. Induction of calcium-dependent nitric oxide synthases by sex hormones. Proc Natl Acad Sci USA 1994; 91:5212–5216.
39. Wang R, Ghahary A, Shen YJ, Scott PG, Tredget EE. Human dermal fibroblasts produce nitric oxide and express both constitutive and inducible nitric oxide synthase isoforms. J Invest Dermatol 1996; 106:419–427.
40. Liu SF, Adcock IM, Old RW, Barnes PJ, Evans TW. Differential regulation of the constitutive and inducible nitric oxide synthase mRNA by lipopolysaccharide treatment in vivo in the rat. Crit Care Med 1996; 24:1219–1225.
41. Lu J-L, Schmiege LM III, Kuo L, Liao JC. Downregulation of endothelial constitutive nitric oxide synthase expression by lipopolysaccharide. Biochem Biophys Res Commun 1996; 225:1–5.
42. Lee S, Barbanel G, Rivier C. Systemic endotoxin increases steady-state gene expression of hypothalamic nitric oxide synthase: comparison with corticotropin-releasing factor and vasopressin gene transcripts. Brain Res 1995; 705:136–148.
43. MacNaul KL, Hutchinson NI. Differential expression of iNOS and cNOS mRNA in human vascular smooth muscle cells and endothelial cells under normal and inflammatory conditions. Biochem Biophys Res Commun 1993; 196:1330–1334.
44. Nishida K, Harrison DG, Navas JP, Fisher AA, Dockery SP, Uematsu M, Nerem RM, Alexander RW, Murphy TJ. Molecular cloning and characterization of the constitutive bovine aortic endothelial cell nitric oxide synthase. J Clin Invest 1992; 90:2092–2096.

45. Yoshizumi M, Perrella MA, Burnett JC, Jr, Lee M-E. Tumor necrosis factor downregulates an endothelial nitric oxide synthase mRNA by shortening its half-life. Circ Res 1993; 73:205–209.
46. Inoue N, Venema RC, Sayegh HS, Ohara Y, Murphy TJ, Harrison DG. Molecular regulation of the bovine endothelial cell nitric oxide synthase by transforming growth factor-β_1. Arterioscler Thromb Vasc Biol 1995; 15:1255–1261.
47. Kostyk SK, Kourembanas S, Wheeler EL, Medeiros D, McQuillan LP, D'Amore PA, Braunhut SJ. Basic fibroblast growth factor increases nitric oxide synthase production in bovine endothelial cells. Am J Physiol 1995; 269:H1583–H1589.
48. Figueroa JP, Massmann GA. Estrogen increases nitric oxide synthase activity in the uterus of nonpregnant sheep. Am J Obstet Gynecol 1995; 173:1539–1545.
49. Hishikawa K, Nakaki T, Marumo T, Suzuki H, Kato R, Saruta T. Up-regulation of nitric oxide synthase by estradiol in human aortic endothelial cells. FEBS Lett 1995; 360:291–293.
50. Goetz RM, Morano I, Calovini T, Studer R, Holtz J. Increased expression of endothelial constitutive nitric oxide synthase in rat aorta during pregnancy. Biochem Biophys Res Commun 1994; 205:905–910.
51. Van Voorhis BJ, Moore K, Strijbos PJLM, Nelson S, Baylis SA, Grzybicki D, Weiner CP. Expression and localization of inducible and endothelial nitric oxide synthase in the rat ovary. Effects of gonadotropin stimulation in vivo. J Clin Invest 1995; 96:2719–2726.
52. Ceccatelli S, Hulting A-L, Zhang X, Gustafsson L, Villar M, Hökfelt T. Nitric oxide synthase in the rat anterior pituitary gland and the role of nitric oxide in regulation of luteinizing hormone secretion. Proc Natl Acad Sci USA 1993; 90:11292–11296.
53. Ueta Y, Levy A, Chowdrey HS, Lightman SL. Hypothalamic nitric oxide synthase gene expression is regulated by thyroid hormones. Endocrinology 1995; 136:4182–4187.
54. Benrath J, Eschenfelder C, Zimmerman M, Gillardon F. Calcitonin gene-related peptide, substance P and nitric oxide are involved in cutaneous inflammation following ultraviolet irradiation. Eur J Pharmacol 1995; 293:87–96.
55. North AJ, Lau KS, Brannon TS, Wu LC, Wells LB, German Z, Shaul PW. Oxygen upregulates nitric oxide synthase gene expression in ovine fetal pulmonary artery endothelial cells. Am J Physiol 1996; 270:L643–L649.
56. Liao JK, Zulueta JJ, Yu F-S, Peng H-B, Cote CG, Hassoun PM. Regulation of bovine endothelial constitutive nitric oxide synthase by oxygen. J Clin Invest 1995; 96:2661–2666.
57. McQuillan LP, Leung GK, Marsden PA, Kostyk SK, Kourembanas S. Hypoxia inhibits expression of eNOS via transcriptional and posttranscriptional mechanisms. Am J Physiol 1994; 267:H1921–H1927.
58. Phelan MW, Faller DV. Hypoxia decreases constitutive nitric oxide synthase transcript and protein in cultured endothelial cells. J Cell Physiol 1996; 167:469–476.
59. Arnet UA, McMillan A, Dinerman JL, Ballermann B, Lowenstein CJ. Regulation of endothelial nitric-oxide synthase during hypoxia. J Biol Chem 1996; 271:15069–15073.

60. Le Cras TD, Xue C, Rengasamy A, Johns RA. Chronic hypoxia upregulates endothelial and inducible NO synthase gene and protein expression in rat lung. Am J Physiol 1996; 270:L164–L170.
61. Bosse HM, Böhm R, Resch S, Bachmann S. Parallel regulation of constitutive NO synthase and renin at JGA of rat kidney under various stimuli. Am J Physiol 1995; 269:F793–F805.
62. Yokoyama M, Hirata K-i, Miyake R, Akita H, Ishikawa Y, Fukuzaki H. Lysophosphatidylcholine: essential role in the inhibition of endothelium-dependent vasorelaxation by oxidized low density lipoprotein. Biochem Biophys Res Commun 1990; 168:301–308.
63. Liao JK, Shin WS, Lee WY, Clark SL. Oxidized low-density lipoprotein decreases the expression of endothelial nitric oxide synthase. J Biol Chem 1995; 270:319–324.
64. Zembowicz A, Tang J-l, Wu KK. Transcriptional induction of endothelial nitric oxide synthase type III by lysophosphatidylcholine. J Biol Chem 1995; 270:17006–17010.
65. Baader SL, Schilling K. Glutamate receptors mediate dynamic regulation of nitric oxide synthase expression in cerebellar granule cells. J Neurosci 1996; 16:1440–1449.
66. Belhassen L, Kelly RA, Smith TW, Balligand J-L. Nitric oxide synthase (NOS3) and contractile responsiveness to adrenergic and cholinergic agonists in the heart. Regulation of NOS3 transcription in vitro and in vivo by cyclic adenosine monophosphate in rat cardiac myocytes. J Clin Invest 1996; 97:1908–1915.
67. Ohara Y, Sayegh HS, Yamin JJ, Harrison DG. Regulation of endothelial constitutive nitric oxide synthase by protein kinase C. Hypertension 1995; 25:415–420.
68. Uematsu M, Ohara Y, Navas JP, Nishida K, Murphy TJ, Alexander RW, Nerem RM, Harrison DG. Regulation of endothelial cell nitric oxide synthase mRNA expression by shear stress. Am J Physiol 1995; 269:C1371–C1378.
69. Awolesi MA, Sessa WC, Sumpio BE. Cyclic strain upregulates nitric oxide synthase in cultured bovine aortic endothelial cells. J Clin Invest 1995; 96:1449–1454.
70. Norris M, Morigi M, Donadelli R, Aiello S, Foppolo M, Todeschini M, Orisio S, Remuzzi G, Remuzzi A. Nitric oxide synthesis by cultured endothelial cells is modulated by flow conditions. Circ Res 1995; 76:536–543.
71. Topper JN, Cai JX, Falb D, Gimbrone MA, Jr. Identification of vascular endothelial genes differentially responsive to fluid mechanical stimuli: cyclooxygenase-2, manganese superoxide dismutase, and endothelial cell nitric oxide synthase are selectively up-regulated by steady laminar shear stress. Proc Natl Acad Sci USA 1996; 93:10417–10422.
72. Nadaud S, Philippe M, Arnal J-F, Michel J-B, Soubrier F. Sustained increase in aortic endothelial nitric oxide synthase expression in vivo in a model of chronic high blood flow. Circ Res 1996; 79:857–863.
73. Spessert R, Layes E, Schollmayer A, Reuss S, Vollrath L. In the rat pineal gland, but not in the suprachiasmatic nucleus, the amount of constitutive neuronal nitric oxide synthase is regulated by environmental lighting conditions. Biochem Biophys Res Commun 1995; 212:70–76.

74. Kishimoto J, Tsuchiya T, Emson PC, Nakayama Y. Immobilization-induced stress activates neuronal nitric oxide synthase (nNOS) mRNA and protein in hypothalamic-pituitary-adrenal axis in rats. Brain Res 1996; 720:159–171.
75. Terada H, Nagai T, Kimura H, Kitahama K, Okada S. Distribution of nitric oxide synthase-immunoreactive neurons in fetal rat brains at embryonic day 15 and day 19. J Chem Neuroanat 1996; 10:273–278.
76. Ogilvie P, Schilling K, Billingsley ML, Schmidt HHHW. Induction and variants of neuronal nitric oxide synthase type I during synaptogenesis. FASEB J 1995; 9:799–806.
77. North AJ, Star RA, Brannon TS, Ujiie K, Wells LB, Lowenstein CJ, Snyder SH, Shaul PW. Nitric oxide synthase type I and type III gene expression are developmentally regulated in rat lung. Am J Physiol 1994; 266:L635–L641.
78. Herdegen T, Brecht S, Mayer B, Leah J, Kummer W, Bravo R, Zimmermann M. Long-lasting expression of JUN and KROX transcription factors and nitric oxide synthase in intrinsic neurons of the rat brain following axotomy. J Neurosci 1993; 13:4130–4145.
79. Wu W, Liuzzi FJ, Schinco FP, Depto AS, Li Y, Mong JA, Dawson TM, Snyder SH. Neuronal nitric oxide synthase is induced in spinal neurons by traumatic injury. Neuroscience 1994; 61:719–726.
80. Kristensson K, Aldskogius M, Peng Z-C, Olsson T, Aldskogius H, Bentivoglio M. Co-induction of neuronal interferon-γ and nitric oxide synthase in rat motor neurons after axotomy: a role in nerve repair or death? J Neurocytol 1994; 23:453–459.
81. Vizzard MA, Erdman SL, de Groat WC. Increased expression of neuronal nitric oxide synthase (NOS) in visceral neurons after nerve injury. J Neurosci 1995; 15:4033–4045.
82. Wu Y, Li Y, Liu H, Wu W. Induction of nitric oxide synthase and motoneuron death in newborn and early postnatal rats following spinal root avulsion. Neurosci Lett 1995; 194:109–112.
83. Wu W. Potential roles of gene expression change in adult rat spinal motoneurons following axonal injury: a comparison among c-jun, low-affinity nerve growth factor receptor (LNGFR), and nitric oxide synthase (NOS). Exp Neurol 1996; 141:190–200.
84. Verge VMK, Xu Z, Xu X-J, Wiesenfeld-Hallin Z, Hökfelt T. Marked increase in nitric oxide synthase mRNA in rat dorsal root ganglia after peripheral axotomy: in situ hybridization and functional studies. Proc Natl Acad Sci USA 1992; 89:11617–11621.
85. O'Hearn E, Zhang P, Molliver ME. Excitotoxic insult due to ibogaine leads to delayed induction of neuronal NOS in Purkinje cells. Neuroreport 1995; 6:1611–1616.
86. Vizzard MA, Erdman SL, de Groat WC. Increased expression of neuronal nitric oxide synthase in dorsal root ganglion neurons after systemic capsaicin administration. Neuroscience 1995; 67:1–5.
87. Krukoff TL, Gehlen F, Ganten D, Wagner J. Gene expression of brain nitric oxide synthase and soluble guanylyl cyclase in hypothalamus and medulla of two-kidney, one clip hypertensive rats. Hypertension 1995; 26:171–176.

88. Kadowaki K, Kishimoto J, Leng G, Emson PC. Up-regulation of nitric oxide synthase (NOS) gene expression together with NOS activity in the rat hypothalamo-hypophysial system after chronic salt loading: evidence of a neuromodulatory role of nitric oxide in arginine vasopressin and oxytocin secretion. Endocrinology 1994; 134:1011–1017.
89. Villar MJ, Ceccatelli S, Rönnqvist M, Hökfelt T. Nitric oxide synthase increases in hypothalamic magnocellular neurons after salt loading in the rat. An immunohistochemical and in situ hybridization study. Brain Res 1994; 644:273–281.
90. Singh I, Grams M, Wang WH, Yang T, Killen P, Smart A, Schnermann J, Briggs JP. Coordinate regulation of renal expression of nitric oxide synthase, renin, and angiotensinogen mRNA by dietary salt. Am J Physiol 1996; 270:F1027–F1037.
91. Ueta Y, Levy A, Chowdrey HS, Lightman SL. Water deprivation in the rat induces nitric oxide synthase (NOS) gene expression in the hypothalamic paraventricular and supraoptic nuclei. Neurosci Res 1995; 23:317–319.
92. O'Shea RD, Gundlach AL. Food or water deprivation modulate nitric oxide synthase (NOS) activity and gene expression in rat hypothalamic neurones: correlation with neurosecretory activity? J Neuroendocrinol 1996; 8:417–425.
93. Ueta Y, Levy A, Chowdrey HS, Lightman SL. Inhibition of hypothalamic nitric oxide synthase gene expression in the rat paraventricular nucleus by food deprivation is independent of serotonin depletion. J Neuroendocrinol 1995; 7:861–865.
94. Ceccatelli S, Eriksson M. The effect of lactation on nitric oxide synthase gene expression. Brain Res 1993; 625:177–179.
95. Sessa WC, Pritchard K, Seyedi N, Wang J, Hintze TH. Chronic exercise in dogs increases coronary vascular nitric oxide production and endothelial cell nitric oxide synthase gene expression. Circ Res 1994; 74:349–353.
96. Martin PY, Xu DL, Niederberger M, Weigert A, Tsai P, St John J, Gines P, Schrier RW. Upregulation of endothelial constitutive NOS: a major role in the increased NO production in cirrhotic rats. Am J Physiol 1996; 270:F494–F499.
97. Tanoue K, Ohta M, Tarnawski AS, Wahlstrom KJ, Sugimachi K, Sarfeh IJ. Portal hypertension activates the nitric oxide synthase genes in the esophageal mucosa of rats. Gastroenterology 1996; 110:549–557.
98. Kanazawa K, Kawashima S, Mikami S, Miwa Y, Hirata K, Suematsu M, Hayashi Y, Itoh H, Yokoyama M. Endothelial constitutive nitric oxide synthase protein and mRNA increased in rabbit atherosclerotic aorta despite impaired endothelium-dependent vascular relaxation. Am J Pathol 1996; 148:1949–1956.
99. Yagihashi N, Nishida N, Seo HG, Taniguchi N, Yagihashi S. Expression of nitric oxide synthase in macula densa in streptozotocin diabetic rats. Diabetologia 1996; 39:793–799.
100. Patel KP, Zhang K, Zucker IH, Krukoff TL. Dccreased gene expression of neuronal nitric oxide synthase in hypothalamus and brainstem of rats in heart failure. Brain Res 1996; 734:109–115.
101. Tang J-L, Zembowicz A, Xu X-M, Wu KK. Role of Sp1 in transcriptional activation of human nitric oxide synthase type III gene. Biochem Biophys Res Commun 1995; 213:673–680.
102. Wariishi S, Miyahara K, Toda K, Ogoshi S, Doi Y, Ohnishi S, Mitsui Y, Yui Y,

Kawai C, Shizuta Y. A SP1 binding site in the GC-rich region is essential for a core promoter activity of the human endothelial nitric oxide synthase gene. Biochem Biophys Res Commun 1995; 216:729–735.

103. Zhang R, Min W, Sessa WC. Functional analysis of the human endothelial nitric oxide synthase promoter. Sp1 and GATA factors are necessary for basal transcription in endothelial cells. J Biol Chem 1995; 270:15320–15326.

104. Pugh BF, Tjian R. Mechanism of transcriptional activation by Sp1: evidence for coactivators. Cell 1990; 61:1187–1197.

105. Lee M-E, Temizer DH, Clifford JA, Quertermous T. Cloning of the GATA-binding protein that regulates endothelin-1 gene expression in endothelial cells. J Biol Chem 1991; 266:16188–16192.

106. Dorfman DM, Wilson DB, Bruns GAP, Orkin SH. Human transcription factor GATA-2. Evidence for regulation of preproendothelin-1 gene expression in endothelial cells. J Biol Chem 1992; 267:1279–1285.

107. Pan J, McEver RP. Characterization of the promoter for the human P-selectin gene. J Biol Chem 1993; 268:22600–22608.

108. Jahroudi N, Lynch DC. Endothelial-cell-specific regulation of von Willebrand factor gene expression. Mol Cell Biol 1994; 14:999–1008.

109. Fischer K-D, Haese A, Nowock J. Cooperation of GATA-1 and Sp1 can result in synergistic transcriptional activation or interference. J Biol Chem 1993; 268:23915–23923.

110. Lecointe N, Bernard O, Naert K, Joulin V, Larsen CJ, Romeo PH, Mathieu-Mahul D. GATA- and SP1-binding sites are required for the full activity of the tissue-specific promoter of the tal-1 gene. Oncogene 1994; 9:2623–2632.

111. Merika M, Orkin SH. Functional synergy and physical interactions of the erythroid transcription factor GATA-1 with the Krüppel family proteins Sp1 and EKLF. Mol Cell Biol 1995; 15:2437–2447.

112. Gregory RC, Taxman DJ, Seshasayee D, Kensinger MH, Bieker JJ, Wojchowski DM. Functional interaction of GATA1 with erythroid Krüppel-like factor and Sp1 at defined erythroid promoters. Blood 1996; 87:1793–1801.

113. Riccio A, Pedone PV, Lund LR, Olesen T, Olsen HS, Andreasen PA. Transforming growth factor β1-responsive element: closely associated binding sites for USF and CCAAT-binding transcription factor-nuclear factor I in the type 1 plasminogen activator inhibitor gene. Mol Cell Biol 1992; 12:1846–1855.

114. Xie J, Roddy P, Rife TK, Murad F, Young AP. Two closely linked but separable promoters for human neuronal nitric oxide synthase gene transcription. Proc Natl Acad Sci USA 1995; 92:1242–1246.

115. Deans Z, Dawson SJ, Xie J, Young AP, Wallace D, Latchman DS. Differential regulation of the two neuronal nitric-oxide synthase gene promoters by the Oct-2 transcription factor. J Biol Chem 1996; 271:32153–32158.

116. He X, Treacy MN, Simmons DM, Ingraham HA, Swanson LW, Rosenfeld MG. Expression of a large family of POU-domain regulatory genes in mammalian brain development. Nature 1989; 340:35–41.

117. Schöler HR, Hatzopoulos AK, Balling R, Suzuki N, Gruss P. A family of octamer-specific proteins present during mouse embryogenesis: evidence for germline-specific expression of an Oct factor. EMBO J 1989; 8:2543–2550.

118. Stuehr DJ, Kwon NS, Nathan CF. FAD and GSH participate in macrophage synthesis of nitric oxide. Biochem Biophys Res Commun 1990; 168:558–565.
119. Yui Y, Hattori R, Kosuga K, Eizawa H, Hiki K, Kawai C. Purification of nitric oxide synthase from rat macrophages. J Biol Chem 1991; 266:12544–12547.
120. Yui Y, Hattori R, Kosuga K, Eizawa H, Hiki K, Ohkawa S, Ohnishi K, Terao S, Kawai C. Calmodulin-independent nitric oxide synthase from rat polymorphonuclear neutrophils. J Biol Chem 1991; 266:3369–3371.
121. Huang J, Roby KF, Pace JL, Russell SW, Hunt JS. Cellular localization and hormonal regulation of inducible nitric oxide synthase in cycling mouse uterus. J Leukoc Biol 1995; 57:27–35.
122. Bandaletova T, Brouet I, Bartsch H, Sugimura T, Esumi H, Ohshima H. Immunohistochemical localization of an inducible form of nitric oxide synthase in various organs of rats treated with *Propionibacterium acnes* and lipopolysaccharide. APMIS 1993; 101:330–336.
123. Spolarics Z, Spitzer JJ, Wang JF, Xie J, Kolls J, Greenberg S. Alcohol administration attenuates LPS-induced expression of inducible nitric oxide synthase in Kupffer and hepatic endothelial cells. Biochem Biophys Res Commun 1993; 197:606–611.
124. Geller DA, Nussler AK, Di Silvio M, Lowenstein CJ, Shapiro RA, Wang SC, Simmons RL, Billiar TR. Cytokines, endotoxin, and glucocorticoids regulate the expression of inducible nitric oxide synthase in hepatocytes. Proc Natl Acad Sci USA 1993; 90:522–526.
125. Palmer RMJ, Hickery MS, Charles IG, Moncada S, Bayliss MT. Induction of nitric oxide synthase in human chondrocytes. Biochem Biophys Res Commun 1993; 193:398–405.
126. Nakayama DK, Geller DA, Lowenstein CJ, Chern HD, Davies P, Pitt BR, Simmons RL, Billiar TR. Cytokines and lipopolysaccharide induce nitric oxide synthase in cultured rat pulmonary artery smooth muscle. Am J Respir Cell Mol Biol 1992; 7:471–476.
127. Galea E, Feinstein DL, Reis DJ. Induction of calcium-independent nitric oxide synthase activity in primary rat glial cultures. Proc Natl Acad Sci USA 1992; 89:10945–10949.
128. Rockey DC, Chung JJ. Inducible nitric oxide synthase in rat hepatic lipocytes and the effect of nitric oxide on lipocyte contractility. J Clin Invest 1995; 95:1199–1206.
129. Yang X, Chowdhury N, Cai B, Brett J, Marboe C, Sciacca RR, Michler RE, Cannon PJ. Induction of myocardial nitric oxide synthase by cardiac allograft rejection. J Clin Invest 1994; 94:714–721.
130. Arany I, Brysk MM, Brysk H, Tyring SK. Regulation of inducible nitric oxide synthase mRNA levels by differentiation and cytokines in human keratinocytes. Biochem Biophys Res Commun 1996; 220:618–622.
131. Eizirik DL, Cagliero E, Björklund A, Welsh N. Interleukin-1b induces the expression of an isoform of nitric oxide synthase in insulin-producing cells, which is similar to that observed in activated macrophages. FEBS Lett 1992; 308:249–252.
132. Markewitz BA, Michael JR, Kohan DE. Cytokine-induced expression of a nitric oxide synthase in rat renal tubule cells. J Clin Invest 1993; 91:2138–2143.

133. Jin Y, Heck DE, DeGeorge G, Tian Y, Laskin JD. 5-Fluorouracil suppresses nitric oxide biosynthesis in colon carcinoma cells. Cancer Res 1996; 56:1978–1982.
134. Rosbe KW, Prazma J, Petrusz P, Mims W, Ball SS, Weissler MC. Immunohistochemical characterization of nitric oxide synthase activity in squamous cell carcinoma of the head and neck. Otolaryngol Head Neck Surg 1995; 113:541–549.
135. Gilbert RS, Herschman HR. "Macrophage" nitric oxide synthase is a glucocorticoid-inhibitable primary response gene in 3T3 cells. J Cell Physiol 1993; 157:128–132.
136. Mordan LJ, Burnett TS, Zhang L-X, Tom J, Cooney RV. Inhibitors of endogenous nitrogen oxide formation block the promotion of neoplastic transformation in C3H 10T1/2 fibroblasts. Carcinogenesis 1993; 14:1555–1559.
137. Stéphan J-P, Guillemois C, Jégou B, Bauché F. Nitric oxide production by Sertoli cells in response to cytokines and lipopolysaccharide. Biochem Biophys Res Commun 1995; 213:218–224.
138. Dong Z, Staroselsky AH, Qi X, Xie K, Fidler IJ. Inverse correlation between expression of inducible nitric oxide synthase activity and production of metastasis in K-1735 murine melanoma cells. Cancer Res 1994; 54:789–793.
139. Asano K, Chee CBE, Gaston B, Lilly CM, Gerard C, Drazen JM, Stamler JS. Constitutive and inducible nitric oxide synthase gene expression, regulation, and activity in human lung epithelial cells. Proc Natl Acad Sci USA 1994; 91:10089–10093.
140. Sato I, Kim Y, Himi T, Murota S. Induction of calcium-independent nitric oxide synthase activity in cultured cerebellar granule neurons. Neurosci Lett 1995; 184:145–148.
141. Luss H, Watkins SC, Freeswick PD, Imro AK, Nussler AK, Billiar TR, Simmons RL, del Nido PJ, McGowan FX Jr. Characterization of inducible nitric oxide synthase expression in endotoxemic rat cardiac myocytes in vivo and following cytokine exposure in vitro. J Mol Cell Cardiol 1995; 27:2015–2029.
142. Suschek C, Fehsel K, Kröncke KD, Sommer A, Kolb-Bachofen V. Primary cultures of rat islet capillary endothelial cells. Constitutive and cytokine-inducible macrophagelike nitric oxide synthases are expressed and activities regulated by glucose concentration. Am J Pathol 1994; 145:685–695.
143. Willis RA, Nussler AK, Fries KM, Geller DA, Phipps RP. Induction of nitric oxide synthase in subsets of murine pulmonary fibroblasts: effect on fibroblast interleukin-6 production. Clin Immunol Immunopathol 1994; 71:231–239.
144. Eizirik DL, Sandler S, Welsh N, Cetkovic-Cvrlje M, Nieman A, Geller DA, Pipeleers DG, Bendtzen K, Hellerström C. Cytokines suppress human islet function irrespective of their effects on nitric oxide generation. J Clin Invest 1994; 93:1968–1974.
145. Robbins RA, Springall DR, Warren JB, Kwon OJ, Buttery LDK, Wilson AJ, Adcock IM, Riveros-Moreno V, Moncada S, Polak J, Barnes PJ. Inducible nitric oxide synthase is increased in murine lung epithelial cells by cytokine stimulation. Biochem Biophys Res Commun 1994; 198:835–843.
146. Kutty RK, Kutty G, Hooks JJ, Wiggert B, Nagineni CN. Transforming growth factor-β inhibits the cytokine-mediated expression of the inducible nitric oxide

synthase mRNA in human retinal pigment epithelial cells. Biochem Biophys Res Commun 1995; 215:386–393.

147. Hukkanen M, Hughes FJ, Buttery LDK, Gross SS, Evans TJ, Seddon S, Riveros-Moreno V, MacIntyre I, Polak JM. Cytokine-stimulated expression of inducible nitric oxide synthase by mouse, rat, and human osteoblast-like cells and its functional role in osteoblast metabolic activity. Endocrinology 1995; 136:5445–5453.

148. Evans TJ, Buttery LDK, Carpenter A, Springall DR, Polak JM, Cohen J. Cytokine-treated human neutrophils contain inducible nitric oxide synthase that produces nitration of ingested bacteria. Proc Natl Acad Sci USA 1996; 93:9553–9558.

149. Laubach VE, Zhang CX, Russell SW, Murphy WJ, Sherman PA. Analysis of expression and promoter function of the human inducible nitric oxide synthase gene in DLD-1 cells and monkey hepatocytes. Biochim Biophys Acta 1997; 1351:287-295.

150. Chu SC, Wu H-P, Banks TC, Eissa NT, Moss J. Structural diversity in the 5′-untranslated region of cytokine-stimulated human inducible nitric oxide synthase mRNA. J Biol Chem 1995; 270:10625–10630.

151. Chizzonite R, Truitt T, Kilian PL, Stern AS, Nunes P, Parker KP, Kaffka KL, Chua AO, Lugg DK, Gubler U. Two high-affinity interleukin 1 receptors represent separate gene products. Proc Natl Acad Sci USA 1989; 86:8029–8033.

152. Zhang X, Alley EW, Russell SW, Morrison DC. Necessity and sufficiency of beta interferon for nitric oxide production in mouse peritoneal macrophages. Infect Immun 1994; 62:33–40.

153. Fujihara M, Ito N, Pace JL, Watanabe Y, Russell SW, Suzuki T. Role of endogenous interferon-β in lipopolysaccharide-triggered activation of the inducible nitric-oxide synthase gene in a mouse macrophage cell line, J774. J Biol Chem 1994; 269:12773–12778.

154. Yamada K, Otabe S, Inada C, Takane N, Nonaka K. Nitric oxide and nitric oxide synthase mRNA induction in mouse islet cells by interferon-γ plus tumor necrosis factor-α. Biochem Biophys Res Commun 1993; 197:22–27.

155. Corbett JA, Kwon G, Misko TP, Rodi CP, McDaniel ML. Tyrosine kinase involvement in IL-1-β-induced expression of iNOS by b-cells purified from islets of Langerhans. Am J Physiol 1994; 267:C48–C54.

156. Koide M, Kawahara Y, Nakayama I, Tsuda T, Yokoyama M. Cyclic AMP-elevating agents induce an inducible type of nitric oxide synthase in cultured vascular smooth muscle cells. Synergism with the induction elicited by inflammatory cytokines. J Biol Chem 1993; 268:24959–24966.

157. Iwashina M, Hirata Y, Imai T, Sato K, Marumo F. Molecular cloning of endothelial, inducible nitric oxide synthase gene from rat aortic endothelial cell. Eur J Biochem 1996; 237:668–673.

158. Liew FY, Li Y, Moss D, Parkinson C, Rogers MV, Moncada S. Resistance to *Leishmania major* infection correlates with the induction of nitric oxide synthase in murine macrophages. Eur J Immunol 1991; 21:3009–3014.

159. Lorsbach RB, Murphy WJ, Lowenstein CJ, Snyder SH, Russell SW. Expression of the nitric oxide synthase gene in mouse macrophages activated for tumor cell

killing. Molecular basis for the synergy between interferon-γ and lipopolysaccharide. J Biol Chem 1993; 268:1908–1913.

160. Nunokawa Y, Ishida N, Tanaka S. Cloning of inducible nitric oxide synthase in rat vascular smooth muscle cells. Biochem Biophys Res Commun 1993; 191:89–94.

161. Koide M, Kawahara Y, Tsuda T, Yokoyama M. Cytokine-induced expression of an inducible type of nitric oxide synthase gene in cultured vascular smooth muscle cells. FEBS Lett 1993; 318:213–217.

162. Punjabi CJ, Laskin JD, Pendino KJ, Goller NL, Durham SK, Laskin DL. Production of nitric oxide by rat type II pneumocytes: increased expression of inducible nitric oxide synthase following inhalation of a pulmonary irritant. Am J Respir Cell Mol Biol 1994; 11:165–172.

163. Bereta J, Cohen MC, Bereta M. Stimulatory effect of ouabain on VCAM-1 and iNOS expression in murine endothelial cells: involvement of NF-κB. FEBS Lett 1995; 377:21–25.

164. Menegazzi M, Guerriero C, Carcereri de Prati A, Cardinale C, Suzuki H, Armato U. TPA and cycloheximide modulate the activation of NF–kB and the induction and stability of nitric oxide synthase transcript in primary neonatal rat hepatocytes. FEBS Lett 1996; 379:279–285.

165. Hoffmann G, Schobersberger W, Frede S, Pelzer L, Fandrey J, Wachter H, Fuchs D, Grote J. Neopterin activates transcription factor nuclear factor-κB in vascular smooth muscle cells. FEBS Lett 1996; 391:181–184.

166. Adcock IM, Brown CR, Kwon O, Barnes PJ. Oxidative stress induces NFkB DNA binding and inducible NOS mRNA in human epithelial cells. Biochem Biophys Res Commun 1994; 199:1518–1524.

167. Mustafa SB, Howard KM, Olson MS. Platelet-activating factor augments lipopolysaccharide-induced nitric oxide formation by rat Kupffer cells. Hepatology 1996; 23:1622–1630.

168. Chen F, Sun S-C, Kuh DC, Gaydos LJ, Demers LM. Essential role of NF-κB activation in silica–induced inflammatory mediator production in macrophages. Biochem Biophys Res Commun 1995; 214:985–992.

169. Melillo G, Cox GW, Radzioch D, Varesio L. Picolinic acid, a catabolite of L-tryptophan, is a costimulus for the induction of reactive nitrogen intermediate production in murine macrophages. J Immunol 1993; 150:4031–4040.

170. LaPointe MC, Sitkins JR. Mechanisms of interleukin-1β regulation of nitric oxide synthase in cardiac myocytes. Hypertension 1996; 27:709–714.

171. Gaillard T, Mülsch A, Klein H, Decker K. Regulation by prostaglandin E_2 of cytokine-elicited nitric oxide synthesis in rat liver macrophages. Biol Chem Hoppe-Seyler 1992; 373:897–902.

172. Sowa G, Przewlocki R. cAMP analogues and cholera toxin stimulate the accumulation of nitrite in rat peritoneal macrophage cultures. Eur J Pharmacol 1994; 266:125–129.

173. Bulut V, Severn A, Liew FY. Nitric oxide production by murine macrophages is inhibited by prolonged elevation of cyclic AMP. Biochem Biophys Res Commun 1993; 195:1134–1138.

174. Harbrecht BG, Kim YM, Wirant EM, Shapiro RA, Billiar TR. PGE_2 and LTB_4 inhibit cytokine-stimulated nitric oxide synthase type 2 expression in isolated rat hepatocytes. Prostaglandins 1996; 52:103–116.
175. McKenna TM, Li S, Tao S. PKC mediates LPS- and phorbol-induced cardiac cell nitric oxide synthase activity and hypocontractility. Am J Physiol 1995; 269:H1891–H1898.
176. Jun C-D, Choi B-M, Hoon-Ryu, Um J-Y, Kwak H-J, Lee B-S, Paik S-G, Kim H-M, Chung H-T. Synergistic cooperation between phorbol ester and IFN-γ for induction of nitric oxide synthesis in murine peritoneal macrophages. J Immunol 1994; 153:3684–3690.
177. Park YC, Jun CD, Kang HS, Kim HD, Kim HM, Chung HT. Role of intracellular calcium as a priming signal for the induction of nitric oxide synthesis in murine peritoneal macrophages. Immunology 1996; 87:296–302.
178. Fujihara M, Connolly N, Ito N, Suzuki T. Properties of protein kinase C isoforms (βII, ε, and ζ) in a macrophage cell line (J774) and their roles in LPS-induced nitric oxide production. J Immunol 1994; 152:1898–1906.
179. Nakayama I, Kawahara Y, Tsuda T, Okuda M, Yokoyama M. Angiotensin II inhibits cytokine-stimulated inducible nitric oxide synthase expression in vascular smooth muscle cells. J Biol Chem 1994; 269:11628–11633.
180. Geng Y-j, Wu Q, Hansson GK. Protein kinase C activation inhibits cytokine-induced nitric oxide synthesis in vascular smooth muscle cells. Biochim Biophys Acta 1994; 1223:125–132.
181. Mühl H, Pfeilschifter J. Possible role of protein kinase C-ε isoenzyme in inhibition of interleukin 1β induction of nitric oxide synthase in rat renal mesangial cells. Biochem J 1994; 303:607–612.
182. Förstermann U, Schmidt HHHW, Kohlhaas KL, Murad F. Induced RAW 264.7 macrophages express soluble and particulate nitric oxide synthase: inhibition by transforming growth factor-β. Eur J Pharmacol 1992; 225:161–165.
183. Cunha FQ, Moncada S, Liew FY. Interleukin-10 (IL-10) inhibits the induction of nitric oxide synthase by interferon-γ in murine macrophages. Biochem Biophys Res Commun 1992; 182:1155–1159.
184. McCall TB, Palmer RM, Moncada S. Interleukin-8 inhibits the induction of nitric oxide synthase in rat peritoneal neutrophils. Biochem Biophys Res Commun 1992; 186:680–685.
185. Sunyer T, Rothe L, Jiang X, Osdoby P, Collin-Osdoby P. Proinflammatory agents, IL-8 and IL-10, upregulate inducible nitric oxide synthase expression and nitric oxide production in avian osteoclast-like cells. J Cell Biochem 1996; 60:469–483.
186. Xie Q, Whisnant R, Nathan C. Promoter of the mouse gene encoding calcium-independent nitric oxide synthase confers inducibility by interferon g and bacterial lipopolysaccharide. J Exp Med 1993; 177:1779–1784.
187. Alley EW, Murphy WJ, Russell SW. A classical enhancer element responsive to both lipopolysaccharide and interferon-γ augments induction of the iNOS gene in mouse macrophages. Gene 1995; 158:247–251.
188. Spink J, Cohen J, Evans TJ. The cytokine responsive vascular smooth muscle cell

enhancer of inducible nitric oxide synthase. Activation by nuclear factor-κB. J Biol Chem 1995; 270:29541–29547.
189. Perrella MA, Patterson C, Tan L, Yet S-F, Hsieh C-M, Yoshizumi M, Lee M-E. Suppression of interleukin-1β-induced nitric-oxide synthase promoter/enhancer activity by transforming growth factor-β1 in vascular smooth muscle cells. Evidence for mechanisms other than NF-κB. J Biol Chem 1996; 271:13776–13780.
190. Amoah-Apraku B, Chandler LJ, Harrison JK, Tang S-S, Ingelfinger JR, Guzman NJ. NF-κB and transcriptional control of renal epithelial-inducible nitric oxide synthase. Kidney Int 1995; 48:674–682.
191. Fujisawa H, Ogura T, Hokari A, Weisz A, Yamashita J, Esumi H. Inducible nitric oxide synthase in a human glioblastoma cell line. J Neurochem 1995; 64:85–91.
192. Sharma K, Danoff TM, DePiero A, Ziyadeh FN. Enhanced expression of inducible nitric oxide synthase in murine macrophages and glomerular mesangial cells by elevated glucose levels: possible mediation via protein kinase C. Biochem Biophys Res Commun 1995; 207:80–88.
193. Kleinert H, Euchenhofer C, Ihrig-Biedert I, Förstermann U. Glucocorticoids inhibit the induction of nitric oxide synthase II by down-regulating cytokine-induced activity of transcription factor nuclear factor-κB. Mol Pharmacol 1996; 49:15–21.
194. Baeuerle PA, Baltimore D. IκB: a specific inhibitor of the NF-κB transcription factor. Science 1988; 242:540–546.
195. Lin Y-C, Brown K, Siebenlist U. Activation of NF-κB requires proteolysis of the inhibitor IκB-α: signal-induced phosphorylation of IκB-a alone does not release active NF-κB. Proc Natl Acad Sci USA 1995; 92:552–556.
196. DiDonato JA, Mercurio F, Karin M. Phosphorylation of IkBa precedes but is not sufficient for its dissociation from NF-κB. Mol Cell Biol 1995; 15:1302–1311.
197. Palombella VJ, Rando OJ, Goldberg AL, Maniatis T. The ubiquitin-proteasome pathway is required for processing the NF-κB1 precursor protein and the activation of NF-κB. Cell 1994; 78:773–785.
198. Grilli M, Chiu JJ-S, Lenardo MJ. NF-κB and Rel: participants in a multiform transcriptional regulatory system. Int Rev Cytol 1993; 143:1–62.
199. Xie Q, Kashiwabara Y, Nathan C. Role of transcription factor NF-κB/Rel in induction of nitric oxide synthase. J Biol Chem 1994; 269:4705–4708.
200. Murphy WJ, Muroi M, Zhang CX, Suzuki T, Russell SW. Both basal and enhancer κB elements are required for full induction of the mouse inducible nitric oxide synthase gene. J Endotoxin Res 1996; 3:381–393.
201. Goldring CEP, Narayanan R, Lagadec P, Jeannin J-F. Transcriptional inhibition of the inducible nitric oxide synthase gene by competitive binding of NF-κB/Rel proteins. Biochem Biophys Res Commun 1995; 209:73–79.
202. Kunz D, Walker G, Eberhardt W, Nitsch D, Pfeilschifter J. Interleukin 1β-induced expression of nitric oxide synthase in rat renal mesangial cells is suppressed by cyclosporin A. Biochem Biophys Res Commun 1995; 216:438–446.
203. Nakayama K, Shimizu H, Mitomo K, Watanabe T, Okamoto S-I, Yamamoto K-I. A lymphoid cell-specific nuclear factor containing c-Rel-like proteins preferentially interacts with interleukin-6 κB-related motifs whose activities are repressed in lymphoid cells. Mol Cell Biol 1992; 12:1736–1746.

204. Matsusaka T, Fujikawa K, Nishio Y, Mukaida N, Matsushima K, Kishimoto T, Akira S. Transcription factors NF-IL6 and NF-κB synergistically activate transcription of the inflammatory cytokines, interleukin 6 and interleukin 8. Proc Natl Acad Sci USA 1993; 90:10193–10197.
205. Grigoriadis G, Zhan YF, Grumont RJ, Metcalf D, Handman E, Cheers C, Gerondakis S. The Rel subunit of NF-κB-like transcription factors is a positive and negative regulator of macrophage gene expression: distinct roles for Rel in different macrophage populations. EMBO J 1996; 15:7099–7107.
206. Goldring CEP, Reveneau S, Algarté M, Jeannin JF. In vivo footprinting of the mouse inducible nitric oxide synthase gene: inducible protein occupation of numerous sites including Oct and NF-IL6. Nucleic Acids Res 1996; 24:1682–1687.
207. Kleinert H, Euchenhofer C, Ihrig-Biedert I, Förstermann U. In murine 3T3 fibroblasts, different second messenger pathways resulting in the induction of NO synthase II (iNOS) converge in the activation of transcription factor NF-κB. J Biol Chem 1996; 271:6039–6044.
208. Busam K, Gieringer C, Freudenberg M, Hohmann H-P. *Staphylococcus aureus* and derived exotoxins induce nuclear factor κB-like activity in murine bone marrow macrophages. Infect Immun 1992; 60:2008–2015.
209. Muroi M, Muroi Y, Suzuki T. The binding of immobilized IgG2a to Fcγ2a receptor activates NF-κB via reactive oxygen intermediates and tumor necrosis factor-α. J Biol Chem 1994; 269:30561–30568.
210. Lin J-Y, Seguin R, Keller K, Chadee K. Tumor necrosis factor alpha augments nitric oxide-dependent macrophage cytotoxicity against *Entamoeba histolytica* by enhanced expression of the nitric oxide synthase gene. Infect Immun 1994; 62:1534–1541.
211. Weisz A, Oguchi S, Cicatiello L, Esumi H. Dual mechanism for the control of inducible-type NO synthase gene expression in macrophages during activation by interferon-γ and bacterial lipopolysaccharide. Transcriptional and post-transcriptional regulation. J Biol Chem 1994; 269:8324–8333.
212. Duval DL, Miller DR, Collier J, Billings RE. Characterization of hepatic nitric oxide synthase: identification as the cytokine-inducible form primarily regulated by oxidants. Mol Pharmacol 1996; 50:277–284.
213. Lawrence R, Chang L-J, Siebenlist U, Bressler P, Sonenshein GE. Vascular smooth muscle cells express a constitutive NF-κB-like activity. J Biol Chem 1994; 269:28913–28918.
214. Bellas RE, Lee JS, Sonenshein GE. Expression of a constitutive NF-κB-like activity is essential for proliferation of cultured bovine vascular smooth muscle cells. J Clin Invest 1995; 96:2521–2527.
215. Shin WS, Hong Y-H, Peng H-B, De Caterina R, Libby P, Liao JK. Nitric oxide attenuates vascular smooth muscle cell activation by interferon-γ. The role of constitutive NF-κB activity. J Biol Chem 1996; 271:11317–11324.
216. Kamijo R, Harada H, Matsuyama T, Bosland M, Gerecitano J, Shapiro D, Le J, Koh SI, Kimura T, Green SJ, Mak TW, Taniguchi T, Vilček J. Requirement for transcription factor IRF-1 in NO synthase induction in macrophages. Science 1994; 263:1612–1615.

217. Williams BRG. Transcriptional regulation of interferon-stimulated genes. Eur J Biochem 1991; 200:1–11.
218. Imam AMA, Ackrill AM, Dale TC, Kerr IM, Stark GR. Transcription factors induced by interferons a and g. Nucleic Acids Res 1990; 18:6573–6580.
219. Harada H, Willison K, Sakakibara J, Miyamoto M, Fujita T, Taniguchi T. Absence of the type I IFN system in EC cells: transcriptional activator (IRF-1) and repressor (IRF-2) genes are developmentally regulated. Cell 1990; 63:303–312.
220. Harada H, Fujita T, Miyamoto M, Kimura Y, Maruyama M, Furia A, Miyata T, Taniguchi T. Structurally similar but functionally distinct factors, IRF-1 and IRF-2, bind to the same regulatory elements of IFN and IFN-inducible genes. Cell 1989; 58:729–739.
221. Geller DA, Nguyen D, Shapiro RA, Nussler A, Di Silvio M, Freeswick P, Wang SC, Tweardy DJ, Simmons RL, Billiar TR. Cytokine induction of interferon regulatory factor-1 in hepatocytes. Surgery 1993; 114:235–242.
222. Neish AS, Read MA, Thanos D, Pine R, Maniatis T, Collins T. Endothelial interferon regulatory factor 1 cooperates with NF-κB as a transcriptional activator of vascular cell adhesion molecule 1. Mol Cell Biol 1995; 15:2558–2569.
223. Karmann K, Min W, Fanslow WC, Pober JS. Activation and homologous desensitization of human endothelial cells by CD40 ligand, tumor necrosis factor, and interleukin 1. J Exp Med 1996; 184:173–182.
224. Abdollahi A, Lord KA, Hoffman-Liebermann B, Liebermann DA. Interferon regulatory factor 1 is a myeloid differentiation primary response gene induced by interleukin 6 and leukemia inhibitory factor: role in growth inhibition. Cell Growth Differ 1991; 2:401–407.
225. Barber SA, Fultz MJ, Salkowski CA, Vogel SN. Differential expression of interferon regulatory factor 1 (IRF-1), IRF-2, and interferon consensus sequence binding protein genes in lipopolysaccharide (LPS)-responsive and LPS-hyporesponsive macrophages. Infect Immun 1995; 63:601–608.
226. Salkowski CA, Barber SA, Detore GR, Vogel SN. Differential dysregulation of nitric oxide production in macrophages with targeted disruptions in IFN regulatory factor-1 and -2 genes. J Immunol 1996; 156:3107–3110.
227. Martin E, Nathan C, Xie Q. Role of interferon regulatory factor 1 in induction of nitric oxide synthase. J Exp Med 1994; 180:977–984.
228. Gao J, Morrison DC, Parmely TJ, Russell SW, Murphy WJ. An interferon-γ-activated site (GAS) is necessary for full expression of the mouse iNOS gene in response to interferon-γ and LPS. J Biol Chem 1997; 272:1256–1260.
229. Horiuchi M, Koike G, Yamada T, Mukoyama M, Nakajima M, Dzau VJ. The growth-dependent expression of angiotensin II type 2 receptor is regulated by transcription factors interferon regulatory factor-1 and -2. J Biol Chem 1995; 270:20225–20230.
230. Kwon G, Corbett JA, McDaniel ML. Interleukin 1-induced Fos and Jun do not regulate inducible nitric oxide synthase in rat islets of Langerhans and RINm5F cells. Endocrinology 1996; 137:825–830.
231. Koide M, Kawahara Y, Tsuda T, Nakayama I, Yokoyama M. Expression of nitric oxide synthase by cytokines in vascular smooth muscle cells. Hypertension 1994; 23:I45–I48.

232. Durante W, Kroll MH, Orloff GJ, Cunningham JM, Scott-Burden T, Vanhoutte PM, Schafer AI. Regulation of interleukin-1β-stimulated inducible nitric oxide synthase expression in cultured vascular smooth muscle cells by hemostatic proteins. Biochem Pharmacol 1996; 51:847–853.
233. Shultz PJ, Archer SL, Rosenberg ME. Inducible nitric oxide synthase mRNA and activity in glomerular mesangial cells. Kidney Int 1994; 46:683–689.
234. Imai T, Hirata Y, Kanno K, Marumo F. Induction of nitric oxide synthase by cyclic AMP in rat vascular smooth muscle cells. J Clin Invest 1994; 93:543–549.
235. Hattori Y, Gross SS. Cycloheximide induces nitric oxide synthase mRNA in vascular smooth muscle cells by prolonging mRNA lifetime. Biochem Mol Biol Int 1995; 37:439–445.
236. Akabane A, Kato I, Takasawa S, Unno M, Yonekura H, Yoshimoto T, Okamoto H. Nicotinamide inhibits IRF-1 mRNA induction and prevents IL-β-induced nitric oxide synthase expression in pancreatic β cells. Biochem Biophys Res Commun 1995; 215:524–530.
237. Hecker M, Preisse C, Klemm P, Busse R. Inhibition by antioxidants of nitric oxide synthase expression in murine macrophages: role of nuclear factor κB and interferon regulatory factor 1. Br J Pharmacol 1996; 118:2178–2184.
238. Lew DJ, Decker T, Strehlow I, Darnell JE. Overlapping elements in the guanylate-binding protein gene promoter mediate transcriptional induction by alpha and gamma interferons. Mol Cell Biol 1991; 11:182–191.
239. Decker T, Lew DJ, Mirkovitch J, Darnell JE Jr. Cytoplasmic activation of GAF, an IFN-γ-regulated DNA-binding factor. EMBO J 1991; 10:927–932.
240. Shuai K, Stark GR, Kerr IM, Darnell JE Jr. A single phosphotyrosine residue of Stat91 required for gene activation by interferon-γ. Science 1993; 261:1744–1746.
241. Shuai K, Schindler C, Prezioso VR, Darnell JE Jr. Activation of transcription by IFN-γ: tyrosine phosphorylation of a 91-κD DNA binding protein. Science 1992; 258:1808–1812.
242. Veals SA, Santa Maria T, Levy DE. Two domains of ISGF3γ that mediate protein-DNA and protein-protein interactions during transcription factor assembly contribute to DNA-binding specificity. Mol Cell Biol 1993; 13:196–206.
243. Decker T, Lew DJ, Darnell JE Jr. Two distinct alpha-interferon-dependent signal transduction pathways may contribute to activation of transcription of the guanylate-binding protein gene. Mol Cell Biol 1991; 11:5147–5153.
244. Silvennoinen O, Schindler C, Schlessinger J, Levy DE. Ras-independent growth factor signaling by transcription factor tyrosine phosphorylation. Science 1993; 261:1736–1739.
245. Seegert D, Strehlow I, Klose B, Levy DE, Schindler C, Decker T. A novel interferon-α-regulated, DNA-binding protein participates in the regulation of the IFP53/tryptophanyl-tRNA synthetase gene. J Biol Chem 1994; 269:8590–8595.
246. Pine R, Canova A, Schindler C. Tyrosine phosphorylated p91 binds to a single element in the ISGF2/IRF-1 promoter to mediate induction by IFNα and IFNγ, and is likely to autoregulate the p91 gene. EMBO J 1994; 13:158–167.
247. Rein T, Schreck R, Willenbrink W, Neubert WJ, Zorbas H, Bäuerle PA. Conserved high-affinity NF-κB binding site in the interferon regulatory factor-1 promoter is

not occupied by NF-κB in vivo and is transcriptionally inactive. J Inflamm 1995; 45:269–282.
248. Lakin ND, Palmer R, Lillycrop KA, Howard MK, Burke LC, Thomas NSB, Latchman DS. Down regulation of the octamer binding protein Oct-1 during growth arrest and differentiation of a neuronal cell line. Brain Res Mol Brain Res 1995; 28:47–54.
249. Ruvkun G, Finney M. Regulation of transcription and cell identity by POU domain proteins. Cell 1991; 64:475–478.
250. Cockerill PN, Klinken SP. Octamer-binding proteins in diverse hemopoietic cells. Mol Cell Biol 1990; 10:1293–1296.
251. Suzuki N, Peter W, Ciesiolka T, Gruss P, Schöler HR. Mouse Oct-1 contains a composite homeodomain of human Oct-1 and Oct-2. Nucleic Acids Res 1993; 21:245–252.
252. Wirth T, Priess A, Annweiler A, Zwilling S, Oeler B. Multiple Oct2 isoforms are generated by alternative splicing. Nucleic Acids Res 1991; 19:43–51.
253. Grenfell SJ, Latchman DS, Thomas NSB. CT-1 and Oct-2 DNA-binding site specificity is regulated in vitro by different kinases. Biochem J 1996; 315:889–893.
254. Herr W, Cleary MA. The POU domain: versatility in transcriptional regulation by a flexible two-in-one DNA-binding domain. Genes Dev 1995; 9:1679–1693.
255. Eaton S, Calame K. Multiple DNA sequence elements are necessary for the function of an immunoglobulin heavy chain promoter. Proc Natl Acad Sci USA 1987; 84:7634–7638.
256. Müller-Immerglück MM, Schaffner W, Matthias P. Transcription factor Oct-2A contains functionally redundant activating domains and works selectively from a promoter but not from a remote enhancer position in non-lymphoid (HeLa) cells. EMBO J 1990; 9:1625–1634.
257. Sawada T, Falk LA, Rao P, Murphy WJ, Pluznik DH. IL-6 induction of protein-DNA complexes via a novel regulatory region of the inducible nitric oxide synthase gene promoter: role of octamer binding proteins. J Immunol 1997; 158:5267-5276.
258. Bergman Y, Strich B, Sharir H, Ber R, Laskov R. Extinction of Ig genes expression in myeloma x fibroblast somatic cell hybrids is accompanied by repression of the oct-2 gene encoding a B-cell specific transcription factor. EMBO J 1990; 9:849–855.
259. Junker S, Pedersen S, Schreiber E, Matthias P. Extinction of an immunoglobulin κ promoter in cell hybrids is mediated by the octamer motif and correlates with suppression of Oct-2 expression. Cell 1990; 61:467–474.
260. Thomson JAF, Parsons PG, Sturm RA. In vivo and in vitro expression of octamer binding proteins in human melanoma metastases, brain tissue, and fibroblasts. Pigment Cell Res 1993; 6:13–22.
261. Maxwell PH, Pugh CW, Ratcliffe PJ. Inducible operation of the erythropoietin 3′ enhancer in multiple cell lines: evidence for a widespread oxygen-sensing mechanism. Proc Natl Acad Sci USA 1993; 90:2423–2427.
262. Semenza GL, Wang GL. A nuclear factor induced by hypoxia via de novo pro-

tein synthesis binds to the human erythropoietin gene enhancer at a site required for transcriptional activation. Mol Cell Biol 1992; 12:5447–5454.

263. Wang GL, Semenza GL. Characterization of hypoxia-inducible factor 1 and regulation of DNA binding activity by hypoxia. J Biol Chem 1993; 268:21513–21518.

264. Wang GL, Semenza GL. Purification and characterization of hypoxia-inducible factor 1. J Biol Chem 1995; 270:1230–1237.

265. Wang GL, Jiang B-H, Rue EA, Semenza GL. Hypoxia-inducible factor 1 is a basic-helix-loop-helix-PAS heterodimer regulated by cellular O_2 tension. Proc Natl Acad Sci USA 1995; 92:5510–5514.

266. Bunn HF, Poyton RO. Oxygen sensing and molecular adaptation to hypoxia. Physiol Rev 1996; 76:839–885.

267. Melillo G, Musso T, Sica A, Taylor LS, Cox GW, Varesio L. A hypoxia-responsive element mediates a novel pathway of activation of the inducible nitric oxide synthase promoter. J Exp Med 1995; 182:1683–1693.

268. Kvietikova I, Wenger RH, Marti HH, Gassmann M. The transcription factors ATF-1 and CREB-1 bind constitutively to the hypoxia-inducible factor-1 (HIF-1) DNA recognition site. Nucleic Acids Res 1995; 23:4542–4550.

269. Gessani S, Di Marzio P, Rizza P, Belardelli F, Baglioni C. Posttranscriptional regulation of interferon mRNA levels in peritoneal macrophages. J Virol 1991; 65:989–991.

270. Darnell JE Jr, Kerr IM, Stark GR. Jak-STAT pathways and transcriptional activation in response to IFNs and other extracellular signaling proteins. Science 1994; 264:1415–1421.

271. Nunokawa Y, Oikawa S, Tanaka S. Human inducible nitric oxide synthase gene is transcriptionally regulated by nuclear factor-κB dependent mechanism. Biochem Biophys Res Commun 1996; 223:347–352.

272. de Vera ME, Shapiro RA, Nussler AK, Mudgett JS, Simmons RL, Morris SM Jr, Billiar TR, Geller DA. Transcriptional regulation of human inducible nitric oxide synthase (NOS2) gene by cytokines: initial analysis of the human NOS2 promoter. Proc Natl Acad Sci USA 1996; 93:1054–1059.

273. Kolyada AY, Savikovsky N, Madias NE. Transcriptional regulation of the human iNOS gene in vascular-smooth-muscle cells and macrophages: evidence for tissue specificity. Biochem Biophys Res Commun 1996; 220:600–605.

274. Thanos D, Maniatis T. Virus induction of human IFNb gene expression requires the assembly of an enhanceosome. Cell 1995; 83:1091–1100.

275. Heck DE, Laskin DL, Gardner CR, Laskin JD. Epidermal growth factor suppresses nitric oxide and hydrogen peroxide production by keratinocytes. Potential role for nitric oxide in the regulation of wound healing. J Biol Chem 1992; 267:21277–21280.

276. Scott-Burden T, Schini VB, Elizondo E, Junquero DC, Vanhoutte PM. Platelet-derived growth factor suppresses and fibroblast growth factor enhances cytokine-induced production of nitric oxide by cultured smooth muscle cells. Effects on cell proliferation. Circ Res 1992; 71:1088–1100.

277. Goureau O, Lepoivre M, Becquet F, Courtois Y. Differential regulation of induc-

ible nitric oxide synthase by fibroblast growth factors and transforming growth factor β in bovine retinal pigmented epithelial cells: inverse correlation with cellular proliferation. Proc Natl Acad Sci USA 1993; 90:4276–4280.
278. Finbloom DS, Larner AC. Induction of early response genes by interferons, interleukins, and growth factors by the tyrosine phosphorylation of latent transcription factors. Implications for chronic inflammatory diseases. Arthritis Rheum 1995; 38:877–889.
279. Gao JJ, Filla MB, Fultz MJ, Vogel SN, Russell SW, Murphy WJ. Autocrine/paracrine IFN-αβ mediates the lipopolysaccharide-induced activation of transcription factor Stat1α in mouse macrophages: pivotal role of Stat1α in induction of the inducible nitric oxide synthase gene. J Immunol 1998; 161:4803–4810.

2

Arginine Synthesis, Metabolism, and Transport:
Regulators of Nitric Oxide Synthesis

Sidney M. Morris, Jr.
University of Pittsburgh School of Medicine, Pittsburgh, Pennsylvania

I. INTRODUCTION

NO has a multiplicity of physiologic and pathophysiologic effects which may be either beneficial or detrimental, depending largely on the level of NO synthesis and various additional factors such as the concentration and type of reactive oxygen species at the local site of NO synthesis. Accordingly, factors that regulate cellular NO synthesis rates play a major role in determining the physiologic impact of NO production. Maximal cellular capacity for NO synthesis is determined by the intracellular content of NOS. In the case of eNOS and nNOS, expression is largely constitutive and enzyme activity is dynamically regulated by changes in intracellular calcium flux. In the case of iNOS, it is the expression of the enzyme itself that is regulated, and its activity is independent of changes in intracellular calcium. Nonetheless, it is important to recognize that actual rates of cellular NO production by any of the NOS isoforms are determined by additional factors, such as availability of the various substrates and cofactors required by NOS (reviewed in 1).

Because arginine is the only physiologically significant nitrogen donor for NO synthesis, this discussion will focus on mechanisms regulating its availability, primarily in cells expressing iNOS. The impact of varying arginine availability on NO production and NO-dependent effects can be quite dramatic, as illustrated in Figure 1. Although concentration-dependent effects of arginine on cellular NO synthesis are generally assumed to reflect changes only in substrate availability, arginine has several other effects which also might affect rates of NO synthesis: It promotes dimerization of iNOS monomers to form the enzymatically active

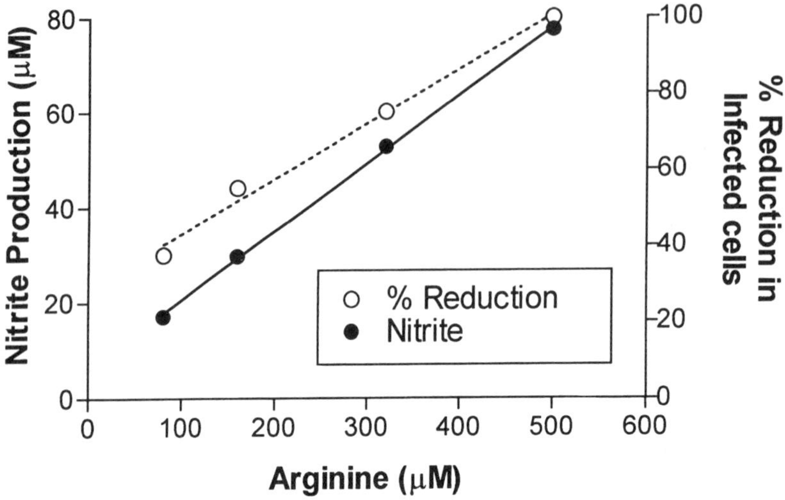

Figure 1 Dependence of NO production and trypanocidal activity on arginine concentration in murine peritoneal macrophages. Lipopolysaccharide- and interferon-γ-stimulated peritoneal macrophages were infected with Trypanosoma cruzi and then incubated for 72 h in RPMI medium containing varying concentrations of arginine. Trypanocidal activity was measured as the percent reduction in infected cells relative to parallel cultures incubated in the presence of sufficient N^G-monomethyl-L-arginine to completely inhibit NO synthesis. (Data from Table 1 of Ref. 20.)

dimer, increases NADPH oxidation, and alters the heme iron spin equilibrium (2,3). Additional physiologic effects of arginine not discussed here can be found in other reviews (4–8).

Regulation of arginine availability is complex, as this amino acid has multiple sources and metabolic fates, and attempts to modulate arginine supply, especially by dietary means, must take these factors into account. Intracellular arginine can be derived from extracellular arginine via specific amino acid transport systems, from intracellular protein degradation, and from endogenous arginine synthesis. Catabolism of arginine by any of several enzymes in addition to NOS could limit availability of extracellular and/or intracellular arginine for NO synthesis. Activities of arginine transport systems vary among different cell types and in response to specific stimuli. Although rates of NO synthesis could potentially be regulated by altering any of these processes, their impact on NO synthesis can easily be misjudged in studies using standard tissue culture media, which contain concentrations of arginine up to 10 times higher than that found in plasma. Nonetheless, evidence of their effects has been obtained for most of these processes, as described in this chapter. The exception is protein degradation, as correlations between rates of intracellular protein degradation, arginine pools, and NO synthesis have not been determined.

II. ARGININE SYNTHESIS

Based on nitrogen balance studies, endogenous rates of arginine synthesis in many mammalian species, including man and rodents, are sufficient to result in the classification of arginine as a nonessential dietary amino acid for healthy adults. Consequently, placing normal adult rodents or humans on an arginine-free or arginine-deficient diet will have little effect on plasma arginine levels. In young animals and during recovery from injury or disease, however, dietary arginine is required for optimal growth and recovery so that under these circumstances it is classified as a conditionally essential or semiessential amino acid (reviewed in 5). Although some textbooks state that arginine is a nonessential amino acid because it is made in the liver, this is not correct. Arginine is synthesized in the liver only as a transient intermediate in the urea cycle and is not exported into the circulation. This is supported by several lines of evidence, including observations that (a) arginine remains an essential dietary amino acid in individuals who have received liver transplants to correct inherited defects in the urea cycle (9); (b) little or no ^{13}N-arginine could be detected in livers of rats following infusion of ^{13}N-ammonia in vivo (10); and (c) ^{15}N-arginine was not detected in either the liver or the perfusate of livers perfused with excess $^{15}NH_4Cl$ for urea synthesis (11).

The primary site of endogenous arginine synthesis is in the proximal tubules of the kidney, where citrulline synthesized and released by epithelial cells of the small intestine is extracted from the blood, efficiently converted to arginine, and then released into the systemic circulation (reviewed in 12). The conversion of citrulline to arginine is accomplished by the sequential action of the cytosolic enzymes argininosuccinate synthetase (AS) and argininosuccinate lyase (AL) (Fig. 2), which are expressed at high levels in both kidney and liver (as part of the urea cycle). However, it has long been known that many other cell types also express AS and AL (13), albeit at very low levels, but the physiologic significance of this widely distributed expression was unclear. Following the elucidation of the NOS-catalyzed reaction that occurs in many cell types, it became apparent that NO and arginine synthesis could potentially be coupled in a recycling pathway which was originally termed the arginine-citrulline cycle (14). However, as this pathway is analogous to the ornithine-urea cycle, we have referred to it as the citrulline-NO cycle, which identifies both the recycled nitrogen carrier (citrulline) and the end product NO (Fig. 3) (15,16).

According to this scheme, the nitrogen atom that is lost as NO is replenished from the amino group of an amino acid by transamination. Evidence to support this pathway was provided by experiments demonstrating that citrulline, which

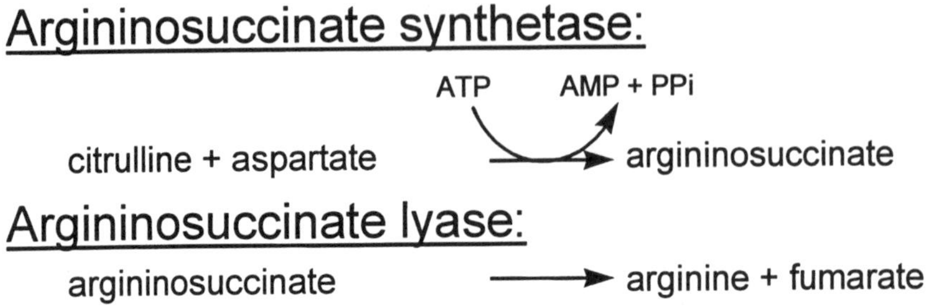

Figure 2 Arginine biosynthetic enzymes.

cannot be utilized as substrate by NOS, can substitute, at least in part, for arginine in supporting NO synthesis by intact cells (16–22). It should be emphasized, however, that the citrulline-NO cycle differs from the urea cycle with regard to coupling efficiency of the different enzymatic reactions. Coupling of reactions within the urea cycle is highly efficient (23), so urea cycle intermediates normally do not accumulate and are not released into the circulation. In contrast, coupling between NO synthesis and arginine synthesis in many cells is inefficient, reflected by the accumulation of citrulline in the medium of NO-producing cells. Although net production of both citrulline and NO often is not reported, total citrulline production can be less than total NO production in some cell types (24), consistent with recycling of citrulline via the citrulline-NO cycle, whereas total citrulline and NO production may be equivalent in other cell types (25).

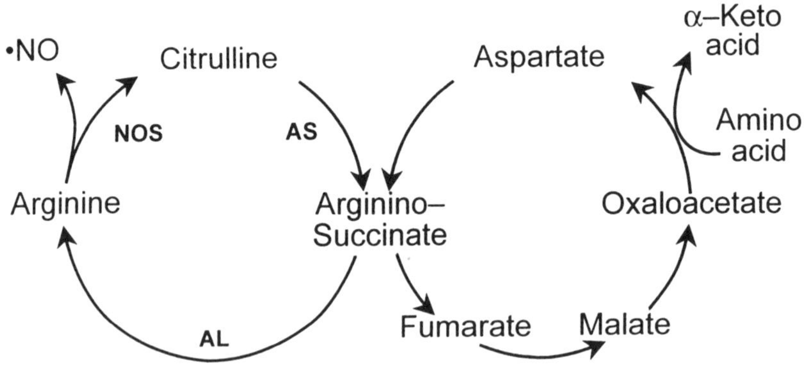

Figure 3 The citrulline-NO cycle. Fumarate is converted to aspartate by the sequential action of fumarase, malate dehydrogenase, and a transaminase (usually glutamate-oxaloacetate transaminase). For the sake of simplicity, not all reactants are shown. (From Ref. 16.)

A. Nonhepatic Cells

The demonstration of the conversion of citrulline to arginine in NO-producing bovine endothelial cells (14) and rat peritoneal macrophages (26) prompted the initial suggestions that endogenous arginine synthesis may play a role in regulating NO synthesis. These suggestions were supported by subsequent findings that citrulline can at least partially substitute for arginine in supporting NO synthesis by a wide variety of cell types, including rodent macrophages (16,18,20), vascular smooth muscle cells (15,17), rat and human islet cells (19,22), and human tumor cell lines (21). Moreover, reports that LPS-stimulated macrophages (26) and macrophages isolated from diabetic BB rats (27) were more active than unstimulated macrophages with regard to both NO and arginine synthesis suggested that iNOS and arginine biosynthetic enzymes were coinduced. At about the same time, our group reported that activity and mRNA abundance for AS, the rate-limiting enzyme in arginine biosynthesis, was coinduced with iNOS expression in both murine macrophages and rat vascular smooth muscle cells (15,16,28). Similar results have since been obtained for a variety of rodent and human cells in culture (17,19–22,29,30), and in tissues of rats treated with lipopolysaccharide in vivo (31,32). For all cells studied to date, it has been observed that whenever iNOS is induced, AS also is induced. The converse, however, is not true; agents such as cAMP or dexamethasone, which induce AS expression in several cell types (12), do not also induce iNOS. The magnitude of AS induction in different cells varies considerably, owing largely to differences in basal levels of AS expression. Although there is little or no coinduction of AL and iNOS in any of the cultured cell models studied, significant induction of both AL and iNOS does occur in tissues of rats treated with lipopolysaccharide (31,32), perhaps suggesting the presence of permissive or stimulatory factors for AL expression in vivo that are not present in cultured cells.

Both glutamine (33,34) and hypoxia (35) inhibited synthesis of arginine from citrulline by intact endothelial cells. Glutamine not only suppressed endothelial arginine synthesis by competitively inhibiting uptake of extracellular citrulline (34), but the activity of AS was modestly reduced in a time- and glutamine concentration-dependent fashion (35). The inhibitory effect of glutamine may be cell-specific, as glutamine did not affect the conversion of citrulline to arginine by unstimulated J774 cells (33). Although hypoxia did not affect citrulline uptake, it resulted in a reduction of endothelial AS activity, with little or no effect on AL activity (35). The inhibitory effect of hypoxia on AS activity was correlated with changes in intracellular glutamine concentration, but the basis for the reduced AS activity has not been identified. Surprisingly, hypoxia for up to 24 hours not only did not result in a decline in intracellular arginine content even though endogenous arginine synthesis declined, but actually resulted in an increase in intracellular arginine concentration (35). This suggests an effect of hypoxia on other pathways involved in the generation, degradation, or uptake of arginine.

It has been speculated that glutamine may inhibit endothelial NO release by limiting availability of arginine, possibly via its effects on endogenous arginine synthesis (33). However, this notion has been challenged by a recent study showing that glutamine inhibited receptor-dependent NO release by endothelial cells but did not inhibit receptor-independent NO release (36). If glutamine were simply limiting intracellular arginine availability, its effect would have been expected to be the same in both cases.

Arginine biosynthetic enzymes also are present in some neural cells (37–41), but there is no information regarding regulation of their activity or their impact on NO synthesis in neurons, other than the fact that citrulline can substitute for arginine in supporting NO synthesis (e.g., 41).

B. Liver

As noted above, the liver is distinct from virtually all other tissues in that it has a highly active urea cycle, with efficient coupling between arginine synthesis and degradation. However, hepatocytes also can be induced to produce NO (reviewed in 42). Under these circumstances, can the urea cycle provide arginine for NO synthesis? This question has been examined by perfusing various nitrogenous substrates into livers isolated from rats treated with inflammatory stimuli to induce high-level hepatic NO synthesis (43–45). In our study using *C. parvum*-treated rats, perfusion of isolated livers with glutamine (Fig. 4) or NH_4Cl resulted in large increases in urea synthesis without any change in NO synthesis (measured as nitrite + nitrate), indicating that arginine made within the urea cycle was not available for NO synthesis (43). Using livers from endotoxin-treated rats, however, Wettstein et al. did find an increase in NO output when livers were perfused with NH_4Cl (44). Although there is no obvious explanation for the difference in results between the two groups, owing to the great difference in magnitude of the rates of hepatic urea and NO synthesis, the results of both studies are consistent with the notion that if the urea cycle does provide any arginine for hepatic NO synthesis, it must represent only a tiny fraction of the total arginine synthesized within the urea cycle. A third group reported that the urea cycle and hepatic iNOS compete for a common pool of arginine, but these investigators were simply measuring hydrolysis of arginine by arginase rather than de novo urea synthesis and thus did not determine whether arginine synthesized within the urea cycle also was available to iNOS (45).

C. Role of Arginine Synthesis in NO-Producing Cells

For a number of cell types, arginine concentrations normally found in plasma (averaging approximately 100 μM for mice and men) are limiting for induced NO synthesis (e.g., 20). In sepsis, a condition in which NO production is elevated (42), plasma arginine levels may be as low as 40 μM (46,47). The impact of the decline

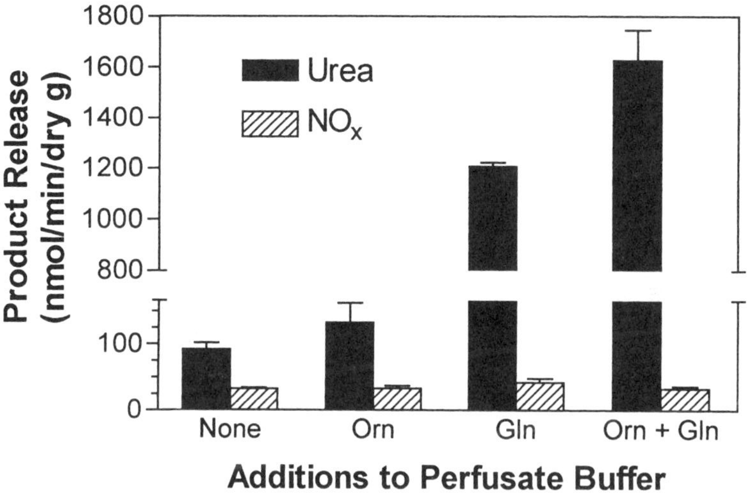

Figure 4 The urea cycle does not provide arginine for hepatic NO synthesis. Hepatic iNOS expression was induced by treating rats with killed Cornybacterium parvum. Isolated livers were then perfused with Krebs-Henseleit bicarbonate buffer containing no added nitrogenous substrate (None), ornithine (Orn), glutamine (Gln), or both ornithine and glutamine. Total NO (NO_x = nitrate + nitrite) and urea production during 20-min perfusion periods were measured. (Figure based on data from Ref. 43.)

in plasma arginine concentration during sepsis may be offset to some extent by increased arginine uptake in response to inflammatory cytokines, as discussed in Section IV. In contrast to plasma, arginine concentrations in standard tissue culture media are in the 0.5 to 1 mM range, thereby giving rise to cellular NO synthesis rates that are not representative of in vivo conditions. As plasma contains both citrulline and arginine, the capability to convert citrulline to arginine should permit a higher output of NO than if arginine alone were present. In fact, such effects have been observed (e.g., 16,20). The extent to which the NO output is increased will depend upon both citrulline concentration and cellular arginine biosynthetic capacity, which varies among different cell types. Normal plasma values of citrulline are approximately 30 to 40 μM (47,48), so NO synthesis in vivo would be expected to be about 30% to 40% higher than in the presence of arginine alone, without taking into account preferential use of arginine recycled via the citrulline-NO pathway. At localized anatomic sites where extracellular arginine may become depleted by the action of arginase (see Sect. III), the presence of citrulline, which is not a substrate for arginase, may sustain a low level of NO synthesis via its intracellular conversion to arginine.

Although the capacity for endogenous arginine synthesis is co-induced with iNOS and arginine synthesis also occurs in cells expressing eNOS or nNOS, whether endogenous arginine synthesis regulates NO synthesis rates in vivo is unclear. For example, as discussed in Section II.A, the results obtained with hypoxia (35) and glutamine (36) suggest that endogenous arginine synthesis may not play a major role in regulating NO synthesis in endothelial cells. On the other hand, endogenous arginine synthesis not only may be rate-limiting in vascular smooth muscle cells but also may preferentially provide arginine to iNOS (49). Whether this is generally true or is limited to specific cell types remains to be determined. Because no highly selective, potent inhibitors for AS or AL are currently available, recombinant DNA techniques using the cloned cDNAs or genes for AS and AL to alter expression of these enzymes in NO-producing cells will be required to more fully explore this problem.

III. ARGININE CATABOLISM

In addition to NOS, there are three known mammalian enzymes that catabolize arginine and could, in principle, regulate NO synthesis by limiting availability of arginine: arginase, L-arginine:glycine amidinotransferase, and arginine decarboxylase (Fig. 5). Thus, arginine is a precursor not only for the synthesis of NO but also for creatine, agmatine, urea, and ornithine, which in turn is a precursor for synthesis of polyamines, proline, and glutamate. The role of arginine catabolic enzymes in regulating arginine availability has been strengthened by recent studies indicating that modulation of arginine degradation, rather than of arginine synthesis, plays a major role in regulating arginine homeostasis in humans (50,51). Although the impact of its activity on NO synthesis has not been studied, the restricted expression and specialized metabolic function of L-arginine:glycine amidinotransferase, which is expressed in liver and kidney and catalyzes the first step in creatine biosynthesis, make it unlikely that this enzyme plays a significant role in regulating NO synthesis. Thus, only arginase and arginine decarboxylase will be discussed here.

A. Arginase

There are at least two isoforms of mammalian arginase, both of which catalyze the conversion of arginine to ornithine and urea (reviewed in 52,53). Although both isoforms require Mn^{2+} for activity, they differ with regard to tissue distribution, subcellular localization, immunologic reactivity, and certain enzymatic properties, reflecting the fact that they are encoded by different genes (52–55). Type I arginase (Arg I) is highly expressed in liver cytoplasm as a component of the urea cycle,

Nitric oxide synthase:

arginine + O_2 → citrulline + NO

Arginase:

arginine → ornithine + urea

Arginine decarboxylase:

arginine → agmatine + CO_2

Arginine: glycine amidinotransferase:

arginine + glycine → ornithine + guanidinoacetate

Figure 5 Arginine catabolic enzymes.

and to a limited extent in a few other tissues. In contrast, type II arginase (Arg II), a mitochondrial enzyme, is expressed in kidney, brain, small intestine, mammary gland, and macrophages, but there is little or no expression in liver (52,53,55). The differential subcellular localization of Arg I and II, as well of other metabolic enzymes (Fig. 6), may facilitate independent regulation of the synthesis of NO, polyamines, proline, and glutamate. Interest in arginase has increased greatly in recent years because of the possibility that it may play a role in regulating arginine availability for NO synthesis, particularly in cells in which arginase and iNOS are coexpressed. Much of the existing information regarding the possible role of arginase in NO synthesis comes from studies of macrophages and endothelial cells, primarily as measurements of arginase activity in whole cells or cell lysates. Little information is currently available regarding the identity of the arginase isoforms expressed or the regulatory mechanisms which underlie changes in arginase activity. As the cDNAs for mammalian Arg II have been cloned recently (55–57), however, more information on the roles and regulation of the arginase isoforms should become available soon.

1. Macrophages

Most studies of the regulation of arginase activity and NO synthesis have used macrophages. Unlike argininosuccinate synthetase, arginase is not consistently coinduced with iNOS. In murine macrophages or macrophage cell lines, arginase

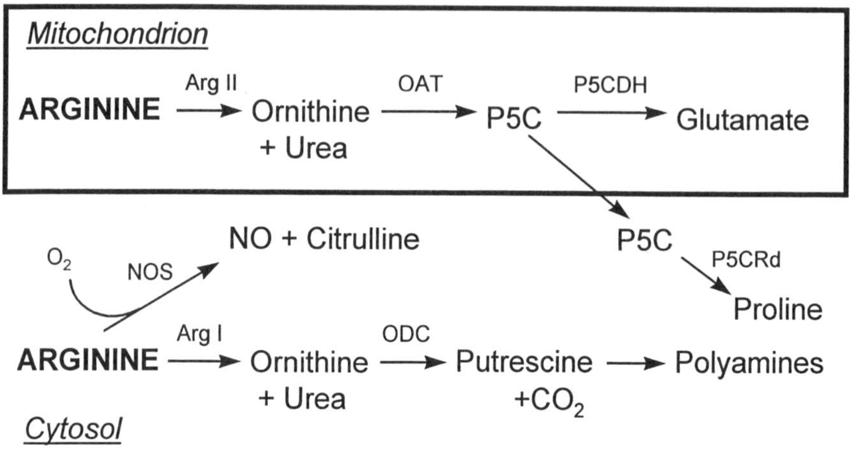

Figure 6 Relationship among arginases, NOS, and ornithine metabolic enzymes. Abbreviations: Arg, arginase; ODC, ornithine decarboxylase; OAT, ornithine aminotransferase; P5C, Δ^1-pyrroline-5-carboxylate; P5CDH, P5C dehydrogenase; P5CRd; P5C reductase. For the sake of simplicity, not all reactants are shown.

activity is induced by lipopolysaccharide (LPS), IL-4, IL-10, cAMP, and prostaglandin E_2 (58–66). With the exception of LPS, these agents inhibit induction of iNOS in macrophages. Unlike iNOS, LPS-dependent induction of arginase in RAW 264.7 cells was not inhibited by pyrrolidine dithiocarbamate, an inhibitor of NFκB activation (66). Furthermore, induction of arginase activity lagged behind that of cellular NO synthesis (66).

Taken together, these observations clearly indicate that the mechanisms regulating expression of arginase and iNOS in macrophages are not identical. Although the identity of the isoforms responsible for arginase activity was not determined in most studies, the arginase induced by LPS in RAW 264.7 cells is solely the type II isoform (66). This correlates with induction of mRNA encoding Arg II (56,67), indicating that LPS-dependent induction probably reflects increased transcription of the Arg II gene. On the other hand, rat peritoneal macrophages treated with LPS express only Arg I (68). This may reflect a species difference because we have found strong expression of Arg II in murine peritoneal macrophages.

In two studies where arginase activity was found to be induced by either tumor necrosis factor-α or LPS (61, 69), arginase induction was suppressed by indomethacin but restored by addition of prostaglandin E_2 or cholera toxin, suggesting that secondary elevation of cAMP levels was required for the induction (69). This contrasts with LPS induction of macrophage iNOS expression, which is not significantly affected by indomethacin (70–72). We have found that cAMP strongly induces arginase activity in RAW 264.7 cells, reflecting induction of mRNAs encoding both Arg I and II. Thus, as described below for endothelial cells, Arg I and II can be expressed and independently regulated within the same cell. Interferon-γ, a potent inducer of iNOS, not only fails to induce arginase (63,66) but also strongly antagonizes the induction of arginase by cAMP in RAW 264.7 cells.

In addition to LPS treatment, culturing wound-derived rat macrophages in an anoxic environment is one of the few treatments in which both apparent arginase activity and iNOS expression in macrophages were increased, even though flux of arginine through iNOS was inhibited (73). Interestingly, this effect depended upon the site from which the macrophages were isolated, as it was not observed with either resident or C. parvum-elicited peritoneal macrophages (73). In a separate study also using rat peritoneal macrophages, concentrations of transforming growth factor-β that reduced basal NO production resulted in significantly increased arginase activity (74). As in the studies that measured arginase activity in cell extracts, these reports provide further evidence that regulation of arginase expression in macrophages is distinct from that of iNOS.

Some investigators have referred to the arginase activity found in supernatants of conditioned medium from cultures of stimulated murine macrophages as "secreted" arginase. However, it should be noted that other investigators have found no arginase activity in conditioned medium from similarly stimulated macrophage cultures (e.g., 75,76), and no secretory process for release of arginase has ever been described. In addition, the known arginase isoforms are intracellular enzymes. Thus, it is likely that any arginase activity found in conditioned media is due to cell lysis.

The preceding results were obtained with rodent macrophages. As in the case of iNOS expression, human macrophages have very little constitutive arginase activity, and arginase is not inducible by inflammatory stimuli which induce iNOS in rodent macrophages (77). However, it should be appreciated that most inducers of iNOS do not induce arginase, and effects of most arginase-inducing agents in human macrophages have not been reported. Consequently, the question of whether there is regulated arginase activity in human macrophages remains open.

2. Endothelial Cells

Two groups have reported that NO-producing rat aortic endothelial cells also express arginase activity (78,79). The arginase constitutively expressed in the absence of added LPS or cytokines is type I, as indicated by its cytosolic localization (78) and immunoreactivity (79). Interestingly, basal arginase activity in endothelial cells from diabetic BB rats was only 8% of that in endothelial cells of nondiabetic BBn rats (78). However, owing to the fact that the arginase specific activity reported for endothelial cells of normal Sprague-Dawley rats (79) is equal to that reported for the diabetic BB rat, it is not clear whether the arginase activity in endothelial cells of diabetic BB rats is unusually low or the arginase activity in endothelial cells of nondiabetic BBn rats is unusually high. Addition of LPS alone to endothelial cells of normal rats resulted in induction of both iNOS and Arg II (79). Addition of interferon-γ together with LPS significantly enhanced induction of iNOS compared to LPS alone, but induction of arginase activity by LPS was blocked by the coaddition of interferon-γ. Thus, as in macrophages, arginase activity in endothelial cells is not consistently coinduced with iNOS expression. Nevertheless, arginase activity of intact cells was inhibited when they were producing NO, apparently as a consequence of producing N-hydroxyarginine, a potent arginase inhibitor (79). This not only demonstrates that inhibitory effects of iNOS and arginase activity can be mutual, but also underscores the importance of measuring arginase activity in intact cells as well as in cell lysates.

3. Liver and Enterocytes

Hepatocytes produce NO in response to inflammatory stimuli (42), despite the fact that these cells contain the highest levels of arginase in the body. There are several possible explanations for this: (a) as noted below in Section A.4, the K_m for arginine is much greater for arginase than for iNOS so that arginine will not become limiting for iNOS; (b) there may be multiple intracellular pools of arginine which are not equally accessible to both enzymes, or (c) hepatic uptake of arginine, which increases during inflammatory conditions such as sepsis (80), may preferentially deliver arginine to iNOS. Possibilities b and c imply that arginase and iNOS have different subcellular localizations. In support of possibility b is our finding that rates of NO synthesis in isolated perfused liver did not correlate with rates of urea synthesis (Fig. 4).

NO-producing pig enterocytes express arginase that is predominantly type I, based on its localization in the cytosolic fraction of cell lysates (81). Its activity is developmentally regulated (82), but little is known regarding regulation of its activity in response to LPS and cytokines or the functional consequences of coexpression of arginase and NOS in enterocytes.

4. Relationship Between Arginase Activity and NO Synthesis

What are the consequences of expressing both arginase and NOS in the same cells? Arginase would not be expected to effectively compete with NOS for a common intracellular pool of arginine because the K_m for arginine is in the 2 to 20 mM range for mammalian arginases (52,53) whereas it is in the 1 to 20 μM range for the various NOS isoforms (83). The relationship between these enzymes is even more complicated in the case of Arg II, which is located in mitochondria, whereas the NOS isoforms are cytosolic. Nevertheless, it is important to remember that enzymes do function at substrate concentrations below the K_m, and sufficient quantities of enzyme can effectively limit substrate availability for other enzymes which utilize the same substrate. Indeed, a recent study found that inhibition of arginase in LPS-treated rat alveolar macrophages resulted in enhanced conversion of arginine to citrulline, indicating that arginase and iNOS do compete for intracellular arginine (84). Also, in wounds (75,76,85) and culture dishes containing macrophages (86,87) the extracellular fluid becomes almost completely depleted of arginine whereas ornithine increases, due to extracellular arginase which is probably released by autolysis of macrophages. Consequently, NO synthesis is markedly reduced by arginase under these conditions.

As it would not be desirable to immediately block NO synthesis at sites of infection, this may provide a rationale for the observation that induction of arginase activity lags induction of NO synthesis in macrophages (66). Although the delayed induction and release of arginase may help to ensure that NO synthesis is not prolonged, thus sparing host tissue from damage by NO and its reactive metabolites, the primary role of arginase activity in stimulated macrophages may be to generate ornithine (Fig. 6). Ornithine is a substrate for synthesis of polyamines, which are required for cell proliferation, and of proline, which is required for collagen biosynthesis. Thus, enhanced polyamine and proline biosynthesis could promote wound healing and tissue remodeling during recovery from infection. This notion is supported by the fact that LPS and cAMP, which are potent inducers of arginase in murine macrophages (Sect. III.A.1), also strongly induce ornithine decarboxylase activity in these cells (88–90). A role for Arg II also in proline synthesis would be consistent with the fact that proline is a better inhibitor of Arg II than of Arg I (91).

The necessity for allowing transient high-level NO synthesis at sites of infection also could provide a rationale for the finding that induced NO production correlates with partial inhibition of cellular arginase activity in rat and murine macrophages (73,92,93). This inhibition apparently is due to extracellular accumulation of N^G-hydroxyarginine (93), an intermediate in NO synthesis that also is a potent inhibitor of arginase (see Sect. 5 below).

Partial inhibition of cellular arginase activity by accumulation of N^G-hydroxyarginine also has been found in LPS- and cytokine-stimulated rat endothelial cells (79). Thus, urea production by NO-producing endothelial cells did not match arginase activity measured in cell lysates. Because endothelial cells in intact animals are constantly perfused whereas cultured cells are not, it seems unlikely that they would be exposed to sufficient N^G-hydroxyarginine in vivo to inhibit cellular arginase activity. The fact that N^G-hydroxyarginine can be oxidized to citrulline and NO by a variety of heme proteins such as peroxidases, cytochromes P450, hemoglobin, and catalase (94,95) also may prevent accumulation of this compound in some tissues. Despite these caveats, N^G-hydroxyarginine may nonetheless regulate arginase activity in vivo, particularly at sites of poor perfusion.

As noted in earlier sections, some investigators have evaluated changes in arginase activity by measuring changes in concentrations of ornithine or urea in the culture medium or by rates of conversion of labeled arginine to ornithine or urea by intact cells. Although it is essential to monitor the metabolic activity of intact cells because this may reveal important phenomena that are not apparent from enzyme assays of cell extracts, the recently discovered inhibitory effect of N^G-hydroxyarginine underscores the point that it may not be valid to equate ornithine or urea production by intact cells with cellular arginase content (e.g., 79). Moreover, inflammatory stimuli also can alter cellular arginine transport (Sect. IV), so changes in ornithine output may simply reflect changes in uptake of arginine. Also, ornithine itself can be metabolized by several enzymes, including ornithine decarboxylase and ornithine aminotransferase, potentially resulting in underestimates of arginase activity. Finally, citrulline can interfere with some colorimetric assays for urea (e.g., 96), which could result in inaccurate estimates of arginase activity in NO-producing cells unless appropriate controls are used.

5. Arginase Inhibitors

To dissect the interplay between arginase and iNOS, it would be desirable to have compounds that effectively inhibit one enzyme but not the other. Because arginase and NOS utilize the same substrate, it might be expected that arginine analogs that inhibit NOS would also inhibit arginase. However, this is not the case. Fortuitously, commonly used NOS inhibitors such as N^G-monomethyl-L-arginine (L-NMMA), N^G-nitro-L-arginine (L-NNA), N^G-nitro-L-arginine methyl ester (L-NAME), and aminoguanidine are weak inhibitors of arginase, even at concentrations much greater than those used for inhibiting NOS (97–99). On the other hand, NOS inhibitors do not fully discriminate among the various NOS isoforms, and they can have side effects independent of their inhibition of NOS. For example, L-NMMA is a competitive inhibitor of the arginine transport system y^+ (Sect. IV.E), and both L-NMMA and L-NNA can inhibit responses mediated by ATP-sensitive K^+ channels (100).

The branched-chain amino acids valine, leucine, and isoleucine, as well as the reaction product ornithine, are effective inhibitors of Arg I, but the branched-chain amino acids are relatively poor inhibitors of Arg II (52,91,101). Conversely, proline is a better inhibitor of Arg II than of Arg I (91). In any case, concentrations of these amino acids required for effective inhibition of the arginases are too high ($\geq$ 10 mM) for physiologic studies and discrimination between the arginase isoforms is modest. No effective isoform-specific inhibitors of arginase have been identified. Nitrite, an end product of NO, is moderately effective in inhibiting arginase whereas nitrate is ineffective (102), but it is unlikely that nitrite concentrations in vivo become sufficiently elevated (>10 mM) to significantly affect arginase activity. Interestingly, N^G-hydroxyarginine, an intermediate in the synthesis of

NO from arginine, is one of the most potent inhibitors of arginase yet identified, particularly at physiologic pH ($K_i \approx$ 10-12 μM at pH 7.4) (79), and both arginase isoforms are inhibited by this intermediate (93,99,103). In fact, inhibition of arginase by this compound may be physiologically significant, as noted in Section 4. Major drawbacks of N^G-hydroxyarginine as an arginase inhibitor are that it is itself a NO donor and has a limited half-life in cells (79). However, these disadvantages are largely absent in the recently described N^{ω}-hydroxyamino-α-amino acid derivatives which are potent arginase inhibitors (104–106). Combining the recently-described crystal structure of rat Arg I (107) with comparative sequence analyses of the arginase family (108) may help in designing arginase inhibitors of even greater potency and selectivity.

B. Arginine Decarboxylase

Agmatine, produced by decarboxylation of arginine, is a ligand for both imidazoline and α_2 adrenergic receptors (109). It also is of interest because it is one of the inhibitors of NOS known to be endogenously produced by mammalian cells (110). Its role as a physiologic regulator of NO synthesis is not immediately obvious as it is a relatively weak competitive inhibitor, with approximate K_i values of 220 μM (iNOS), 660μM (nNOS), and 7.5 mM (eNOS) (110). However, arginine decarboxylase is expressed in many cell types that produce NO (111), and it is possible that highly localized concentrations of agmatine may become sufficiently elevated that significant inhibition of NOS will occur, particularly in the case of iNOS. Whether arginine decarboxylase and its product agmatine play a role in regulating NO synthesis in vivo remains to be determined.

IV. ARGININE TRANSPORT

As extracellular arginine is a primary source of arginine for most mammalian cells, arginine transport systems represent a major control point for regulating cellular NO synthesis rates. Uptake of arginine can occur via several transport systems, designated as systems y^+, $b^{o,+}$, $B^{o,+}$, or y^+L (reviewed in 112,113). Not all transporters are expressed in every cell type. The Na^+-independent system y^+ is involved in transport of arginine, as well as of lysine and ornithine, in all NO-producing cells studied to date.

cDNAs for three cationic amino acid transporters (CATs) have been cloned from mice, rats, and humans and are identified as mCATs (murine) or hCATs (human) (reviewed in 114, 115). As noted below, changes in abundance of the corresponding mRNAs have been found in a variety of cell types following induction of NO synthesis, indicating regulation of the synthesis of transporter molecules (see refs. 114 and 116 for reviews of CAT gene expression). CAT-1 and CAT-2 (also known as CAT-2B or Tea, for T-cell early-activation gene) exhibit the characteristics of system y^+, whereas CAT-2A (a splice variant of CAT-2 originally isolated from murine hepatocytes) has a 10-fold lower affinity for substrate than the other two CATs and does not exhibit other properties characteristic of system y^+. K_ms of CAT-1 and -2 for arginine are in the 0.1 to 0.3 mM range (117), approximately the normal concentration of arginine in plasma (about 0.1 mM). Thus, these two transporters will not be saturated with arginine in vivo but will be saturated in cultured cells, owing to the high arginine concentrations (0.4 to 1.0 mM) in standard tissue culture media. The high arginine concentrations in tissue culture media also will facilitate arginine uptake by low-affinity transport systems which probably play minor roles in vivo.

A. Macrophages

Both arginine transport and NO synthesis were induced by LPS in murine macrophages and macrophage cell lines (118–120). Interferon-γ further enhanced the NO production elicited by LPS but did not enhance arginine uptake (119). Dexamethasone markedly inhibited induction of NO synthesis without affecting arginine uptake (120), demonstrating that mechanisms for regulating arginine transport activity and iNOS expression are not identical. Similarly, LPS or LPS + IFN-γ induced both NO production and arginine uptake in rodent astrocytes (121–123), corresponding with induction of iNOS and CAT-2 mRNAs (123).

B. Endothelial Cells

Three arginine transport systems are present in porcine pulmonary artery endothelial cells: y^+, $b^{o,+}$, and $B^{O,+}$ (124). System y^+ accounts for 60% to 70% of arginine transport, whereas system $B^{O,+}$ accounts for 30% to 40% of the transport. Transport by system $b^{o,+}$ at physiologic levels of arginine is negligible, owing to its very low affinity for arginine. Arginine uptake by these pulmonary artery endothelial cells could be induced several fold by LPS (125), an effect apparently mediated by an LPS-stimulated autocrine loop involving both IL-1 and TNF-α (126). Induction by LPS was blocked by both actinomycin D and cycloheximide, suggesting that increased transport reflects increased synthesis of transporters (125). Activities of both systems y^+ and $B^{O,+}$ were induced by these stimuli, though not equally well by each stimulus (126). Conversely, arginine transport by both systems in pulmonary artery endothelial cells was reduced by approximately 30% to 40% in response to hypoxia (127). This reduction was time-dependent, with reductions in system y^+ becoming apparent within 4 hours, whereas the reduction in system $B^{O,+}$ became apparent at 20 hours.

Several possible mechanisms have been proposed to account for the inhibition of arginine transport by hypoxia, including reduction in the number of transporter molecules, reduction in transporter activity without any change in transporter number, a decrease in endothelial cell membrane potential, which could affect system y^+ activity, and dissipation of the extracellular-to-intracellular sodium gradient that maintains sodium-dependent transport through system $B^{O,+}$. However, the fact that system y^+ activity was reduced by 32% after only 4 hours of hypoxia (127) but was unchanged after 8 or 12 hours of exposure to cycloheximide or actinomycin D, respectively (125), suggests that hypoxia may enhance turnover or inactivation of transporter molecules.

Stimulation of arginine transport in bovine aortic endothelial cells was similar to that of porcine pulmonary artery endothelial cells: transport was enhanced by TNF-α and LPS but not by IFN-γ or IL-1β (128). Unlike the pulmonary artery endothelial cells, however, none of these stimuli affected NO production, demonstrating that NO production and arginine transport are not linked in these cells. Western blot analysis confirmed that lack of enhanced NO production in TNF-α-stimulated bovine aortic endothelial cells reflected lack of iNOS expression rather than a simple lack of essential cofactors such as tetrahydrobiopterin.

C. Muscle

1. Vascular Smooth Muscle

Arginine transport in vascular smooth muscle cells occurs via system y^+. Both NO production and arginine transport activity in rat vascular smooth muscle cells were induced by TNF-α, LPS, and IL-1β, whereas IFN-γ and cAMP induced NO synthesis without affecting arginine transport activity (129–131). Moreover, dexamethasone inhibited induction of NO production by LPS + IFN-γ without inhibiting induction of arginine transport (129), whereas angiotensin II inhibited the stimulation of NO production by IL-1β + TNF-α while further enhancing the stimulation of arginine transport by these cytokines (131). Whereas unstimulated cells expressed both CAT-1 and CAT-2B mRNAs, induction of arginine transport by IL-1β + TNF-α correlated with induction only of CAT-2B mRNA (131). Thus, as in other cell types, arginine transport in vascular smooth muscle cells is not obligatorily coinduced with enhanced NO production.

Results obtained with bovine vascular smooth muscle cells were similar but not identical to those obtained with the rat vascular smooth muscle cells: TNF-α and IL-1β induced NO production along with a modest increase in arginine transport but IFN-γ and LPS enhanced NO production without affecting arginine transport (128).

2. Cardiac Myocytes

Both NO production and arginine transport were induced by IL-1β + IFN-γ in neonatal rat cardiac myocytes (29). CAT-1 mRNA was constitutively expressed in resting myocytes and was induced 10-fold by the cytokine combination. In addition, expression of CAT-2B and CAT-2A mRNAs appeared following exposure to the cytokines. Interestingly, insulin, which enhanced cytokine-induced NO production by the myocytes without changing NOS activity, further increased arginine transport and abundance of CAT-1, but not of CAT-2A or CAT-2B, mRNA.

D. Liver

Basal activity of system y^+ in normal adult hepatocytes is barely detectable (132), reflecting the fact that there is little uptake of plasma arginine by this organ. This may prevent depletion of plasma arginine by the high levels of arginase present in liver. Currently, there is little information correlating hepatic arginine transport activity with hepatic NO production. However, activity of hepatic system y^+ is stimulated by a variety of agents associated with inflammation, including TNFα, LPS, interleukin-6, and prostaglandin E_2 (80,133–136), indicating that arginine transport activity is enhanced whenever hepatic NO production is induced. This appears to be the case also during liver regeneration, when both hepatic NO production and mCAT-1 mRNA are induced (137–139). However, as system y^+ also is responsible for uptake of ornithine, which is the precursor of polyamines, the induction of system y^+ during liver regeneration also may be linked to the enhanced polyamine synthesis required for cell proliferation. Induction of arginine transport and NO synthesis is not uniformly coordinate, as hepatic arginine transport is enhanced by glucocorticoids (135), whereas glucocorticoids inhibit hepatic iNOS induction (140,141).

E. Regulation of Arginine Transport and NO Synthesis

The strong dependence of NO synthesis—particularly in iNOS-expressing cells—on extracellular arginine concentration points to the arginine transport system as an important control point for physiologic and pharmacologic regulation. For example, induced NO synthesis in stimulated rat astrocytes is strictly dependent on co-induction of CAT-2 gene expression (142), underscoring the significance of the co-induction of iNOS and CATs described in preceding sections. Because system y^+ transports lysine and ornithine as well as arginine, arginine uptake—particularly in cell culture models—can be reduced simply by competitive inhibition with these amino acids, and this can result in reduced NO synthesis (e.g., 119). Similarly, certain arginine analogs that are NOS inhibitors also can competitively inhibit arginine uptake. These include L-NMMA, N^G-nitro-L-ornithine (L-NIO), and asymmetric N^G, N^G-dimethyl-L-arginine (ADMA); other NOS inhibitors such as aminoguanidine, L-NAME, and L-NNA are not effective inhibitors of arginine uptake (143–145). Thus, administration of L-NMMA or L-NIO to cultured cells or intact animals will inhibit NO synthesis not only by inhibiting NOS but also by inhibiting arginine uptake. This raises the possibility that these compounds may have side effects as a consequence of reducing substrate availability for other arginine-requiring pathways.

The fact that CATs can exchange lysine for arginine has been exploited to determine the dependence of cellular NO synthesis on extracellular versus intracellular arginine pools (146). Incubation of an activated murine macrophage cell line in 2 mM lysine in the absence of arginine for 2 hours reduced intracellular arginine pools from 2 mM to 0.3 mM and totally abolished NO production, indicating a virtually complete reliance on extracellular arginine to support NO synthesis. In contrast, incubating a human endothelial cell line in 2 mM lysine in the absence of arginine for up to 48 hours reduced intracellular arginine pools from 1.5 mM to about 0.68 mM without reducing NO production, suggesting the presence of an intracellular arginine pool that is accessible to eNOS but not freely exhangeable with extracellular arginine. These results suggest that inhibitors of arginine transport could selectively inhibit NO synthesis by iNOS-expressing cells without reducing NO synthesis in cells expressing eNOS.

V. CONCLUDING REMARKS

The many studies summarized here indicate that the relative contributions of arginine synthesis, degradation, and transport in regulating cellular NO synthesis at specific anatomic sites in vivo are a complex function of cell type and vascular perfusion. As these processes are regulated by many of the same stimuli that regulate expression of iNOS, the concentrations of various inflammatory and other stimuli also are critical in determining the relative impact of each of these processes. At present, there is relatively little quantitative information regarding the impact of each of these processes on NO synthesis in specific cell types in vivo or in culture, and it is important that research efforts be directed toward obtaining such information. The strong dependence of cellular NO synthesis on extracellular arginine suggests that among the processes of arginine synthesis, degradation and transport, arginine transport is likely to be, as a general rule, the principal step regulating delivery of arginine to iNOS, as well to other NOS isoforms. In healthy men, for example, plasma arginine provides at least half the argininc used in whole body NO synthesis (147).

Essentially all the key proteins involved in arginine synthesis, degradation, and transport have been cloned. Thus, it is now possible not only to more precisely characterize expression of these proteins in health and disease but also to manipulate these processes by recombinant DNA techniques to directly test their impact on synthesis of NO by the various NOS isoforms and to conduct structure-function studies on purified recombinant proteins in order to design highly specific inhibitors. The next few years should provide a wealth of exciting new information on regulation of arginine metabolism and its relationship to NO synthesis.

ACKNOWLEDGMENTS

Because of space limitations, the author regrets that not all relevant publications could be cited. Work in the author's laboratory was supported in part by NIH grants DK33144 and GM50897.

REFERENCES

1. Morris SM Jr, Billiar TR. New insights into the regulation of induced nitric oxide synthesis. Am J Physiol 1994; 266:E829–E839.
2. Baek KJ, Thiel BA, Lucas S, Stuehr DJ. Macrophage nitric oxide subunits. Purification, characterization, and role of prosthetic groups and substrate in regulating their association into a dimeric enzyme. J Biol Chem 1993; 268:21120–21129.
3. Sennequier N, Stuehr DJ. Analysis of substrate-induced electronic, catalytic, and structural changes in inducible NO synthase. Biochemistry 1996; 35:5883–5892.
4. Visek WJ. Arginine and disease states. J Nutr 1985; 115:532–541.
5. Barbul A. Arginine: biochemistry, physiology, and therapeutic implications. JPEN 1986; 10:227–238.
6. Barbul A, Dawson H. Arginine and immunity, In: Fourse RA, Editor.Diet, Nutrition, and Immunity. Boca Raton, FL:CRC Press, 1994:199–216.
7. Ketteler M, Border WA, Noble NA. Cytokines and L-arginine in renal injury and repair. Am J Physiol 1994; 267:F197–F207.
8. Reyes AA, Karl IE, Klahr S. Role of arginine in health and in renal disease. Am J Physiol 1994; 267:F331–F346.
9. Rabier D, Narcy C, Bardet J, Parvy P, Saudubray JM, Kamoun P. Arginine remains an essential amino acid after liver transplantation in urea cycle deficiencies. J Inher Metab Dis 1991; 13:277–280.
10. Cooper AJL, Nieves E, Coleman AE, Filc-DeRicco S, Gelbard AS. Short-term metabolic fate of [^{13}N]ammonia in rat liver in vivo. J Biol Chem 1987; 262:1073–1080.
11. Geissler A, Kanamori K, Ross BD. Real-time study of the urea cycle using ^{15}N n.m.r. in the isolated perfused rat liver. Biochem J 1992; 287:813–820.
12. Morris SM Jr. Regulation of enzymes of urea and arginine synthesis. Annu Rev Nutr 1992; 12:81–101.
13. Beaudet AL, O'Brien WE, Bock HGO, Freytag SO, Su TS. The human argininosuccinate synthetase locus and citrullinemia. Adv Hum Genet 1986; 15:161–196.
14. Hecker M, Sessa WC, Harris HJ, Anggard EE, Vane JR. The metabolism of

L-arginine and its significance for the biosynthesis of endothelium-derived relaxing factor: Cultured endothelial cells recycle L-citrulline to L-arginine. Proc Natl Acad Sci USA 1990; 87:8612–8616.

15. Morris SM Jr, Nakayama DK, Nussler AK, Liu ZZ, Davies P, Pitt BR, Simmons RL, Billiar TR. Co-induction of nitric oxide synthase and argininosuccinate synthetase gene expression: implications for regulation of nitric oxide synthesis, In: Moncada S, Freelisch M, Busse R, and Higgs EA, Editor.The Biology of Nitric Oxide. Vol. 4. Enzymology, Biochemistry, and Immunology. London: Portland Press Ltd., 1994:301–303.
16. Nussler AK, Billiar TR, Liu ZZ, Morris SM Jr. Coinduction of nitric oxide synthase and argininosuccinate synthetase in a murine macrophage cell line. Implications for regulation of nitric oxide production. J Biol Chem 1994; 269:1257–1261.
17. Hattori Y, Campbell EB, Gross SS. Argininosuccinate synthetase mRNA and activity are induced by immunostimulants in vascular smooth muscle. J Biol Chem 1994; 269:9405–9408.
18. Baydoun AR, Bogle RG, Pearson JD, Mann GE. Discrimination between citrulline and arginine transport in activated macrophages: inefficient synthesis of NO from recycling of citrulline to arginine. Br J Pharmacol 1994; 112:487–492.
19. Flodstrom M, Nieman A, Bedoya FJ, Morris SM Jr, Eizirik DL. Expression of the citrulline-nitric oxide cycle in rodent and human pancreatic β-cells. Induction of argininosuccinate synthetase by cytokines. Endocrinology 1995; 136:3200–3206.
20. Norris KA, Schrimpf JE, Flynn JL, Morris SM Jr. Enhancement of macrophage microbicidal activity: Supplemental arginine and citrulline augment nitric oxide production in murine peritoneal macrophages and promote intracellular killing of *Trypanosoma cruzi*. Infect Immun 1995; 63:2793–2796.
21. Nussler AK, Liu ZZ, Hatakeyama K, Geller DA, Billiar TR, Morris SM Jr. A cohort of supporting metabolic enzymes is coinduced with nitric oxide synthase in human tumor cell lines. Cancer Lett 1996; 103:79–84.
22. Flodstrom M, Morris SM Jr, Eizirik DL. Role of the citrulline-nitric oxide cycle in the functional response of adult human and rat pancreatic islets to cytokines. Cytokine 1996; 8:642–650.
23. Cheung CW, Cohen NS, Raijman L. Channeling of urea cycle intermediates in situ in permeabilized hepatocytes. J Biol Chem 1989; 264:4038–4044.
24. Marzinzig M, Nussler AK, Stadler J, Marzinzig E, Bartlen W, Morris SM Jr, Nussler NC, Beger HG, Bruckner UB. Improved methods to measure end products of nitric oxide in biological fluids: nitrite, nitrate, and S-nitrosothiols. Nitric Oxide Biol Chem 1997; 1:177–189.
25. Granger DL, Hibbs JB Jr, Perfect JR, Durack DT. Metabolic fate of L-arginine in relation to microbiostatic capability of murine macrophages. J Clin Invest 1990; 85:264–273.
26. Wu G, Brosnan JT. Macrophages can convert citrulline into arginine. Biochem J 1992; 281:45–48.
27. Wu G, Flynn NE. The activation of the arginine-citrulline cycle in macrophages from the spontaneously diabetic BB rat. Biochem J 1993; 294:113–118.

28. Billiar TR, Nussler AK, Liu Z, Toczek T, Geller DA, Simmons RL, Morris SM Jr. Coinduction of argininosuccinate synthetase and nitric oxide synthase by LPS and interferon-gamma in murine macrophages. J Leuko Biol 1992; 52:18.
29. Simmons WW, Closs EI, Cunningham JM, Smith TW, Kelly RA. Cytokines and insulin induce cationic transporter (CAT) expression in cardiac myocytes. J Biol Chem 1996; 271:11694–11702.
30. Simmons WW, Ungureanu-Longrois D, Smith GK, Smith TW, Kelly RA. Glucocorticoids regulate inducible nitric oxide synthase by inhibiting tetrahydrobiopterin synthesis and L-arginine transport. J Biol Chem 1996; 271:23928–23937.
31. Hattori Y, Shimoda SI, Gross SS. Effect of lipopolysaccharide treatment in vivo on tissue expression of argininosuccinate synthetase and argininosuccinate lyase mRNAs: relationship to nitric oxide synthase. Biochem Biophys Res Commun 1995; 215:148–153.
32. Nagasaki A, Gotoh T, Takeya M, Yu Y, Takiguchi M, Matsuzaki H, Takasuki K, Mori M. Coinduction of nitric oxide synthase, argininosuccinate synthetase, and argininosuccinate lyase in lipopolysaccharide-treated rats. RNA blot, immunoblot, and immunohistochemical analyses. J Biol Chem 1996; 271:2658–2662.
33. Sessa WC, Hecker M, Mitchell JA, Vane JR. The metabolism of L-arginine and its significance for the biosynthesis of endothelium-derived relaxing factor: L-glutamine inhibits the generation of L-arginine by cultured endothelial cells. Proc Natl Acad Sci USA 1990; 87:8607–8611.
34. Wu G, Meininger CJ. Regulation of L-arginine synthesis from L-citrulline by L-glutamine in endothelial cells. Am J Physiol 1993; 265:H1965–H1971.
35. Su Y, Block ER. Hypoxia inhibits L-arginine synthesis from L-citrulline in porcine pulmonary artery endothelial cells. Am J Physiol 1995; 269:L581–L587.
36. Arnal JF, Munzel T, Venema RC, James NL, Bai C, Mitch WE, Harrison DG. Interactions between L-arginine and L-glutamine change endothelial NO production. An effect independent of NO synthase substrate availability. J Clin Invest 1995; 95:2565–2572.
37. Nakamura H, Saheki T, Nakagawa S. Differential cellular localization of enzymes of L-arginine metabolism in the rat brain. Brain Res 1990; 530:108–112.
38. Nakamura H, Saheki T, Ichiki H, Nakata K, Nakagawa S. Immunocytochemical localization of argininosuccinate synthetase in the rat brain. J Comp Neurol 1991; 312:652–679.
39. Arnt-Ramos LR, O'Brien WE, Vincent SR. Immunohistochemical localization of argininosuccinate synthetase in the rat brain in relation to nitric oxide synthase-containing neurons. Neuroscience 1992; 51:773–789.
40. Chen FY, Lee TJF. Arginine synthesis from citrulline in perivascular nerves of cerebral artery. J Pharmacol Exp Ther 1995; 273:895–901.
41. Shuttleworth CWR, Burns AJ, Ward SM, O'Brien WE, Sanders KM. Recycling of L-citrulline to sustain nitric oxide-dependent enteric neurotransmission. Neuroscience 1995; 68:1295–1304.
42. Nussler AK, Billiar TR. Inflammation, immunoregulation, and inducible nitric oxide synthase. J Leukoc Biol 1993; 54:171–178.
43. Pastor CM, Morris SM Jr, Billiar TR. Sources of arginine for induced nitric oxide synthesis in the isolated perfused liver. Am J Physiol 1995; 269:G861–G866.

44. Wettstein M, Gerok W, Haussinger D. Endotoxin-induced nitric oxide synthesis in the perfused rat liver: effects of L-arginine and ammonium chloride. Hepatology 1994; 19:641–647.
45. Stadler J, Barton D, Beil-Moeller H, Diekmann S, Hierholzer C, Erhard W, Heidecke CD. Hepatocyte nitric oxide inhibits glucose output and competes with urea synthesis for arginine. Am J Physiol 1995; 268:G183–G188.
46. Clowes GHA, Heideman M, Lindberg B, Randall HT, Hirsch EF, Cha CJ, Martin H. Effects of parenteral alimentation on amino acid metabolism in septic patients. Surgery 1980; 88:531–540.
47. Clowes GHA, Randall HT, Cha CJ. Amino acid metabolism in septic and traumatized patients. JPEN 1980; 4:195–205.
48. Christensen ML, Burgess J, Helms RA, Mirro J Jr, Kalwinsky DK, Storm MC. Plasma amino acids in patients with acute nonlymphocytic leukemia receiving parenteral nutrition. Ann Pharmacother 1993; 27:146–150.
49. Xie L, Gross SS. Argininosuccinate synthetase overexpression in vascular smooth muscle cells potentiates immunostimulant-induced NO production. J Biol Chem 1997; 272:16624–16630.
50. Castillo L, Chapman TE, Sanchez M, Yu YM, Burke JF, Ajami AM, Vogt J, Young VR. Plasma arginine and citrulline kinetics in adults given adequate and arginine-free diets. Proc Natl Acad Sci USA 1993; 90:7749–7753.
51. Castillo L, Sanchez M, Chapman TE, Ajami A, Burke JF, Young VR. The plasma flux and oxidation rate of ornithine adaptively decline with restricted arginine intake. Proc Natl Acad Sci USA 1994; 91:6393–6397.
52. Grody WW, Dizikes GJ, Cederbaum SD. Human arginase isozymes. Isozymes Curr Topics Biol Med Res 1987; 13:181–214.
53. Jenkinson CP, Grody WW, Cederbaum SD. Comparative properties of arginases. Comp Biochem Physiol 1996; 114B:107–132.
54. Jenkinson CP, Vockley JG, Kern RM, Cederbaum SD. Isolation and characterization of a novel human arginase. FASEB J 1996; 10:A1372.
55. Morris SM Jr, Bhamidipati D, Kepka-Lenhart D. Human type II arginase: sequence analysis and tissue-specific expression. Gene 1997; 193:157-161.
56. Gotoh T, Sonoki T, Nagasaki A, Terada K, Takiguchi M, Mori M. Molecular cloning of cDNA for nonhepatic mitochondrial arginase (arginase II) and comparison of its induction with nitric oxide synthase in a murine macrophage-like cell line. FEBS Lett 1996; 395:119–122.
57. Vockley JG, Jenkinson CP, Shukula H, Kern RM, Grody WW, Cederbaum SC. Cloning and characterization of the human type II arginase gene. Genomics 1996; 38:118–123.
58. Currie GA. Activated macrophages kill tumor cells by releasing arginase. Nature 1978; 273:758–759.
59. Ryan JL, Yohe WB, Morrison DC. Stimulation of peritoneal cell arginase by bacterial lipopolysaccharide. Am J Pathol 1980; 99:451–462.
60. Jakway JP, Morris HG, Blumenthal EJ, Talmage DW. Serum factors required for arginase induction in macrophages. Cell Immunol 1980; 54:253–263.
61. Lehmann V, Benninghoff B, Droge W. Tumor necrosis factor-induced activation

of peritoneal macrophages is regulated by prostaglandin E_2 and cAMP. J Immunol 1988; 141:587–591.
62. Jansen A, Lewis S, Cattell V, Cook TH. Arginase is a major pathway of L-arginine metabolism in nephritic glomeruli. Kidney Int 1992; 42:1107–1112.
63. Corraliza IM, Campo ML, Soler G, Modollel M. Determination of arginase activity in macrophages: a micromethod. J Immunol Meth 1994; 174:231–235.
64. Corraliza IM, Soler G, Eichmann K, Modollel M. Arginase induction by suppressors of nitric oxide synthesis (IL-4, IL-10, and PGE_2) in murine bone marrow-derived macrophages. Biochem Biophys Res Commun 1995; 206:667–673.
65. Modollel M, Corraliza IM, Link F, Soler G, Eichmann K. Reciprocal regulation of the nitric oxide synthase/arginase balance in mouse bone marrow-derived macrophages by TH1 and TH2 cytokines. Eur J Immunol 1995; 25:1101–1104.
66. Wang WW, Jenkinson CP, Griscavage JM, Kern RM, Arabolos NS, Byrns RE, Cederbaum SD, Ignarro LJ. Co-induction of arginase and nitric oxide synthase in murine macrophages activated by lipopolysaccharide. Biochem Biophys Res Commun 1995; 210:1009–1016.
67. Morris SM Jr. Arginine metabolism and nitric oxide production: harmony or discord? Second International Conference on the Biochemistry and Molecular Biology of Nitric Oxide, Los Angeles, CA, July 13–16, 1996.
68. Sonoki T, Nagasaki A, Gotoh T, Takiguchi M, Takeya M, Matsuzakis H, Mori M. Coinduction of nitric-oxide synthase and arginase I in cultured rat peritoneal macrophages and rat tissues *in vivo* by lipopolysaccharide. J Biol Chem 1997; 272:3689–3693.
69. Benninghoff B, Droge W, Lehmann V. The lipopolysaccharide-induced stimulation of peritoneal macrophages involves at least two signal pathways. Eur J Biochem 1989; 179:589–594.
70. Imai Y, Kolb H, Burkart V. Nitric oxide production from macrophages is regulated by arachidonic acid metabolites. Biochem Biophys Res Commun 1993; 197:105–109.
71. Amin AR, Vyas P, Attur M, Leszczynska-Piziak J, Patel IR, Weissmann G, Abramson SB. The mode of action of aspirin-like drugs: effect on nitric oxide synthase. Proc Natl Acad Sci USA 1995; 92:7926–7930.
72. Kepka-Lenhart D, Chen LC, Morris SM Jr. Novel actions of aspirin and sodium salicylate: discordant effects on nitric oxide synthesis and induction of nitric oxide synthase mRNA in a murine macrophage cell line. J Leukoc Biol 1996; 59:840–849.
73. Albina JE, Henry WL Jr, Mastrofrancesco B, Martin BA, Reichner JS. Macrophage activation by culture in an anoxic environment. J Immunol 1995; 155:4391–4396.
74. Boutard V, Havouis R, Fougueray B, Philippe C, Moulinoux JP, Baud L. Transforming growth factor-b stimulates arginase activity in macrophages. J Immunol 1995; 155:2077–2084.
75. Albina JE, Mills CD, Barbul A, Thirkill CE, Henry WL Jr, Mastrofrancesco B, Caldwell MD. Arginine metabolism in wounds. Am J Physiol 1988; 254:E459–E467.

76. Albina JE, Abate JA, Mastrofrancesco B. Role of ornithine as a proline precursor in healing wounds. J Surg Res 1993; 55:97–102.
77. Schneemann M, Schoedon G, Hofer S, Blau N, Guerrero L, Schaffner A. Nitric oxide synthase is not a constituent of the antimicrobial armature of human mononuclear phagocytes. J Infect Dis 1993; 167:1358–1363.
78. Wu G, Meininger CJ. Impaired arginine metabolism and NO synthesis in coronary endothelial cells of the spontaneously diabetic BB rat. Am J Physiol 1995; 269:H1312–H1318.
79. Buga GM, Singh R, Pervin S, Rogers NE, Schmitz DA, Jenkinson CP, Cederbaum SD, Ignarro LJ. Arginase activity in endothelial cells: inhibition by N^G-hydroxyarginine during high-output nitric oxide production. Am J Physiol 1996; 271:H1988–H1998.
80. Inoue Y, Bode BP, Beck DJ, Li AP, Bland KI, Souba WW. Arginine transport in human liver. Characterization and effects of nitric oxide synthase inhibitors. Ann Surg 1993; 218:350–363.
81. M'Rabet-Touil H, Blachier F, Morel MT, Darcy-Vrillon B, Duee PH. Characterization and ontogenesis of nitric oxide synthase activity in pig enterocytes. FEBS Lett 1993; 331:243–247.
82. Wu G. Urea synthesis in enterocytes of developing pigs. Biochem J 1995; 312:717–723.
83. Griffith OW, Stuehr DJ. Nitric oxide synthases: properties and catalytic mechanism. Annu Rev Physiol 1995; 57:707–736.
84. Hey C, Boucher J-L, Vadon-Le Goff S, Ketterer G, Wessler I, Racke K. Inhibition of arginase in rat and rabbit alveolar macrophages by N^w-hydroxy-D,L-indospicine, effects on L-arginine utilization by nitric oxide synthase. Br J Pharmacol 1997; 121:395–400.
85. Currie GA, Gyure L, Cifuentes L. Microenvironmental arginine depletion by macrophages in vivo. Br J Cancer 1979; 39:613–620.
86. Kung JT, Brooks SB, Jakway JP, Leonard LL, Talmage DW. Suppression of in vitro cytotoxic response by macrophages due to induced arginase. J Exp Med 1977; 146:665–672.
87. Vodovotz Y, Kwon NS, Pospischil M, Manning J, Paik J, Nathan C. Inactivation of nitric oxide synthase after prolonged incubation of mouse macrophages with IFN-g and bacterial lipopolysaccharide. J Immunol 1994; 152:4110–4118.
88. Nichols WK, Prosser FH. Induction of ornithine decarboxylase in macrophages by bacterial lipopolysaccharides (LPS) and mycobacterial cell wall material. Life Sci 1980; 27:913–920.
89. Shurtleff SA, McElwain CM, Taffet SM. Rapid expression of ornithine decarboxylase mRNA in a macrophage-like cell line: cAMP repression of the requirement for prior protein synthesis. J Cell Physiol 1988; 134:453-459.
90. Taffet SM, Singhel KJ, Overholtzer JF, Shurtleff SA. Regulation of tumor necrosis factor expression in a macrophage-like cell line by lipopolysaccharide and cyclic AMP. Cell Immunol 1989; 120:291–300.
91. Carvajal N, Cederbaum SD. Kinetics of inhibition of rat liver and kidney arginases by proline and branched-chain amino acids. Biochim Biophys Acta 1986; 870:181–184.

92. Hibbs JB Jr, Taintor RR, Vavrin Z, Rachlin EM. Nitric oxide: a cytotoxic activated macrophage effector molecule. Biochem Biophys Res Commun 1988; 157:87–94.
93. Hecker M, Nematollahi H, Hey C, Busse R, Racke K. Inhibition of arginase by N^G-hydroxy-L-arginine in alveolar macrophages: implications for the utilization of L-arginine for nitric oxide synthesis. FEBS Lett 1995; 359:251–254.
94. Boucher JL, Genet A, Vadon S, Delaforge M, Mansuy D. Formation of nitrogen oxides and citrulline upon oxidation of N^ω-hydroxy-L-arginine by hemeproteins. Biochem Biophys Res Commun 1992; 184:1158–1164.
95. Boucher JL, Genet A, Vadon S, Delaforge M, Henry Y, Mansuy D. Cytochrome P450 catalyses the oxidation of N^ω-hydroxy-L-arginine by NADPH and O_2 to nitric oxide and citrulline. Biochem Biophys Res Commun 1992; 187:880–886.
96. Ratner S. Enzymatic synthesis of arginine (condensing and splitting enzymes). Meth Enzymol 1955; II:356–367.
97. Robertson CA, Green BG, Niedzwiecki L, Harrison RK, Grant SK. Effect of nitric oxide synthase substrate analog inhibitors on rat liver arginase. Biochem Biophys Res Commun 1993; 197:523–528.
98. Hrabak A, Bajor T, Temesi A. Comparison of substrate and inhibitor specificity of arginase and nitric oxide (NO) synthase for arginine analogues and related compounds in murine and rat macrophages. Biochem Biophys Res Commun 1994; 198:206–212.
99. Daghigh F, Fukuto JM, Ash DE. Inhibition of rat liver arginase by an intermediate in NO biosynthesis, N^G-hydroxy-L-arginine: implications for the regulation of nitric oxide biosynthesis by arginase. Biochem Biophys Res Commun 1994; 202:174–180.
100. Kontos HA, Wei EP. Arginine analogues inhibit responses mediated by ATP-sensitive K^+ channels. Am J Physiol 1996; 271:H1498–H1506.
101. Reczkowski R, Ash DE. Rat liver arginase: kinetic mechanism, alternate substrates, and inhibitors. Arch Biochem Biophys 1994; 312:31–37.
102. Hrabak A, Bajor T, Temesi A, Meszaros G. The inhibitory effect of nitrite, a stable product of nitric oxide (NO) formation, on arginase. FEBS Lett 1996; 390:203–206.
103. Boucher JL, Custot J, Vadon S, Delaforge M, Lepoivre M, Tenu JP, Yapo A, Mansuy D. N^ω-hydroxy-L-arginine, an intermediate in the L-arginine to nitric oxide pathway, is a strong inhibitor of liver and macrophage arginase. Biochem Biophys Res Commun 1994; 203:1614–1621.
104. Vadon S, Custot J, Boucher J-L, Mansuy D. Synthesis and effects on arginase and nitric oxide synthase of two novel analogues of N^ω-hydroxyarginine, N^ω-hydroxyindospicine and *p*-hydroxyamidinophenylalanine. J Chem Soc Perkin Trans 1996; 1:645–648.
105. Custot J, Boucher JL, Vadon S, Guedes C, Dijols S, Delaforge M, Mansuy D. N^ω-hydroxyamino-a-amino acids as a new class of very strong inhibitors of arginases. J Biol Inorg Chem 1996; 1:73–82.
106. Custot J, Moali C, Brolio M, Boucher JL, Delaforge M, Mansuy D, Tenu JP, Zimmerman JL. The new a-amino acid N^ω-hydroxy-nor-L-arginine: a high-affinity

inhibitor of arginase well adapted to bind to its manganese cluster. J Am Chem Soc 1997; 119:4086–4087.

107. Kanyo ZF, Scolnick LR, Ash DE, Christianson DW. Structure of a unique binuclear manganese cluster in arginase. Nature 1996; 383:554–557.

108. Perozich J, Hempel J, Morris SM Jr. Roles of conserved residues in the arginase family. Biochim Biophys Acta 1998; 1382:23–37.

109. Li G, Regunathant S, Barrow J, Eshraghi J, Cooper R, Reis DJ. Agmatine: an endogenous clonidine-displacing substance in the brain. Science 1994; 263:966–969.

110. Galea E, Regunathan S, Eliopoulos V, Feinstein DL, Reis DJ. Inhibition of mammalian nitric oxide synthases by agmatine, an endogenous polyamine formed by decarboxylation of arginine. Biochem J 1996; 316:247–249.

111. Lortie MJ, Novotny WF, Peterson OW, Vallon V, Malvey K, Mendonca M, Satriano J, Insel P, Thomson SC, Blantz RC. Agmatine, a bioactive metabolite of arginine. Production, degradation, and functional effects in the kidney of the rat. J Clin Invest 1996; 97:413–420.

112. Kilberg MS, Stevens BR, Novak DA. Recent advances in mammalian amino acid transport. Annu Rev Nutr 1993; 13:137–165.

113. Van Winkle LJ. Endogenous amino acid transport systems and expression of mammalian amino acid transport proteins in *Xenopus* oocytes. Biochim Biophys Acta 1993; 1154:157–172.

114. Malandro MS, Kilberg MS. Molecular biology of mammalian amino acid transporters. Annu Rev Biochem 1996; 65:305–336.

115. Closs EI. CATs, a family of three distinct mammalian cationic amino acid transporters. Amino Acids 1996; 11:193–208.

116. MacLeod CL, Kakuda DK. Regulation of CAT-Cationic amino acid transporter gene expression. Amino Acids 1996; 11:171–191.

117. Closs EI, Lyons CR, Kelly C, Cunningham JM. Characterization of the third member of the MCAT family of cationic amino acid transporters. J Biol Chem 1993; 268:20796–20800.

118. Sato H, Fujiwara M, Bannai S. Effect of lipopolysaccharide on transport and metabolism of arginine in mouse peritoneal macrophages. J Leukoc Biol 1992; 52:161–164.

119. Bogle RG, Baydoun AR, Pearson JD. L-arginine transport is increased in macrophages generating nitric oxide. Biochem J 1992; 284:15–18.

120. Baydoun AR, Bogle RG, Pearson JD, Mann GE. Selective inhibition by dexamethasone of induction of NO synthase, but not of induction of L-arginine transport, in activated murine macrophage J774 cells. Br J Pharmacol 1993; 110:1401–1406.

121. Schmidlin A, Wiesinger H. Transport of L-arginine in cultured glial cells. Glia 1994; 11:262–268.

122. Schmidlin A, Wiesinger H. Stimulation of arginine transport and nitric oxide production by lipopolysaccharide is mediated by different signaling pathways in astrocytes. J Neurochem 1995; 65:590–594.

123. Stevens BR, Kakuda DK, Yu K, Waters M, Vo CB, Raizada MK. Induced nitric

oxide synthesis is dependent on induced alternatively spliced CAT-2 encoding L-arginine transport in brain astrocytes. J Biol Chem 1996; 271:24017–24022.
124. Greene B, Pacitti AJ, Souba WW. Characterization of L-arginine transport by pulmonary artery endothelial cells. Am J Physiol 1993; 264:L351–L356.
125. Lind DS, Copeland EM III, Souba WW. Endotoxin stimulates arginine transport in pulmonary artery endothelial cells. Surgery 1993; 114:199–205.
126. Cendan JC, Souba WW, Copeland EM III, Lind DS. Cytokines regulate endotoxin stimulation of endothelial cell arginine transport. Surgery 1995; 117:213–219.
127. Block ER, Herrera H, Couch M. Hypoxia inhibits L-arginine uptake by pulmonary artery endothelial cells. Am J Physiol 1995; 269:L574–L580.
128. Durante W, Liao L, Iftikhar I, O'Brien WE, Schafer AI. Differential regulation of L-arginine transport and nitric oxide production by vascular smooth muscle and endothelium. Circ Res 1996; 78:1075–1082.
129. Wileman SM, Mann GE, Baydoun AR. Induction of L-arginine transport and nitric oxide synthase in vascular smooth muscle cells: synergistic actions of pro-inflammatory cytokines and bacterial lipopolysaccharide. Br J Pharmacol 1995; 116:3243–3250.
130. Durante W, Liao L, Schafer AI. Differential regulation of L-arginine transport and inducible NOS in cultured vascular smooth muscle cells. Am J Physiol 1995; 268:H1158–H1164.
131. Gill DJ, Low BC, Grigor MR. Interleukin-1b and tumor necrosis factor-a stimulate the cat-2 gene of the L-arginine transporter in cultured vascular smooth muscle cells. J Biol Chem 1996; 271:11280–11283.
132. White MF, Christensen HN. Cationic amino acid transport into cultured animal cells. II. Transport system barely perceptible in ordinary hepatocytes, but active in hepatoma cell lines. J Biol Chem 1982; 257:4450–4457.
133. Pacitti AJ, Copeland EM, Souba WW. Stimulation of hepatocyte system y^+-mediated L-arginine transport by an inflammatory agent. Surgery 1992; 112:403–411.
134. Inoue Y, Bode BP, Souba WW. Hepatic Na^+-independent amino acid transport in endotoxemic rats: evidence for selective stimulation of arginine transport. Shock 1994; 2:164–172.
135. Watkins KT, Dudrick PS, Copeland EM, Souba WW. Interleukin-6 and dexamethasone work coordinately to augment hepatic amino acid transport. J Trauma 1994; 36:523–528.
136. Watkins KT, Copeland EM, Souba WW. Induction of hepatic system y^+-mediated L-arginine transport by prostaglandin E_2. Am J Surg 1994; 167:128–134.
137. Obolenskaya MY, Vanin AF, Mordvintcev PI, Mulsch A, Decker K. EPR evidence of nitric oxide production by the regenerating rat liver. Biochem Biophys Res Commun 1994; 202:571–576.
138. Obolenskaya M, Schulze-Specking A, Plaumann B, Frenzer K, Freudenberg N, Decker K. Nitric oxide production by cells isolated from regenerating rat liver. Biochem Biophys Res Commun 1994; 204:1305–1311.
139. Wu JY, Robinson D, Kung HJ, Hatzoglou M. Hormonal regulation of the gene for the type C ecotropic retrovirus receptor in rat liver cells. J Virol 1994; 68:1615–1623.

140. Geller DA, Nussler AK, Di Silvio M, Lowenstein CJ, Shapiro RA, Wang SC, Simmons RL, Billiar TR. Cytokines, endotoxin, and glucocorticoids regulate the expression of inducible nitric oxide synthase in hepatocytes. Proc Natl Acad Sci USA 1993; 90:522–526.
141. Geller DA, Freeswick PD, Nguyen D, Nussler AK, Di Silvio M, Shapiro RA, Wang SC, Simmons RL, Billiar TR. Differential induction of nitric oxide synthase in hepatocytes during endotoxemia and the acute-phase response. Arch Surg 1994; 129:165–171.
142. Stevens BR, Kakuda DK, Yu K, Waters M, Vo CB, Raizada MK. Induced nitric oxide synthesis is dependent on induced alternatively spliced CAT-2 encoding L-arginine transport in brain astrocytes. J Biol Chem 1996; 271:24017–24022.
143. Schmidt K, Klatt P, Mayer B. Characterization of endothelial cell amino acid transport systems involved in the actions of nitric oxide synthase inhibitors. Mol Pharmacol 1993; 44:615–621.
144. Baydoun AR, Mann GE. Selective targeting of nitric oxide synthase inhibitors to system y+ in activated macrophages. Biochem Biophys Res Commun 1994; 200:726–731.
145. Bogle RG, MacAllister RJ, Whitley GSJ, Vallance P. Induction of N^G-monomethyl-L-arginine uptake: a mechanism for differential inhibition of NO synthases? Am J Physiol 1995; 269:C750–C756.
146. Closs EI, Graf P, Scheld JS, Sharafi M, Forstermann U. The role of cationic amino acid transporters (CATs) in the substrate supply for NOS. Second International Conference on the Biochemistry and Molecular Biology of Nitric Oxide, Los Angeles, CA, July 13–16, 1996.
147. Castillo L, Beaumier L, Ajami AM, Young VR. Whole body nitric oxide synthesis in healthy men determined from [^{15}N]arginine-to-[^{15}N]citrulline labeling. Proc Natl Acad Sci USA 1996; 93:11460–11465.

3

Inactivation of Nitric Oxide Synthase Isoforms by Metalloporphyrins

Donald J. Wolff, Darren A. Saks, Andrew Lubeskie, and Robert A. Naddelman

University of Medicine and Dentistry of New Jersey–Robert Wood Johnson Medical School, Piscataway, New Jersey

I. INTRODUCTION

Metalloporphyrins exert a variety of effects on diverse physiological processes and enzyme systems. Metalloporphyrins inhibit heme oxygenase (1,2), block the effects of metabotropic glutamate receptor activation (3), inhibit long-term potentiation (4,5), inhibit the allosteric activation of soluble guanylate cyclase (6), and repress hepatic δ-aminolevulinate synthase (7). Metalloporphyrins have been administered for the therapy of neonatal hemolytic disease or hyperbilirubinemia (8,9). In lead poisoning, lead interferes with the incorporation of iron into heme and zinc then replaces iron to form zinc protoporphyrin IX (10). Zinc protoporphyrin IX measured in whole blood fluorimetrically is considered a reliable indicator of the severity of lead poisoning. Elevations of zinc protoporphyrin IX concentrations also occur in patients suffering from iron deficiency anemia and erthyropoietic porphyria (10).

Abbreviations used: BH_4, (6R)-5,6,7,8-tetrahydro-L-biopterin; CaM, calmodulin; CHAPSO, 3-(cholamidopropyl dimethylammonio)-2-hydroxy-1-propane sulfonate; DTT, dithiothreitol; eNOS, endothelial nitric oxide synthase; iNOS, cytokine-inducible nitric oxide synthase; nNOS, neuronal nitric oxide synthase; CrMP, chromium mesoporphyrin IX; FePP, ferriprotoporphyrin IX; MnPP, manganese protoporphyrin IX; SnPP, tin protoporphyrin IX; ZnPP, zinc protoporphyrin IX; HEPES, N-(2-hydroxyethyl)piperazine-N′-2-ethane sulfonic acid; MOPS, 3-(N-morpholino) propane sulfonic acid.

Recently Meffert et al. (4) reported that metalloporphyrins inhibit nitric oxide synthase activity in hippocampal homogenates. Our laboratory has observed that zinc protopophyrin IX produces an inhibition of the citrulline-forming capability of all three nitric oxide synthase isoforms (11) that progresses with time of incubation. Given that metalloporphyrins are administered clinically and are elevated in diverse pathological states, and considering the central role of nitric oxide in the regulation of diverse physiological functions, we considered it important to explore in further detail the apparent inactivation of nitric oxide synthase by metalloporphyrins. This chapter reports the results of these studies.

II. MATERIALS AND METHODS

A. Materials

Zinc protoporphyrin IX chloride, hemin (ferriprotoporphyrin IX chloride), chromium mesoporphyrin IX dichloride, tin protoporphyrin IX dichloride, and manganese protoporphyrin IX chloride were obtained from Porphyrin Products, Inc. (Logan, UT). Murine interferon-γ was the kind gift of Dr. Sidney Pestka (Department of Molecular Genetics and Microbiology, Robert Wood Johnson Medical School, Piscataway, NJ).

B. Preparation and Assay of NO Synthase Isoforms

ADP-agarose purified nNOS, iNOS, and eNOS were prepared and characterized from rat GH_3 pituitary cells (12), interferon-γ-treated murine RAW 264.7 cells (13), and bovine pulmonary arterial endothelial cells (14) as described previously. The citrulline-forming activity of the nitric oxide synthases was measured by a modification of the procedure of Bredt and Synder (15) as described previously (12). Standard incubations (150 μL) for the measurement of citrulline formation by either endothelial constitutive nitric oxide synthase or GH_3 constitutive nitric oxide synthase contained 30 mM HEPES, pH 7.4, 1 mM dithiothreitol, 120 nM ^{3}H-arginine (a subsaturating concentration), 1 mM EGTA, 0.85 mM Ca^{2+}, CaM 6 μM, 100 μM NADPH, and 100 μM BH_4. Standard incubations for the measurement of citrulline formation by the interferon-γ-inducible macrophage nitric oxide synthase contained 30 mM HEPES, pH 7.4, 1 mM dithiothreitol, 120 nM ^{3}H-arginine, 1 mM EGTA, 100 μM NADPH, and 300 μM BH_4. Incubations were conducted at 30°C for 30 min in duplicate and the mean values were calculated. Variability of values about the mean routinely averaged ± 3% of the mean. Routinely assays were conducted at dilutions of enzyme that provided 5% to 10% of total substrate consumption. At these conditions of measurement (without added metalloporphyrin) product formation was linear over time. The substrate-independent NADPH oxidase activity of the GH_3 pituitary nNOS was measured in standard reactions mixtures of 1 mL volume containing 50 mM HEPES, pH 7.4, 100 μM NADPH, 6 μM CaM, 1 mM EGTA, and 0.85 mM Ca^{2+}. Reactions conducted in quartz cuvettes were initiated by the addition of enzyme and NADPH consumption measured from the decrease in light absorbance at 340 nm compared to an identical reference cuvette without added enzyme. NADPH consumption was calculated using an extinction coefficient of 6.34 $mM^{-1}cm^{-1}$.

Cytochrome c reductase activity measurements were conducted in disposable polystyrene cuvettes in a volume of 1 mL containing 50 mM HEPES, pH 7.4, 1 mM EGTA, 0.85 mM Ca^{2+}, 6 μM CaM, 100 μM NADPH, and 50 μM cytochrome c. Incubations were initiated by addition of nitric oxide synthase source, and absorbance changes at 550 nm were followed using a Shimadzu 160U spectrophotometer. Calculation of product formation was performed using the published absorption coefficient for reduced cytochrome c of 21 $mM^{-1}cm^{-1}$ (16).

C. Sucrose Density Gradient Centrifugation Analysis

Sucrose gradients of 12 mL volume containing 5% to 20% sucrose in 50 mM MOPS, pH 7.5, 100 mM NaCl, 1 mM EGTA, 1 mM DTT, and 0.3% CHAPSO were generated for analysis of ZnPP and protoporphyrin IX-treated nitric oxide synthase isoforms. The sedimentation marker proteins hemoglobin (4.09 S), catalase (11.9 S), and thyroglobulin (19.4 S) were included in each gradient. The samples were centrifuged in an SW 41 rotor in a Beckman L5-75 ultracentrifuge at 35,000 rpm at 4°C for 16 hours. The gradients were emptied using an ISCO 185 density gradient fractionator with online ultraviolet monitor.

D. Miscellaneous Procedures

Protein concentrations were determined according to the method of Bradford (17) with bovine serum albumin as reference standard.

III. RESULTS

A. Effect of Metalloporphyrins on the Activity of GH_3 Pituitary nNOS

It has been reported by Meffert et al. (4) that chromium mesoporphyrin inhibits the nitric oxide synthase activity of hippocampal homogenates. We decided to examine the effect of chromium mesoporphyrin on the activity of ADP-agarose-purified GH_3 pituitary nNOS in preincubations without added substrate or cofactors. This would allow the subsequent determination of the effect of arginine substrate and cofactors on metalloporphyrin-induced nitric oxide synthase inactivation. Preincubations were constructed containing CrMP at concentrations ranging from 25 to 250 μM (Fig. 1). Samples withdrawn from the preincubations at the specified times over a 15-min period were diluted 50-fold and were used immediately to initiate standard 30-min incubations for measurement of citrulline formation. The activity of the nNOS declined exponentially with time, with inactivation occurring at progressively increased rates as the concentration of CrMP in the preincubations was increased. Half-times of inactivation of 28, 13, 8, 4.8, 2.2, 1.6, and 1 min were observed at concentrations of 25, 50, 75, 100, 150, 200 and 250 μm CrMP, respectively (Fig. 1A). When the half-times of inactivation versus the reciprocal of the CrMP concentration nonsaturating plots (the ordinal intercept was zero) were obtained; that is, a saturating maximal inactivation rate and a concentration of CrMP providing a half-maximal inactivation rate could not be determined.

The effect of increasing concentrations of zinc protoporphyrin IX in a preincubation without substrate or cofactors on nNOS activity was also determined (Fig. 2). In incubations containing ZnPP, nNOS activity declined at exponential rates that increased with increasing ZnPP concentration. In a control preincubation without added metalloporphyrin, only 8% of initial activity was lost over a 15-min period. Half-times of inactivation of 12, 7.2, 4.8, 3.7, and 2.6 min were observed in preincubations containing 1.5, 3, 6, 10, and 20 μM ZnPP. In the preincubations containing 50 and 100 μM ZnPP, the zero time value declined to 65% and 18% of the values obtained in a preincubation either without added ZnPP or at concentrations ranging from 1.5 to 20 μM. This apparent decrease of activity at zero time is presumably due to the transfer of ZnPP into the subsequent incubation at concentrations sufficient despite dilution to exert an inhibitory action. As had been noted for CrMP, ZnPP inactivation rate did not display saturation at increased concentrations (data not depicted). To be assured that enzyme inactivation by ZnPP was not due to the release of zinc from its coordination linkage, nNOS was preincubated with a 30-μM solution of zinc acetate. The enzyme displayed stability equal to that of a control without addition (data not shown).

Metalloporphyrins, particularly tin protoporphyrin IX, have been shown in the presence of light and NADPH to produce CO (1,18), a known inhibitor of nitric oxide synthase (19). Since our preincubations contained NADPH derived from the ADP-agarose purification, we decided to explore whether this mechanism contributes to or is responsible for the inactivation of nitric oxide synthase. A preincubation with ZnPP was conducted in light-impermeant glass vials. The rate of inactivation by ZnPP observed was identical to that observed in the presence of ambient light (data not shown).

The effect of adding arginine substrate and tetrahydrobiopterin cofactor to the preincubation on the rate of ZnPP-induced inactivation was determined (Fig. 3). Without additions, 10 μM ZnPP produced an inactivation half-time of 6.7 min. In preincubations containing arginine substrate the inactivation half-time was

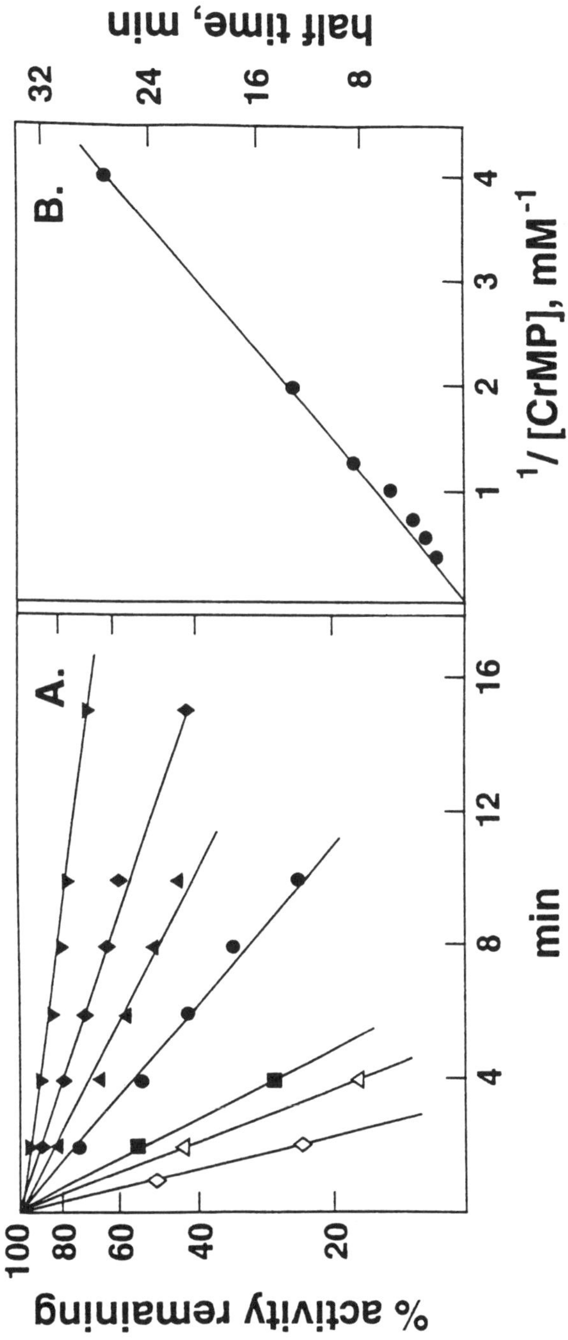

Figure 1 Chromium mesoporphyrin IX concentration dependence of inactivation of GH_3 nNOS. (A) Preincubations (120 μL) conducted at 3C° were constructed containing 3.6 μg of ADP-agarose-purified GH_3 nNOS in 25 mM Mops, pH 7.5, 0.5 mM EGTA, 0.5 mM DTT, 2.5 mM NADPH with 25 (▼), 50 (◆) 75 (▲), 100 (●), 150 (■), 200 (Δ), and 250 (◇) μmM CrMP. Samples (10 μL) were removed at the specified times and were diluted 50-fold in BSA (1 mg/mL). Dilutions were used immediately to initiate standard incubations for measurement of citrulline formation. Samples withdrawn at zero time exhibited an activity of 35 nmol citrulline formed/min-mg protein as assayed at 120 nM arginine substrate. Values of the percent activity remaining at the specified times are plotted on a semilogarithmic axis. (B) Half-times of inactivation are plotted versus the reciprocal of the CrMP concentration.

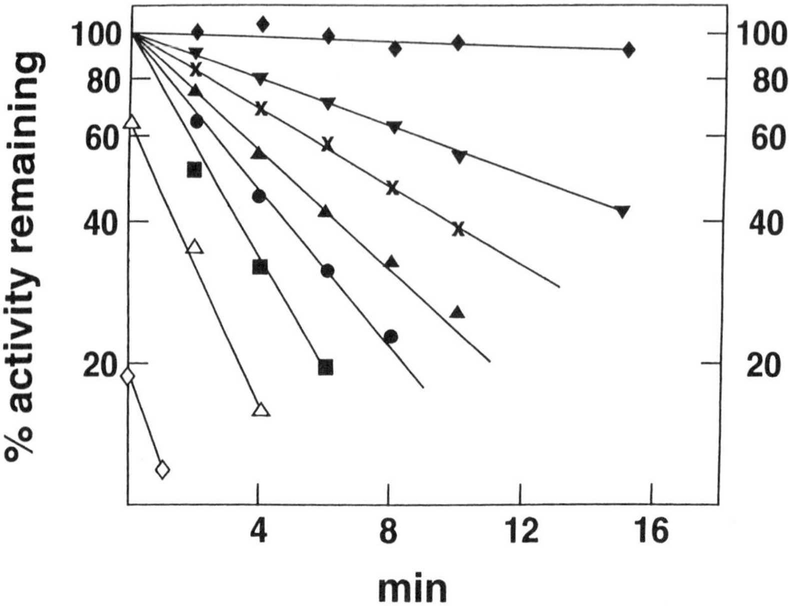

Figure 2 Zinc protoporphyrin IX concentration dependence of inactivation of GH_3 nNOS. Preincubations (120 μL) conducted at 30°C were constructed containing 3.6 μg GH_3 nNOS in 25 mM Mops, pH 7.5, 0.5 mM EGTA, 0.5 mM DTT, 2.5 mM NADPH with 0 (◆), 1.5 (▼), 3 (X), 6 (▲), 10 (●), 20 (■), 50 (Δ), or 100 (◇) μM ZnPP. Samples (10 μL) removed at the specified times were diluted 50-fold and were used immediately to initiate standard incubations for determination of citrulline-forming activity. Samples withdrawn at zero time from preincubations containing 0 to 20 μM added ZnPP exhibited an activity of 37 nmol citrulline formed/min-mg protein as assayed at 120 nM arginine substrate.

extended to 9 min, while in a preincubation containing BH_4 a half-time of inactivation of 14 min was observed. The concurrent presence of both arginine substrate and BH_4 produced in inactivation half-time of 19 min. Thus both arginine substrate and BH_4 stabilized the enzyme to ZnPP-induced inactivation, but BH_4 was particularly effective in this regard.

Imidazole is known to bind to the sixth coordination position of the heme iron of nNOS, producing a type II difference spectrum (20), and to act as a reversible inhibitor of nitric oxide synthase activity (Fig. 4) (21). We examined whether the presence of imidazole in the preincubation at a concentration fivefold above its K_d altered the rate of ZnPP-induced inactivation of nNOS. In the absence of imidazole, 10 μM ZnPP produced an inactivation half-time of 6.5 min that was extended to 16 min by the addition of imidazole.

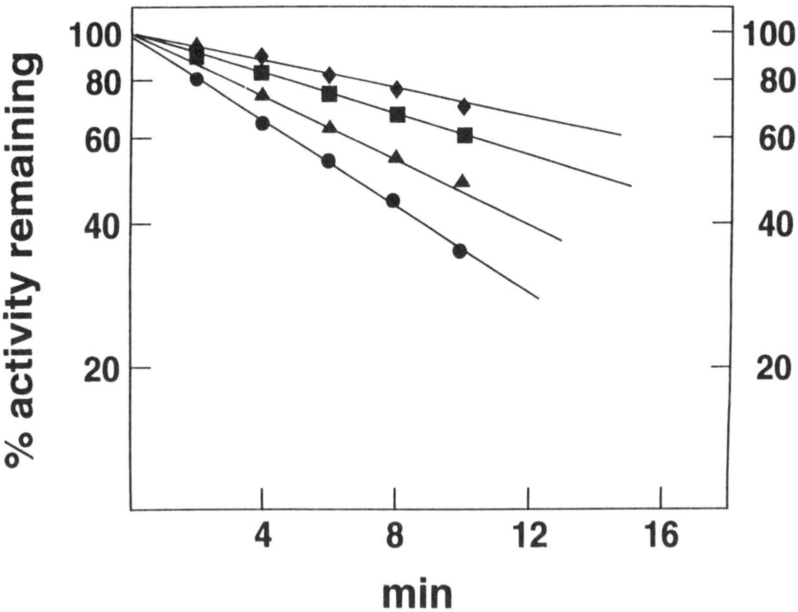

Figure 3 Effect of tetrahydrobiopterin and arginine on the inactivation of nNOS by ZnPP. Preincubations (120 μL) conducted at 30°C were constructed containing 4.2 μg of GH_3 nNOS in 25 mM Mops, pH 7.5, 0.5 mM EGTA, 0.5 mM DTT, 2.5 mM NADPH, 10 μM ZnPP without further additions (●), or with 10 μM arginine (▲), 12 μM BH_4 (■), or both 10 μM arginine and 12 μM BH_4 (◆). Samples (10 μL) removed at the specified times were diluted 50-fold and used immediately to initiate standard incubations for the determination of citrulline-forming activity. Samples withdrawn at zero time from incubations containing no added unlabeled arginine (●, ■) exhibited an activity of 24 nmol citrulline formed/min-mg protein. Samples withdrawn at zero time from preincubations containing 10 μM arginine (▲, ◆) exhibited an activity of 53 nmol/min-mg protein as adjusted based on the decreased specific activity of arginine substrate caused by the transfer of unlabeled arginine from the preincubation into the subsequent incubation. Values expressed as the percentage of the zero time activity remaining at the specified time are plotted on a logarithmic ordinal axis.

B. Effect of ZnPP on the Arginine-Independent Activities of GH_3 Pituitary nNOS

Samples of nNOS were preincubated with 10 μM ZnPP and a portion was removed either immediately or after 4, 8, 12, or 16 min of preincubation (Fig. 5). The samples were used to initiate measurements of the substrate-independent NADPH oxidase activity. NADPH oxidase activity declined exponentially with

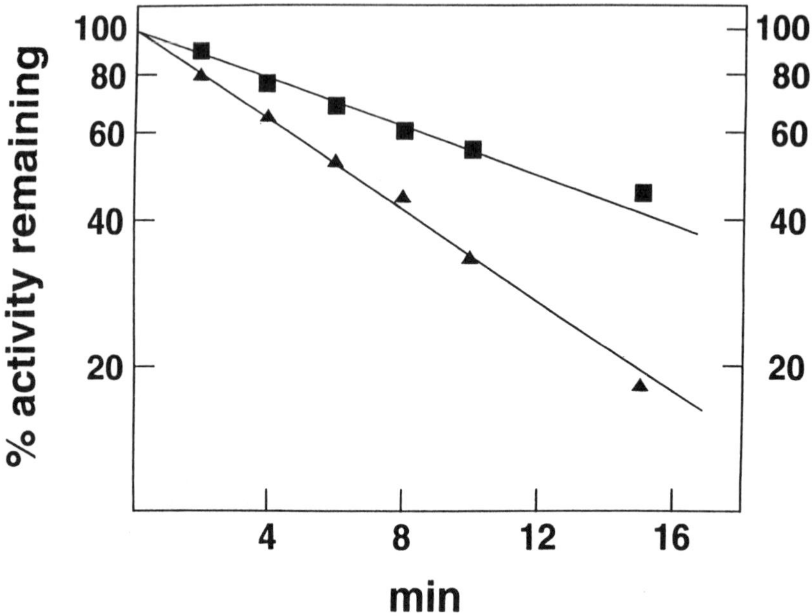

Figure 4 Effect of imidazole on inactivation of nNOS by zinc protoporphyrin IX. Preincubations (120 μL) conducted at 30°C were constructed containing 3.6 μg of ADP-agarose purified GH_3 nNOS, 2.5 mM NADPH, 10 μM ZnPP, without (▲) or containing 1 mM (■) imidazole in 25 mM Mops, pH 7.5, 0.5 mM EGTA, 0.5 mM DTT. Samples (10 μL) withdrawn at the specified times were diluted 50-fold and used immediately to initiate standard incubations for measurement of citrulline formation. Samples withdrawn at zero time exhibited an activity of 40 nmol citrulline formed/min-mg protein as assayed at 120 nM arginine.

time, with a half-time of inactivation of 9 min being observed. By contrast, preincubation of nNOS with a fivefold higher concentration of ZnPP (50 μM) over a 15-min period had no detectable effect on the cytochrome c reductase activity of nNOS.

C. Effects of Diverse Metalloporphyrins on NO Synthase Isoforms

Since the inactivation of nNOS by CrMP and ZnPP had not exhibited a maximal inactivation rate or K_I (concentration producing half-maximal inactivation; Fig. 1B), direct quantitative comparison of their relative effectiveness as inactivators was not readily obtainable. Further, direct comparison of the stability of the nNOS and iNOS isoforms under identical conditions of preincubation

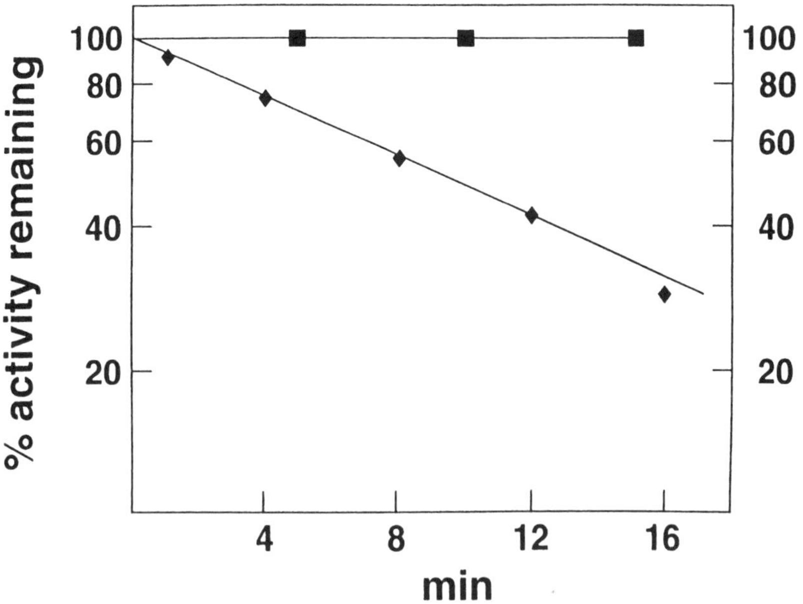

Figure 5 The effect of zinc protoporphyrin IX on the NADPH oxidase and cytochrome c reductase activities of GH_3 nNOS. ADP-agarose-purified nNOS (3.5 μg) was preincubated with 10 μM ZnPP for the specified time at 30°C. The preincubated enzyme was added to a cuvette containing 1 mL of 50 mM HEPES, pH 7.4, 100 μM NADPH, 6 μM CaM, 1 mM EGTA, 0.85 mM Ca^{2+}, and NADPH consumption was measured at 340 nm over a 4-min period (linear initial velocity measurement). Values of NADPH oxidase activity (◆) are expressed as the percentage of the zero time activity (0.14 μmol NADPH consumed/min-mg) that remain following the specified period of preincubation. Alternately, nNOS (0.7 μg) was preincubated at 30°C with 50 μM ZnPP for the specified periods of time and was assayed for cytochrome c reductase activity (■) measured as described in Materials and Methods. Values of cytochrome c reductase activity are expressed as the percentage of the zero time activity (1.2 μmol/min-mg) that remain following the specified period of preincubation.

was not possible since the iNOS isoform was found not to be stable in control preincubations without added metalloporphyrin unless a saturating concentration of BH_4 cofactor was present. To allow a comparison, we decided to determine the concentration of diverse metalloporphyrins that produced a 50% reduction of citrulline formation in standard incubations containing saturating concentrations of BH_4 (Table 1) using the three NOS isoforms. ZnPP was found to be the most potent inactivator among the metalloporphyrins examined, exhibiting IC_{50} values of 2, 6, and 7 μM versus the nNOS, iNOS, and eNOS isoforms, respectively. Preincubations containing the natural ligand ferriprotoporphyrin IX also

Table 1 IC_{50} Values of Diverse Metalloporphyrins for the Three Nitric Oxide Synthase Isoforms

Metalloporphyrin	nNOS (μM)	iNOS (μM)	eNOS (μM)
ZnPP	2	6	7
CrMP	14	20	16
FePP	6	20	40
MnPP	>>30	30	30
SnPP	5	17	15

Standard incubations conducted at 30°C for 30 min were constructed for each NOS isoform as described in Materials and Methods in the absence or presence of the various metalloporphyrins at concentrations ranging from 0.1 to 300 μM. The concentrations were generated from 3-mM stock solutions dissolved in dimethylformamide. The dimethylformamide carried into the incubation did not affect nitric oxide synthase activities at or below 30 μM metalloporphyrin (=1% dimethylformamide).

underwent inactivation, with IC_{50} values of 6, 20, and 40 μM being observed for the nNOS, iNOS, and eNOS isoforms, respectively. Tin protoporphyrin and chromium mesoporphyrin each elicited inactivation of the three NOS isoforms, while manganese protoporphyrin was a very poor inactivator. Among the isoforms examined nNOS appeared consistently to be somewhat more sensitive to metalloporphyrin-induced inactivation.

D. Effect of ZnPP Pretreatment on the Sedimentation

The active form of nitric oxide synthase capable of synthesizing nitric oxide is known for both the nNOS (22) and the iNOS (23) isoforms to have a homodimeric structure. In the case of the iNOS isoform it is known that the monomeric form of enzyme retains cytochrome c reductase activity but is unable both to reduce molecular oxygen (substrate-independent NADPH oxidase activity) and to synthesize citrulline and NO (23). Baek et al. (23) have shown that monomers can be reconstituted to functional dimers by incubation in the presence of arginine, BH_4, and stoichiometric concentrations of heme (ferriprotoporphyrin IX). At suprastoichiometric concentrations of heme, activity was reduced dramatically. Since ZnPP treatment leads to the inactivation of both the NADPH-oxidase- and citrulline-forming activities of nNOS without altered cytochrome c reductase activity, we considered it plausible that metalloporphyrins produced inactivation by promoting dissociation of fully functional dimers to partially functional monomers. To examine this proposal, untreated preparations of iNOS (Fig. 6A) and nNOS (Fig. 6C) or ZnPP-pretreated iNOS (Fig. 6B) and nNOS (Fig. 6D) were subjected to sucrose density gradient centrifugation in the

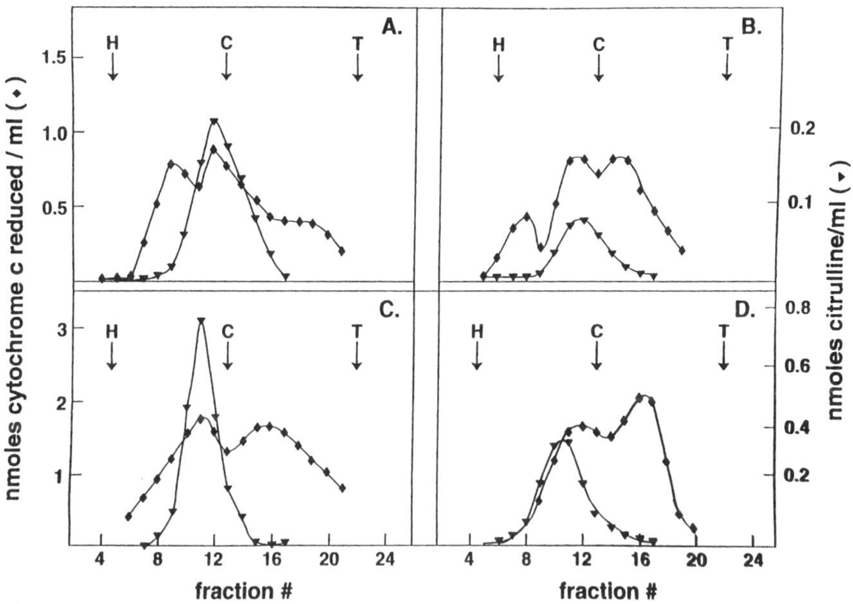

Figure 6 The effect of ZnPP pretreatment on the sedimentation of the iNOS and nNOS isoforms in sucrose density gradients. (A,B) iNOS. ADP-agarose purified RAW 264.7 iNOS (35 μg) either untreated (A) or preincubated with 20 μM ZnPP for 30 min was applied to a 5% to 20% sucrose gradient containing 10 μM BH_4 and was centrifuged at 35,000 rpm in an SW 41 rotor for 16 hours as described in Materials and Methods. Fractions (0.35 mL) were collected and were assayed for either citrulline-forming (▼) or cytochrome c reductase (◆) activity. (C,D) nNOS. ADP-agarose purified GH_3 pituitary nNOS (11 μg) was applied to 5% to 20% sucrose gradients without added BH_4. The gradients were run and analyzed as described above. Gradients contained added hemoglobin (4.09 S), catalase (11.9 S), and thryoglobulin (19.4 S) as markers whose sedimentation in the respective gradients is indicated by the arrows.

presence of sedimentation markers. Samples from across the gradients were assayed for cytochrome c reductase and citrulline forming activities. A control sample of nNOS (Fig. 6C) displayed a peak of citrulline-forming activity corresponding to fraction 11. In comparison to the sedimentation of the internal markers, this corresponded to a sedimentation coefficient of 9.8 S. Gel filtration of the nNOS isoform (data not shown) indicated that the Stokes radius of the citrulline-forming activity of nNOS is 7.7 nm. Using the procedure of Siegel and Monty (24), we calculated a native molecular weight of 308,000 corresponding to that expected for the functional dimer. The peak of activity (fraction 11) displayed a citrulline-forming activity of 0.77 nmol/min-mL and a cytochrome c reductase activity of 1.7 nmol/min-mL. The gradient (Fig. 6D) containing the ZnPP-treated nNOS sample also had a peak of residual citrulline-forming activity in fraction 11 sedimenting as a dimer. Fraction 11 from the gradient containing the ZnPP-treated sample had a citrulline-forming activity of 0.34 nmol/min-mL and a cytochrome c reductase activity of 1.5 nmol/min-mL.

The gradient containing untreated iNOS (Fig. 6A) exhibited a peak of citrulline-forming activity in fraction 12 with a sedimentation coefficient of 10.7 S, corresponding to functional dimer. Cytochrome c reductase exhibited two peaks of activity, one at fraction 9 (7.2 S) and one at fraction 12 (10.7 S). Fraction 12 (dimer) exhibited a citrulline-forming activity of 0.22 nmol/min-mL and a cytochrome c reductase activity of 0.9 nmol/min-mL. The gradient (Fig. 6B) containing the ZnPP-treated iNOS possessed a peak of citrulline-forming and cytochrome c reductase activity also at fractions 11 (10.7 S, functional dimer). This fraction exhibited a citrulline-forming activity of 0.08 nmol citrulline formed/min-mL and a cytochrome c reductase activity of 0.9 nmol/min-mL. Thus for both the nNOS and iNOS isoforms, ZnPP-treated samples possessed a dimeric enzyme form with diminished citrulline-forming capability but unaltered cytochrome c reductase activity. If ZnPP treatment produces the dissociation of dimer to monomer, one would have expected the citrulline-forming and cytochrome c reductase activities of the peak of dimeric enzyme to decline at a constant ratio of activities, with increased cytochrome c reductase activity being recovered at a sedimentation velocity characteristic of the monomer. This was not observed.

We have reported that protoporphyrin IX inhibits nitric oxide synthase isoforms in a manner reversible upon dilution. We therefore constructed gradients containing 10 μM protoporphyrin IX throughout and applied samples of either nNOS or iNOS to the gradients (data not shown). The inclusion of protoporphyrin IX in the gradients did not alter the sedimentation behavior of cytochrome c reductase or citrulline-forming activity for either isoform; that is, the sustained presence of protoporphyrin IX did not produce dissociation of dimeric enzyme.

E. Effect of Protoporphyrin IX on Inactivation of nNOS by Zinc Protoporphyrin

Protoporphyrin IX inhibits reversibly nitric oxide synthase isoforms, while metalloporphyrins produce an irreversible inactivation. We examined whether protoporphyrin IX alters the rate of ZnPP-induced inactivation of nitric oxide synthase. Preincubations were constructed containing nNOS and 30 μM ZnPP without or containing 5, 10, or 25 μM protoporphyrin IX. The rate of inactivation of activity observed was identical in the control preincubation and those containing protoporphyrin IX (data not shown).

F. Effect of ZnPP on NO Formation by Cytokine-Treated RAW 264.7 Cells

We attempted unsuccessfully to measure the effect of ZnPP on NO formation by cytokine-treated RAW 264.7 cells. The assay we employed measures spectrophotometrically the conversion of oxyhemoglobin to methemoglobin by NO (25). We found that ZnPP in the absence of cells underwent a time-dependent formation of a product that interfered with the spectral measurement of methemoglobin formation.

IV. DISCUSSION

We have observed that diverse metalloporphyrins affect all three nitric oxide synthase isoforms. When present in incubations metalloporphyrins produce nonlinear time course curves; that is, the degree of inhibition progresses with time of incubation (11). When added to preincubations containing enzyme, they produce a time- and concentration-dependent loss of activity not reversible upon dilution, consistent with the proposal that they inactivate the enzyme. The rate of inactivation by metalloporphyrins was slowed by the presence of arginine substrate, BH_4 cofactor, and imidazole. The metalloporphyrins inactivate both the citrulline-forming and oxygen-reducing capabilities of nitric oxide synthase but do not alter its ability to catalyze cytochrome c reduction. It is known that the dimeric form of NOS catalyzes the formation of citrulline and nitric oxide, the reduction of oxygen, and the reduction of cytochrome c. The monomeric form of iNOS catalyzes cytochrome c reduction only (23). A reasonable hypothesis for these actions of metalloporphyrins is that metalloporphyrins produce their effects on activity by disrupting fully functional dimers to form partially functional monomers. Sucrose density gradient centrifugation of ZnPP-treated nNOS and iNOS revealed that this treatment resulted in a dimeric enzyme form with unaltered cytochrome c reductase activity but reduced citrulline-forming capability; that is, dimers were not disrupted, rather the dimers exhibited altered catalytic capabilities.

A second plausible hypothesis for the effects of metalloporphyrins on nitric oxide synthase was that the metalloporphyrins had the ability to exchange for the ferriprotoporphyrin IX (heme) residue of the enzyme, resulting in an enzyme with a nonfunctional metalloporphyrin at the catalytic site where oxygen is reduced and citrulline and NO are formed. This similarly seems untenable since preincubation with the natural metalloporphyrin (ferriprotoporphyrin IX) also produced a time- and concentration-dependent loss of activity. Baek et al. (23) have reported that reconstitution of partially functional monomers to fully functional dimers requires stoichiometric quantities of heme to restore citrulline-synthesizing capability. At higher heme concentrations citrulline-forming activity declined; that is, a biphasic response of citrulline-forming activity to added heme was observed. It is possible that the inhibitory phase in response to added heme noted by Baek et al. (23) corresponds to the phenomenon we have described herein for heme as well as the other metalloporphyrins.

The inactivation of nNOS in preincubations containing ZnPP (Fig. 2), CrMP (Fig. 1) and FePP (not shown) followed nonsaturating, first-order inactivation rates; that is, at constant NOS concentration an exponential inactivation rate was observed that increased without apparent limit as the metalloporphyrin concentration was increased. The failure to observe a maximal limiting inactivation rate is consistent with the proposal that the rate of inactivation is limited by the bimolecular reaction between enzyme and metalloporphyrin to form the initial enzyme-metalloporphyrin complex, rather than by a subsequent rate-limiting rearrangement of those complexes to form an inactive enzyme form. We have reported that protoporphyrin IX inhibits NOS activity reversibly (11). When protoporphyrin IX is preincubated with nitric oxide synthase at concentrations > 30-fold higher than its IC_{50} value and is then diluted for assay, full activity is recovered irrespective of the time of preincubation. Given the close structural homology of protoporphyrin IX and zinc protoporphyrin IX, it seemed reasonable that they exerted their actions, though different, by binding to a common binding site. Presumably, protoporphyrin IX by competition could reduce the rate of formation of the inactive zinc protoporphyrin-NOS complex. However, when present in preincubations at a concentration more than 50-fold higher than its IC_{50} value, protoporphyrin IX had no effect on the rate of ZnPP-induced inactivation, failing to confirm a common site of action for porphyrins and metalloporphyrins.

A diverse array of different metalloporphyrins inactivate all three NOS isoforms. Among the metalloporphyrins examined, ZnPP was the most potent, particularly versus the nNOS isoform. Ferriprotoporphyrin IX and tin protoporphyrin IX were also effective inactivators, while MnPP was the least effective agent, particularly versus the nNOS isoform. The metallomesoporphyrin CrMP, whose porphyrin ring differs from protoporphyrin by replacement of vinyl groups at positions 2 and 4 with ethyl groups, was also an inactivator. Some metalloporphyrins, particularly tin protoporphyrin IX, in the presence of light and NADPH generate carbon monoxide (1,26). CO binds tenaciously to the iron of heme proteins including nitric oxide synthases (20,27) in place of oxygen, inhibiting their activity. The nitric oxide synthase preparations employed in these studies were prepared by eluting enzyme adsorbed to ADP-agarose with NADPH and thus contained NADPH. To evaluate whether this phenomenon was responsible for nitric oxide synthase inactivation we conducted preincubations in light-impermeant glass vials and observed an identical rate of inactivation to that seen in ambient light. That a photochemical mechanism is not involved is further supported by the observation of the relative inhibitory effectiveness of the metalloporphyrins. Tin protoporphyrin is highly reactive photochemically while zinc protoporphyrin is largely unreactive (1). However, ZnPP was the more potent inactivator of nitric oxide synthase.

The conduct of these experiments was hampered by several phenomena. Several metalloporphyrins are sensitive to light. Zinc protoporphyrin IX did not appear to be fully stable in air. We observed that ZnPP underwent a time-dependent formation of an unidentified product that possessed an altered spectrum. The product exhibited increased light absorbance at 426 nm, reduced light absorbance at 409 nm, and an isosbestic point at 418 nm. The formation of this product occurred in the absence of cells and interfered with the ability to measure spectrophotometrically NO formation by intact cells using the known ability of NO to convert oxyhemoglobin to methemoglobin (25). Finally, we observed that all of the metalloporphyrins were readily soluble in dimethylformamide at millimolar concentrations. Upon dilution in water, variable, time-dependent formation of colloidal suspensions apparently occurs, making it difficult to assure achieving identical concentrations between experiments. These difficulties were minimized by storing the metalloporphyrins dissolved in dimethylformamide in light-impermeant, sealed glass vials with dilution into neutral aqueous solutions immediately prior to use with vigorous vortexing.

The physiological significance, if any, of the effects of metalloporphyrins on nitric oxide synthase is unclear. Metalloporphyrins have been assumed to be relatively selective inhibitors of heme oxygenase and have been used extensively to implicate CO (the product of heme oxygenase) in diverse physiological processes (4). It is clear that metalloporphyrins have effects on other enzymes important in signal transduction, including soluble guanylate cyclase (6) and nitric oxide synthase. The inhibitory effect of the natural ligand heme on nitric oxide synthase is interesting, but it is uncertain whether such effects would occur at concentrations achieved intracellularly. ZnPP is formed within erythrocytes under pathophysiological conditions. In iron deficiency anemia and lead poisoning the lack of "available" iron leads to the incorporation of zinc into protoporphyrin IX in place of iron to form ZnPP. ZnPP measured in whole blood fluorimetrically is used to screen for undue lead absorption and is considered positive when intracellular concentrations exceed 2 μM, a concentration that inactivates the nNOS isoform. While all cells are believed capable of synthesizing heme, there are dramatic differences in their quantitative synthetic rates. It is doubtful that cells not involved in hemoglobin synthesis are capable of achieving significant ZnPP formation.

V. SUMMARY

Zinc protoporphyrin IX produces a time- and concentration-dependent, nonsaturable, and irreversible inactivation of all three nitric oxide synthase isoforms that follows pseudo-first-order kinetics. The IC_{50} values for inactivation in a 30-min incubation containing zinc protoporphyrin IX and a saturating concentration of tetrahydrobiopterin cofactor are 2, 6, and 7 μM for the nNOS, iNOS, and eNOS isoforms, respectively. Ferriprotoporphyrin IX (heme), chromium mesoporphyrin IX, and tin protoporphyrin IX behave similarly but are less potent inactivators. These inactivations are not light-dependent. The presence of either tetrahydrobiopterin cofactor, arginine substrate, or imidazole in a preincubation slows the rate of zinc protoporphyrin IX-induced inactivation. Preincubation with zinc protoporphyrin IX also produces a time- and concentration-dependent inactivation of the substrate-independent NADPH oxidase activity of nNOS but does not affect its cytochrome c reductase activity. Sucrose density gradient centrifugation analysis reveals that zinc protoporphyrin IX-induced inactivation of both the nNOS and iNOS isoforms is not associated with the dissociation of fully functional dimers into partially functional monomers.

ACKNOWLEDGMENTS

This work was supported by National Institute of Health grant HL 54768

REFERENCES

1. Vreman HK, Ekstrand BC, Stevenson DK. Pediatr Res 1993; 33:195–200.
2. Maines MD. FASEB J 1988; 2:2557–2568.
3. Glaum SR, Miller RJ. Mol Pharmacol 1993; 43:965–969.
4. Meffert MK, Haley JE, Schuman EM, Shulman H, Madison DV. Neuron 1994; 13:1225–1233.
5. Schuman EM, Madison DV. Science 1991; 254:1503–1506.
6. Ignarro LJ, Ballot B, Wood KS. J Biol Chem 1984; 259:6201–6207.
7. Cable EE, Cable JW, Bonkovsky HL. Hepatology 1993; 18, 119–127.
8. Drummond GS, Kappas A. J Exp Med 1982; 156:1878–1883.
9. Vreman HK, Stevenson DK. Am J Dis Child 1990; 144:590–594.
10. Louria DB. Trace metal poisoning. In Wyngaarden JB, Smith LJ, eds. Cecil Textbook of Medicine. 18th ed. Philadelphia: W.B. Saunders, 1988, pp 1182–1187.
11. Wolff DJ, Naddelman RA, Lubeskie A, Saks D. Arch Biochem Biophys 1996; 333:27–34.
12. Wolff DJ, Datto GA. Biochem J. 1992; 258:201–206.
13. Wolff DJ, Gribin BJ. Arch Biochem Biophys 1994; 311:293–299.

14. Wolff DJ, Lubeskie A, Umansky S. Arch Biochem Biophys 1994; 314:360–366.
15. Bredt DS, Snyder SH. Proc Natl Acad Sci USA 1990; 87:682–685.
16. Klatt P, Schmidt K, Uray G, Mayer B. J Biol Chem 1993 268:14781–14787.
17. Bradford MM. Anal Biochem 1976; 72:248–254.
18. Vreman HJ, Gillman MJ, Downum KR, Stevenson DK. Dev Pharmacol Ther 1990; 15:112–124.
19. White KA, Marletta MA. Biochemistry 1993; 31:6627–6631.
20. McMillan K, Masters BSS. Biochemistry 1993; 32:9875–9880.
21. Wolff DJ, Datto GA, Samatovicz RA, Tempsick RA. J Biol Chem 1993; 268:9425–9429.
22. Schmidt HHHW, Pollock JS, Nakane M, Gorsky LD, Forstermann U, Murad F. Proc Natl Acad Sci USA 1991; 88:365–369.
23. Baek KJ, Thiel BA, Lucas S, Stuehr DJ. J Biol Chem 1993; 268:21120–21129.
24. Siegel LM, Monty KJ. Biochim Biophys Acta 1966; 112:346–362.
25. Noack E, Kubitzek D, Kodja G. Nueroprotocols 1992; 1:133–139.
26. Vreman HJ, Gillman MJ, Downum KR, Stevenson DK. Dev Pharmacol Ther 1990; 15:112–124.
27. Wang J, Stuehr DJ, Ikeda-Saito M, Rousseau D. J Biol Chem 1993; 268:22255–22258.

4

Nitric Oxide, Peroxynitrite, and Carbon Monoxide Poisoning

Stephen R. Thom
University of Pennsylvania, Philadelphia, Pennsylvania

Harry Ischiropoulos
Stokes Research Institute, Philadelphia, Pennsylvania

I. PEROXYNITRITE BIOCHEMISTRY

Peroxynitrite is generated by the second-order reaction between nitric oxide (·NO) and superoxide. The rate constant is $6.7 \times 10^9\ M^{-1}s^{-1}$, three times faster than the reaction between superoxide and superoxide dismutase and 30 times as rapid as the reaction between ·NO and oxyhemoglobin (1). Hence, formation of peroxynitrite is kinetically favored over competing reactions in vivo when both superoxide and ·NO are available. The pKa of peroxynitrite ($ONOO^-$) is 6.8, and at pH 7.4 approximately 25% will be protonated to form peroxynitrous acid (ONOOH). Peroxynitrite is a powerful oxidant that can readily react with many cellular components. The peroxynitrite anion is stable at alkaline pH; however, peroxynitrous acid isomerizes to nitrate at the rate of 0.6 sec^{-1} in phosphate buffer at 37°C (2).

$$ONOO^- + H^+ \rightarrow ONOOH \rightarrow NO_3^- \qquad (1)$$

At physiological pH, three pathways are known for reactions involving peroxynitrite (Fig. 1). Peroxynitrous acid does not appear to physically separate to form hydroxyl radical and nitrogen dioxide ($\cdot OH \cdots \cdot NO_2$); however, it will oxidize biological and other organic molecules to yield the same oxidized products as hydroxyl radical (2–5). Although this hydroxyl radical-like reactivity represents a potential threat to cells, it may not be important in vivo because the rate

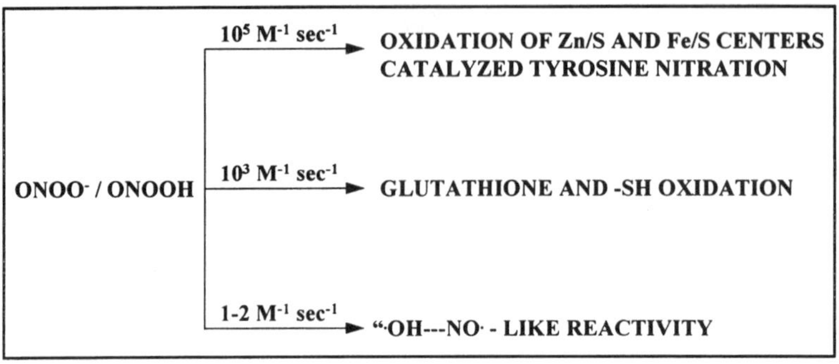

Figure 1 Potential reactive pathways of peroxynitrite based on rate constants.

of ONOOH dissociation to yield hydroxyl radical-like reactivity is slower than the other reactions in biological systems (Fig. 1).

Direct reactions with target molecules that can significantly contribute to the in vivo toxicity of $ONOO^-$ include reactions with zinc-thiolate, iron-thiolate centers, thiols, and formation of strong nitrating species. Crow et al. showed that $ONOO^-$ induced the release of zinc and the oxidation of the thiol moiety of zinc finger-containing proteins (6). Superoxide and $ONOO^-$, but not as ·NO, can inhibit the mitochondrial and cytosolic aconitases (7,8). ·NO-mediated inactivation of mitochondrial enzymes is one mechanism involved with macrophage-derived target cell killing (7,8). In addition, $ONOO^-$ was shown to oxidize glutathione and protein cysteine residues (10), potentially depleting cells of antioxidant defenses. Peroxynitrite and/or peroxynitrous acid will also initiate the oxidation of unsaturated fatty acids to conjugated dienes and TBARS (11). Importantly, the oxidation of lipids and other biomolecules by $ONOO^-$ does not require a metal catalyst (11,12). Peroxynitrite also exhibits reactivity with protein tyrosine residues that has been extremely useful in detecting $ONOO^-$ mediated reactions in vivo. All of the reactive pathways so far described may occur under physiological conditions because $ONOO^-$ is sufficiently stable that it can diffuse from the site of formation and attack critical cellular sites (3,4). However, the proportion of molecules that follow each of the three reaction pathways will vary with experimental conditions, such as pH and the presence of chelated metals. It will also depend on the concentration of dissolved CO_2, which recently was shown to direct the reactivity of peroxynitrite toward nitration reactions and away from thiol oxidations (13–15).

II. PEROXYNITRITE-MEDIATED TYROSINE NITRATION

A major product from the reaction between peroxynitrite and proteins is nitrotyrosine (16,17). Endogenous tyrosine nitration is almost certainly derived via enzymatically produced ·NO, but ·NO itself is not a nitrating agent (15–18). Protein tyrosine residues can be nitrated in vivo by peroxynitrite, nitrogen dioxide, and acidified nitrite solutions such as nitrite plus hypochlorous acid (15,16,19,20). Among these potential nitrating agents, the concentration of peroxynitrite required to produce protein nitrotyrosine is much lower than the other agents. For example, generation of peroxynitrite at approximately 5 μM/min by the simultaneous generation of ·NO and superoxide was sufficient to selectively nitrate tyrosine residues in plasma proteins (15). In contrast, ·NO, nitrogen dioxide and acidified nitrite were not found to cause detectable tyrosine nitration at exposures of 17 μM/min, or with bolus additions in the mM range (15,16,18). Peroxynitrite can react with other components in plasma besides protein tyrosine residues, but these competing reactions may be minimized in vivo because of the presence of CO_2, which catalyzes peroxynitrite-mediated protein nitration (13–15).

The role of peroxynitrite as a nitrating agent in vivo is underscored by detection of nitrotyrosine is cells and tissues that were not exposed to alternative nitrating agents such as nitrite plus hypochlorous acid (18,21–28). Although peroxynitrite may not account for all protein tyrosine nitration in vivo, under pathophysiological conditions it is likely to be the predominant species that forms protein nitrotyrosine. This view is supported by the relative potency of the different nitrating agents (15), and because alternative nitrating agents besides peroxynitrite can be readily scavaged by other pathways (15,19). Using analytical and immunological methods, nitrotyrosine has been detected in a variety of human and animal disease states (Table 1), which implicates the involvement of peroxynitrite in these disorders. Peroxynitrite has been also suggested to play a

Table 1 Detection of Nitrotyrosine in Human and Animal Disease States

Human:
Atherosclerotic plaques of coronary vessels (18)
Lungs of infants with sepsis and/or respiratory disease (29,30)
Synovial fluid of patients with arthritis (31)
Multiple sclerosis plaques (24)
Animal
Mouse striatum following MPTP neurotoxicity (21)
Mouse brain after NMDA-receptor activation or 3-nitropropionic acid administration (23)
Cerebral vasculature following CO exposure (25)
Rabbit lungs following exposure to hyperoxia (29)
Ischemia reperfusion-injured rat lungs (26)
Aorta of septic rats (27)
Lungs of endotoxin-treated rats (28)
Inflammatory bowel disease (32,33)

role in heart and brain ischemia-reperfusion injuries (34–40), inflammatory injury in the lung (41,42), and in cold- and glutamate-induced brain injuries (43,44).

Tyrosine nitration induced using tetranitromethane (TNM) as the nitrating agent has been shown to inactivate over 100 mammalian proteins whose activity is dependent on tyrosine residues (45). Thus far, peroxynitrite-mediated nitration of tyrosine residues has been shown to inactivate the mitochondrial Mn superoxide dismutase (16), the lipid aggregatory activity of surfactant protein A (46), and the activity of glutamine synthase (47). Recently MacMillan et al. showed that Mn superoxide dismutase is nitrated and inactivated in rejected human renal allografts (48).

Tyrosine residues in proteins also serve a significant role in signal transduction. Tyrosine phosphorylation is an important regulator of signal transduction in endothelial cells and has been implicated in cellular responses to growth factors, cytokines, and calcium ionophores. Gow et al. (49) showed that tyrosine nitration inhibits tyrosine phosphorylation by $p43^{v\text{-}abl}$ and $p60^{c\text{-}arc}$ kinases. Kong et al. (50) recently showed that $ONOO^-$-mediated nitration of a single tyrosine residue in the cell cycle kinase CDC2 prevents tyrosine phosphorylation. Therefore, nitrotyrosine formation may also interfere with normal signal transduction pathways.

As with other forms of oxidative stress, it is likely that defensive and repair processes exist to counter the effects of peroxynitrite. Removal of modified proteins is considered one line of defense against oxidative stress. Proteins that have undergone oxidative modification exhibit a higher susceptibility to proteolytic degradation (50,51). After reaction with peroxynitrite, nitrated bovine serum albumin has been found to undergo proteolysis at an increased rate (49). Although no specific protective mechanisms have been identified which deal with peroxynitrite, tissues and plasma appear to possess a pathway whereby nitrotyrosine residues are specifically removed from proteins (49).

III. CELLULAR INJURIES DUE TO PEROXYNITRITE

Reactive oxygen species are normally produced during cellular metabolism (53). Neurons, glial cells, macrophages, neutrophils, and endothelial cells also generate ·NO, and a portion of ·NO is released into the extracellular space (54). When appropriately stimulated, the latter four cell types have been shown to generate peroxynitrite (55–58). At first glance, it may appear that peroxynitrite is similar to hydroxyl radical in that it reacts with many cellular targets and therefore may not reach vital cell structures such as mitochondria or specific proteins. However, the reaction rates of peroxynitrite with thiols, sugars, and low-molecular-weight molecules such as ascorbate are relatively slow, approximately 10^3

$M^{-1}sec^{-1}$. Reactions of peroxynitrite with zinc fingers, iron-sulfur centers, metals, and metaloproteins occur at rates of approximately 10^5 $M^{-1}sec^{-1}$, thus making them more likely targets (12).

The magnitude of cell injury and also the type of cell death have been found to depend upon the concentration of peroxynitrite. Bonfoco et al. (59), using cultured cortical neurons, showed that peroxynitrite was capable of inducing necrotic death at high concentrations and death with apoptotic features at lower concentrations. The outcome is also dependent on the cell type. The inherent ability of cells to withstand oxidative stress is dependent upon their antioxidant capacity, the ability to derive energy from alternate pathways, to repair oxidatively modified biomolecules, and the availability of trophic support to maintain energy requirements, ionic homeostasis, and structural integrity. As such, Bolanos et al. (60), demonstrated that the same concentration of peroxynitrite inhibited components of the mitochondrial electron transport chain in primary cultures of neurons but not in primary cultures of astrocytes, and Estevez et al. (61) demonstrated that neurotrophic factors can differentially regulate peroxynitrite-mediated apoptosis in PC12 cells. These observations may also explain the finding that a bolus addition of chemically synthesized peroxynitrite induced apoptosis to HL-60 cells but not to human umbilical cord endothelial cells (62). Insofar as mitochondrial dysfunction has been linked to the apoptotic process (63,64), it appears from these studies that at low concentrations peroxynitrite may selectively react with mitochondria to induce loss of membrane potential and uncoupling, possibly key events in the initiation of cascades that lead to metabolic dysfunction and death. Endothelial cells exposed to peroxynitrite at 200 nM/min/cm^2 will inactivate mitochondrial respiratory chain components and depolarize mitochondrial membrane (65).

It is now clear that ·NO is an important intercellular messenger. Nitric oxide is involved in neuronal communications; with control of vascular tone, inhibition of platelet aggregation, and neutrophil adherence; and with bactericidal and tumoristatic functions of macrophages (54). Given these facts and also the ubiquitous presence of reactive oxygen species in aerobic metabolism, it is obvious to question whether cells can restrict peroxynitrite synthesis under normal physiological conditions. Nitrated proteins have been detected in normal human plasma and nondiseased or control cell homogenates (65). However, nitrotyrosine synthesis may have been an artifact of isolation and tissue processing. It is as yet unclear whether there is a basal, or "normal," level of peroxynitrite production by cells.

It appears that cells regulate peroxynitrite synthesis by controlling the rate of production of ·NO and superoxide. However, the production of peroxynitrite can also be modified by altering the cellular ability to scavaging superoxide and ·NO. Troy et al. (67) eloquently demonstrated this phenomenon by treating cells with antisense oligonucleotide to Cu/Zn superoxide dismutase. Downregulation of superoxide dismutase caused cell death by peroxynitrite, which was prevented by inhibitors of nitric oxide synthase and augmented by nitric oxide generators. Peroxynitrite synthesis can also be increased by inhibiting scavaging of intercellular ·NO. We have carried out a series of investigations on the pathophysiology of brain injury mediated by carbon monoxide. These studies have led to elucidation of a rather novel mechanism of oxidative stress which appears to be based on an imbalance in ·NO scavaging potential.

IV. NITRIC OXIDE IN CARBON MONOXIDE-MEDIATED BRAIN INJURY

A. Clinical Carbon Monoxide Poisoning

Injuries from CO occur primarily in the cardiovascular and central nervous systems. Traditionally, these injuries have been viewed as occurring due to the hypoxic stress mediated by an elevated carboxyhemoglobin (COHb) level. Data derived from human and animal studies indicate that major cardiac injury may indeed be due primarily to CO-induced hypoxic stress (68-70). However, the predominant morbidity associated with CO poisoning occurs in the central nervous system, and the mechanism of CO-mediated brain injury is incompletely understood. Hypoxia per se does not adequately explain the development of neurological injuries, which have not been found to be correlated with elevated COHb levels (71–75).

Morphological features of CO-mediated brain injuries are most often localized to cerebral white matter, globus pallidus, and hippocampus (76). In this respect, CO poisoning is not fundamentally different from other anoxic encephalopathies (77). Demyelination is common following CO poisoning, although the magnitude of white-matter changes varies widely (78). Bilateral lesions of the globus pallidus are also relatively common, and typically involve the basal ganglia (79). Lesions are found slightly less commonly, in about 50% of cases, in the hippocampus. This injury often requires days or weeks to be manifest, and the delayed neuronal loss is most often localized to the CA1 layer (80).

Early investigators who studied CO pathophysiology remarked that demyelination occurred in a perivascular distribution and that CO poisoning also disturbed the blood-brain barrier (80–82). Acute vascular and perivascular changes have been found with pathological examinations of experimental animals (83). Clinical pathological examiners have also stressed the extent of vascular disorders (79). Recently, brain blood flow abnormalities have been detected by SPECT and PET scanning in patients with symptoms ranging from relatively subtle cognitive impairments to major functional deficits (84–88).

B. Nitric Oxide and Peroxynitrite in Experimental CO Poisoning

We hypothesized that peroxynitrite may be produced in animals during CO poisoning because exposure to CO has been found to enhance the rates of production of both ·NO and partially reduced oxygen species (89,90). Platelets liberate large amounts of ·NO coincident with exposure to CO due to impaired intraplatelet ·NO binding (89). Exposure to CO at concentrations sufficient to disturb cardiovascular function can also perturb mitochondrial electron transport and increase the production of partially reduced oxygen species (90). Oxidative stress, documented as brain lipid peroxidation, occurs in rats when the exposure to CO terminates with events that include systemic hypotension, cerebral hypoperfusion, and impaired mitochondrial function (91–93).

The production of ·NO has been found to be increased by ninefold, and nitrotyrosine by 10-fold, in rat brains immediately after CO poisoning (Fig. 2) (25). Moreover, nitrotyrosine depositions occur predominantly in a perivascular distribution (Fig. 3). Brain oxidative stress in this animal model, documented as lipid peroxidation, is caused by a cascade of biochemical events (Fig. 4). It appears that perivascular oxidative stress from ·NO-derived oxidant(s) is responsible for leukocyte sequestration because when synthesis of ·NO was inhibited by injecting rats with the L-arginine analog L-nitroargine methyl ester (L-NAME), leukocyte sequestration and the other biochemical changes that normally followed CO poisoning failed to occur (25). These results provide evidence of a discrete role for peroxynitrite in this model of CO poisoning.

Relatively high concentrations of nitrotyrosine have been measured in both the brain and aorta of rats exposed to CO (25,94). It is possible that peroxynitrite-mediated nitration reactions are particulary favored in CO poisoning because of a concomitant elevation in the tissue CO_2 concentration (95). Carbon dioxide has been shown to catalyze tyrosine nitration by peroxynitrite and to direct the reactivity of peroxynitrite toward nitration reactions and away from alternative processes such as thiol oxidations (14–16). Under normal conditions, CO is only slowly oxidized in the body to CO_2 (96); however, this process increases in relation to the CO body burden (95). Hence, during CO poisoning, the tissue CO_2 concentration can be expected to increase.

To date, two cell types have been found to release increased amounts of ·NO in the presence of CO. Platelets liberate ·NO when exposed to CO, and this ·NO contributes to perivascular oxidative stress (25,89). However, the activity of nitric oxide synthase in platelets is not enhanced by CO. Instead it appears that CO impairs intraplatelet scavaging of ·NO (89). The data indicate that the normal site of ·NO binding in platelets are heme-containing proteins which, once bound with CO, are not available to scavage the ·NO that is normally generated. Hence, ·NO diffuses out from platelets to the surrounding medium. This phenom-

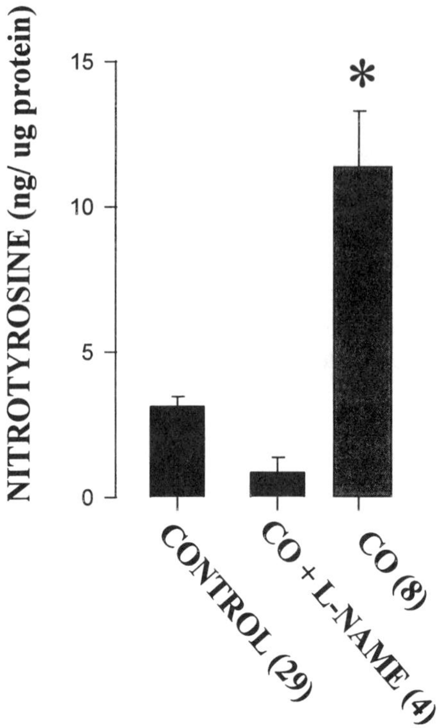

Figure 2 Nitrotyrosine concentration in rat brain homogenates following CO poisoning. Where indicated, rats were injected with 1 mg/ kg L-NAME i.p. 2 hours before CO poisoning. Values are mean±SE. (From Ref. 25.)

enon can be reversed by illuminating CO-exposed platelets with a quartz lamp, which causes photodissociation of CO for hemoproteins (89).

When rats are exposed to CO, platelets contribute to the systemic oxidative stress (25). However, when CO exposure triggers the major cascade of reactions as depicted in Figure 4, additional sources of ·NO appear to be important. Recently, we have found that CO causes endothelial cells to liberate both ·NO and peroxynitrite (97). Cultured vascular endothelial cells were exposed to CO, and liberation of ·NO was quantified in terms of the stable end products nitrite and nitrate. Intracellular production of peroxynitrite was assessed by measurements of nitrotyrosine, and after exposure for 2 hours to 55 nM CO the concentration was 4.7 ± 0.4 (SE, n = 4) pmol/µg cell protein ($P < .05$, ANOVA), versus only 2.5 ± 0.4 (n = 4) for control cells and 3.1 ± 0.2 (n = 3) for cells exposed to CO plus 100 µM L-NAME. Peroxynitrite also was detected in the surrounding medium when cells were exposed to CO. For example, whereas no sig-

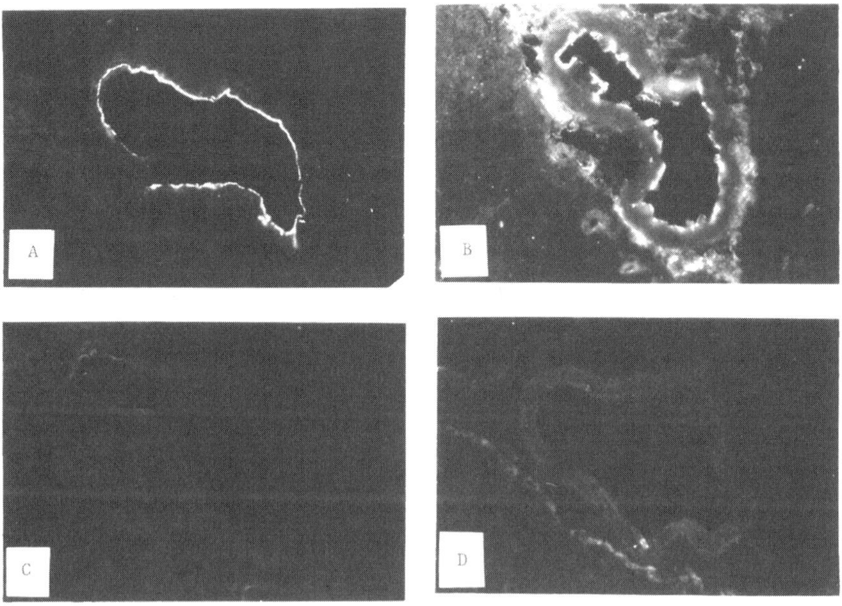

Figure 3 Immunofluorescence photomicrographs from cerebral cortex. Panels show serial thin sections from a rat poisoned with CO. (A) Stained with antinitrotyrosine antibody. (B) Stained with anti-CD31 antibody, which is expressed on endothelial cells. (C) Stained with antigen-competed antinitrotyrosine antibody. (D) Staining with nonspecific IgG. ×875. (From Ref. 25.)

nificant increase in H_2O_2 was detected, enhanced oxidation of the fluorescent marker parahydroxyphenyl acetic acid increased by more than threefold when cells were exposed to 110 nM CO (97).

The mechanism for CO-mediated release of ·NO by endothelial cells appears to be similar to that found with platelets. Endothelial cell synthesis of ·NO is not altered by exposure to CO; however, CO can displace ·NO from intracellular binding sites (97). This phenomenon was observed after cells were incubated with 100 μM L-NAME to inhibit ongoing ·NO synthesis. These cells were exposed to 100 ppm CO, which produced a dissolved CO concentration of 110 nM, and ·NO displacement was detected by measuring the peak height with a ·NO-selective electrochemical sensor (World Precision Instruments, Sarasota, FL) using techniques we have described in a previous publication (89). The concentration of ·NO in the medium was 114 ± 4 (SE, n = 4, $P < .05$) nM, whereas the concentration when cells were exposed to only air was 3 ± 2 (n = 4) nM. It appears that peroxynitrite synthesis is increased due to the increased availability for ·NO, which undergoes reactions with superoxide.

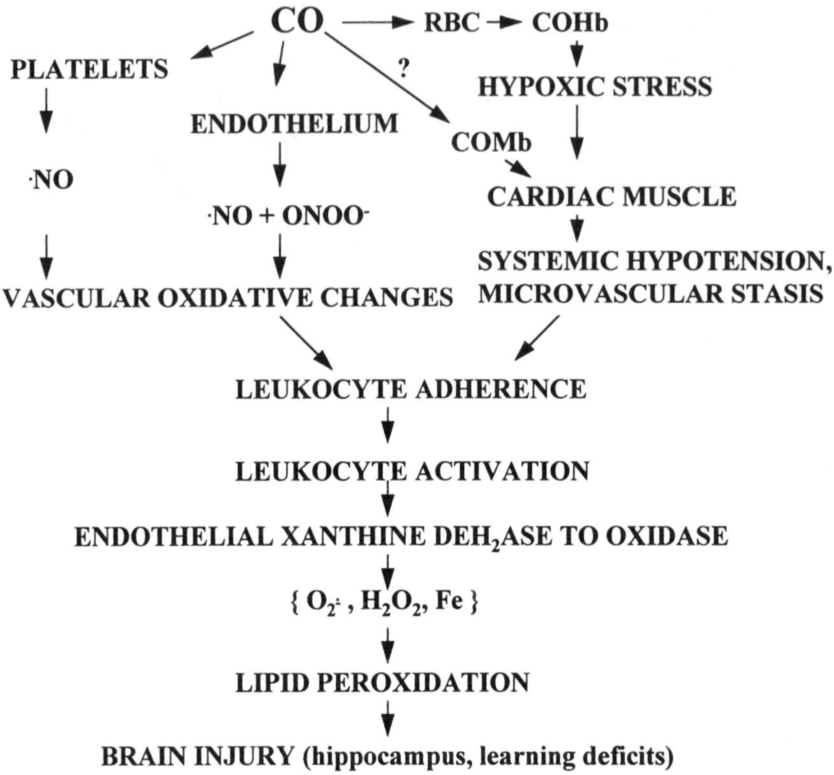

Figure 4 Sequence of biochemical events identified following experimental CO poisoning in the rat. Original data for this flow chart were reported in Refs. 43, 83, 85–87, 93.

C. Hypothesis for CO-Mediated Brain Injuries

Vascular insults appear to precede or coincide with neurological deficits in clinical studies, and the biochemical changes mediated by ·NO in experimental CO poisoning offer a plausible mechanism for CO neuropathology. The linkage between ·NO-derived oxidative stress and experimental neuropathology from CO still needs to be demonstrated. In preliminary studies, CO poisoning has been found to cause lesions in the basal ganglia and hippocampus of rats. Moreover, in a model similar to ours, CO poisoning has been shown to cause learning deficits and neuronal loss in the hippocampus (98). There are a number of examples where others have found that pathology linked to ·NO is involved with CO poisoning (99–101). Clinical neuroimaging studies have caused several investigators to hypothesize that the primary site of injury from CO may be the vasculature, and that focal pathology arises secondary to hemorrhagic necrosis (102,103). In experimental studies, the variability observed in lesions found in the cerebral white matter and globus pallidus of animals has been correlated with the fall in local blood flow and metabolic acidosis (76,104). It is therefore possible that vascular oxidative stress, which appears to be a global phenomenon based on the distribution of nitrotyrosine (25), is the presursor to more traditional evidence of neuropathology.

In addition to the mechanisms we have outlined for CO-mediated oxidative stress (Fig. 4), there are several other potential sources for ·NO and therefore ·NO-mediated oxidative stress that have been elucidated in other models of CO poisoning. Excessive release of excitatory amino acids (EAA), such as glutamate, has been implicated as a component in delayed neuronal death of the hippocampus CA1 subfield following both CO and ischemia/reperfusion (73,105–107). There are three types of receptors activated by EAA: N-methyl-D aspartaic acid (NMDA), D-amino-3-hydroxy-5-methyl-4-isoxazolepropionic acid, and kainic acid (108). Agents that inhibit NMDA receptor activation attenuate CO-mediated delayed neuronal degeneration of pyramidal cells in the hippocampus and cochlear ganglion cells (101,109). When EAA bind to postsynaptic cells, one response elicited is retrograde release of ·NO which causes greater release of neurotransmitter with subsequent depolarizations. The physiological role for this process, termed long-term potentiation, appears to be to strengthen axonal interactions that may be important in certain forms of learning and memory. Neurotoxicity can occur with excessive stimulation by mechanisms involving elevations in intracellular calcium and also generation of superoxide and ·NO (108,110).

NMDA-induced neuronal death can be reduced by scavaging either superoxide or ·NO in some, but not all studies (44,108,111–113). If CO enhanced the availability of ·NO in neurons, as it does in the perivascular space of the central nervous system and in cultured endothelial cells, this would offer a biochemical mechanism for how CO poisoning may augment release of excitatory amino acids. Ishimaru et al. (101) posed the hypothesis that CO-induced damage to the hippocampus was not so much due to excessive activation of NMDA receptors as to increased sensitivity of damaged cells to NMDA receptor activation. If so, then the fundamental mechanism of hippocampal injury with CO may be related to the diffuse oxidative stress as measured by brain concentrations of nitrotyrosine and products of lipid peroxidation. In short, oxidative stress to neurons by CO may predispose these cells to injury in association with NMDA-mediated stimulation.

Monoamine neurotransmitters such as norepinephrine and dopamine are elevated after CO exposure, and enzymatic breakdown as well as auto-oxidation will generate reactive oxygen species (114,115). These agents appear to contribute

Table 2 Oxidizing Agents and Their Possible Sources in CO Poisoning

NO· and ONOO
Endothelium and platelets (25,89,97)
Neutrophils (93)
Excitatory amino acids (99–101)
Reactive oxygen species
Mitochondria (90,116)
Endothelium (xanthine oxidase) (92)
Monoamine neurotransmitters (110,114)

to oxidative stress after CO poisoning because free radical production in brain can be diminished by inhibiting monoamine oxidase-B, an enzyme located principally in microglial cells (116–118). As with EAA, it appears that ·NO plays a role in augmenting presynaptic release of monoamine neurotransmitters by activating NMDA receptors (119,120). Activated microglia can also mediate neuronal injury by generating ·NO-derived oxidants (55). As microglia have been linked to demyelination due to their ability to attack oligodendroglia (121), once again, ·NO-mediated oxidative stress may be a common biochemical link among the varied morphological manifestations of CO poisoning.

In closing, there appear to be a number of potential sources of oxidizing species in CO poisoning (Table 2), and there are also several mechanisms by which ·NO may play a role in CO-mediated neuropathology. There is also a growing body of evidence which suggests that CO has a physiological role as a neurotransmitter, as does ·NO (122,123). The origin of CO in vivo is heme oxygenase, which catalyzes equimolar production of CO and biliverdin for each molecule of heme degraded. Heme oxygenase mRNA colocalizes in the central nervous system with cytosolic guanylate cyclase (122). Exposure to CO has been shown to increase evoked synaptic potentials in hippocampus in the same range of concentrations as does ·NO (123).CO can also activate cytosolic guanylate cyclase by binding to the heme-containing regulatory subunit (124). CO appears to modulate cerebellar α3-Na,K-ATPase by altering cellular cGMP levels in in vitro studies (125), and exposure to CO increases the concentration of cGMP in platelets and smooth muscle cells (124–126). However, CO is less than one-tenth as potent as ·NO for activating guanylate cyclase (124). Heme oxygenase has been shown to be induced by oxidative stress in fibroblasts and in the rat liver (127,128). The physiological rationale for this is speculated to be enhanced production of bilirubin, a potent antioxidant. Hence, there may be still more associations between ·NO and CO with relation to the issue of oxidative stress.

ACKNOWLEDGMENTS

This work was supported in part by funding provided by the National Institute of Environmental Health Sciences, NIH, No. ES-05211 (SRT), from the Council for Tobacco Research (SRT), and from the Pennsylvania Lung and Southeastern Pennsylvania Heart Associations (HI). H. Ischiropoulos is an Established Investigator of the American Heart Association.

REFERENCES

1. Huie RE, Padjama S. The reaction of NO with superoxide. Free Rad Res Comm 1993; 18:195–199.
2. Koppenol WH, Moreno JJ, Pryor WA, Ischiropoulos H, Beckman JS. Peroxynitrite, a cloaked oxidant formed by nitric oxide and superoxide. Chem Res Toxicol 1992; 5:834–842.
3. Beckman JS, Beckman TW, Chen J, Marshall PA, Freeman BA. Apparent hydroxyl radical production by peroxynitrite: implications for endothelial injury from nitric oxide and superoxide. Proc Natl Acad Sci USA 1990; 87:1620–1624.
4. Beckman JS, Chen J, Ischiropoulos H, Crow JP. Examining the oxidative chemistry of peroxynitrite. Methods Enzymol 1994; 233:229–240.
5. Crow JP, Spruell C, Chen J, Gunn C, Ischiropoulos H, Tsai JHM, Smith CD, Radi R, Koppenol WH, Beckman JS. On the pH-dependent yield of hydroxyl radical products from peroxynitrite. Free Radic Biol Med 1994; 16:331–338.
6. Crow JP, Beckman JS, McCord JM. Sensitivity of the essential zinc-thiolate moiety of yeast alcohol dehydrogenase to hypochlorite and peroxynitrite. Biochemistry 1995; 34:3544–3522.
7. Hausladen A, Fridovich I. Superoxide and peroxynitrite inactivate aconitases but nitric oxide does not. J Biol Chem 1994; 269:29405–29408.
8. Castro L, Rodriguez M, Radi R. Aconitase is readily inactivated by peroxynitrite, but not by its precursor, nitric oxide. J Biol Chem 1994; 269:29409–29415.
9. Rubbo H, Denicola A, Radi R. Peroxynitrite inactivates thiol-containing enzymes of Trypanosoma cruzi energetic metabolism and inhibits cell respiration. Arch Biochem Biophys 1994; 308:96–102.
10. Radi R, Beckman JS, Bush KM, Freeman BA. Peroxynitrite oxidation of sulfhydryls. The cytotoxic potential of superoxide and nitric oxide. J Biol Chem 1991; 266:4244–4250.
11. Radi R, Beckman JS, Bush KM, Freeman BA. Peroxynitrite-induced lipid peroxidation: the cytotoxic potential of superoxide and nitric oxide. Arch Biochem Biophys 1991; 288:481–487.
12. Augusto O, Gatti RM, Radi R. Spin-trapping studies of peroxynitrite decomposition and of 3-morpholinosydnonimine N-ethylcarbamide autooxidation: direct evidence for metal-independent formation of free radical intermediates. Arch Biochem Biophys 1994; 310:118–125.

13. Uppu RM, Squadrito GL, Pryor WA. Acceleration of peroxynitrite oxidations by carbon dioxide. Arch Biochem Biophys 1996; 327:335–343.
14. Denicola A, Trugillo M, Freeman BS, Radi R. Peroxynitrite reaction with carbon dioxide/bicarbonate: kinetics and influence on peroxynitrite-mediated oxidation reactions. Arch Biochem Biophys. 1996; 333:49–58.
15. Gow A, Duran D, Thom SR, Ischiropoulos H. Carbon dioxide enhancement of peroxynitrite-mediated protein tyrosine nitration. Arch Biochem Biophys 1996; 333:42–48.
16. Ischiropoulos H, Zhu L, Chen J, Tsai M, Martin JC, Smith CD, Beckman JS. Peroxynitrite-mediated tyrosine nitration catalyzed by superoxide dismutase. Arch Biochem Biophys 1992; 298:431–437.
17. Beckman JS, Ischiropoulos H, Zhu L, van der Woerd M, Smith CD, Harison J, Martin JC, Tsai M. Peroxynitrite-mediated tyrosine nitration catalyzed by superoxide dismutase. Arch Biochem Biophys 1992; 298:438–445.
18. Beckman JS, Ye YZ, Anderson PG, Chen J, Accavitti MA, Tarpey MM, White CR. Extensive nitration of protein tyrosine in human atherosclerosis detected by immunohistochemistry. Biol Chem Hoppe-Seyler 1994; 375:81–88.
19. Prutz WA, Monig H, Butler J, Land EJ. Reactions of nitrogen dioxide in aqueous model systems: oxidation of tyrosine units in peptides and proteins. Arch Biochem Biophys 1985; 243:125–134.
20. Eiserich JP, Cross CE, Jones D, Halliwell B, Van der Vliet A. Formation of nitrating and chloronating species from the reaction of nitrite with hypochlorous acid. J Biol Chem 1996; 271:19199–19208.
21. Schulz JB, Matthews RT, Mugit MK, Browne SE, Beal MF. Inhibition of neuronal nitric oxide synthase by 7-nitroindazole protects against MPTP-induced neurotoxicity in mice. J Neurochem 1995; 64:936–939.
22. Beal MF, Ferrante RJ, Henshaw R, Matthews RT, Chan PH, Kowall NW, Epstein CJ, Schulz JB. 3-Nitropropionic acid neurotoxicity is attenuated in copper/zinc superoxide dismutase transgenic mice. J Neurochem 1995; 65:919–922.
23. Schulz JB, Matthews RT, Jenkins BG, Ferrente RJ, Siwek D, Henshaw DR, Cipolloni PB, Mecocci P, Kowall NW, Rosen BR, Beal MF. Blockade of neuronal nitric oxide synthase protects against excitotoxicity in vivo. J Neurosci 1995; 15:8419–8429.
24. Basarga O, Michaels FH, Zheng YM, Borboski LE, Spitsin SV, Fu ZF, Tawadros R, Koprowski H. Activation of the inducible form of nitric oxide synthase in the brains of patients with multiple sclerosis. Proc Natl Acad Sci USA 1995: 92:12041–12045.
25. Ischiropoulos H, Beers MF, Ohnishi ST, Fisher D, Garner SE, Thom SR. Nitric oxide production and perivascular tyrosine nitration in brain after carbon monoxide poisoning in the rat. J Clin Invest 1996; 97:2260–2267.
26. Ischiropoulos H, Al-Mehdi A, Fisher AB. Reactive species in ischemic rat lung injury: contribution of peroxynitrite. Am J Physiol 1995; 269:L158–L164.
27. Szabo C, Salzman AL, Ischiropoulos H. Endotoxin triggers the expression of an inducible isoform of nitric oxide synthase and the formation of peroxynitrite in the rat aorta in vivo. FEBS Lett 1995; 363:235–238.

28. Wizemann TM, Gardner CR, Laskin JD, Quinones S, Durham SK, Goller NL, Ohnishi ST, Laskin DL. Production of nitric oxide and peroxynitrite in the lung during acute endotoxemia. J Leuko Biol 1994; 56:759–768.
29. Haddad IY, Pataki G, Hu P, Galliani C, Beckman JS, Matalon S. Quantitation of nitrotyrosine levels in lung sections of patients and animals with acute lung injury. J Clin Invest 1994; 94:2407–2413.
30. Kooy NW, Royall JA, Ye YZ, Kelly DR, Beckman JS. Evidence for in vivo peroxynitrite production in human acute lung injury. Am J Respir Crit Care Med 1995; 151:1250–1254.
31. Kaur H, Halliwell B. Evidence for nitric oxide-mediated oxidative damage in chronic inflammation. Nitrotyrosine in serum and synovial fluid from rheumatoid patients. FEBS Lett 1994; 350:9–12.
32. Seago ND, Thompson JH, Zhang X-J, Eloby-Childress S, Sadowska-Krowicka H, Rossi JL, Currie MG, Manning PT, Clark DA, Miller MJS. Inducible nitric oxide synthase and guinea-pig ileitis induced by adjuvant. Mediators Inflamm 1995; 4:19–24.
33. Miller MJS, Thompson JH, Zhang X-J, Sadowska-Krowicka H, Kakkis JL, Munshi UK, Sandoval M, Rossi JL, Eloby-Childress S, Beckman J, Ye YZ, Rodi CP, Manning PT, Currie MG, Clark DA. Role of inducible nitric oxide synthase expression and peroxynitrite formation in the guinea pig ileitis. Gastroenterology 1995; 109:1475–1483.
34. Nowicki JP, Duval D, Poignet H, Scatton B. Nitric oxide mediates neuronal death after focal cerebral ischemia in the mouse. Eur J Pharmacol 1991; 204:339–340.
35. Buisson A, Plotkine M, Boulu RG. The neuroprotective effect of a nitric oxide inhibitor in a rat model of focal ischemia. Br J Pharmacol 1992; 106:766–767.
36. Cazevielle C, Muller A, Meynier F, Bonne C. Superoxide and nitric oxide cooperation in hypoxia/reoxygenation-induced neuronal injury. Free Radic Biol Med 1993; 14:389–395.
37. Ashwal S, Cole DJ, Osborne TN, Pearce WJ. Dual effects of L-NAME during transient focal cerebral ischemia in spontaneously hypertensive rats. Am J Physiol 1994; 267:H276–H284.
38. Huang Z, Huang PL, Panahian N, Dalkara T, Fishman MC, Moskowitz MA. Effects of cerebral ischemia in mice deficient in neuronal nitric oxide synthase. 1994; 265:1883–1885.
39. Matheis G, Sherman MP, Buckberg GD, Haybron DP, Young HH, Ignarro LJ. Role of L-arginine-nitric oxide pathway in myocardial reoxygenation injury. Am J Physiol 1992; 262 (Heart Circ Physiol 31):H616–H620.
40. Patel VC, Yellon DM, Singh KJ, Neild GH, Woolfson RG. Inhibition of nitric oxide limits infarct size in the in situ rabbit heart. Biochem Biophys Res Commun 1993; 194:234–238.
41. Mulligan MS, Hevel JM, Marletta MA, Ward PA. Tissue injury caused by deposition of immune complexes is L-arginine dependent. Proc Natl Acad Sci USA 1991; 88:6338–6342.
42. Mulligan MS, Varani J, Warren JS, Till GO, Smith CW, Anderson DC, Todd RF III, Ward PA. Roles of B2 integrins of rat neutrophils in complement- and oxygen radical-mediated acute inflammatory injury. J Immunol 1992; 148:1847–1857.

43. Oury TD, Piantadosi CA, Crapo JD. Cold-induced brain edema. Involvement of extracellular superoxide dismutase and nitric oxide. J Biol Chem 1993; 268:15394–15398.
44. Dawson VL, Dawson TM, London ED, Bredt DS, Snyder SH. Nitric oxide mediates glutamate neurotoxicity in primary cortical cultures. Proc Natl Acad Sci USA 1991; 88:6368–6371.
45. Nielsen AT. Nitrocarbons. VCH Publishers. N.Y., N.Y. 1995, pp 40–41.
46. Haddad IY, Zhu S, Ischiropoulos H, Matalon S. Nitration of surfactant protein A results in decreased ability to aggregate lipids. Am J Physiol 1996; 270:L281–L288.
47. Berlett BS, Friguet B, Yim MB, Chock PB, Stadtman ER. Peroxynitrite-mediated nitration of tyrosine residues in Escherichia coli glutamine synthetase mimics adenylylation: relevance to signal transduction. Proc Natl Acad Sci USA 1996; 93:1776–1780.
48. MacMillan-Crow LA, Crow JP, Kerby JD Beckman JS, Thompson JA. Nitration and inactivation of Mn superoxide dismutase in chronic rejection of human renal allografts. Proc Natl Acad Sci USA. 1996; 93:11853–11858.
49. Gow A, Duran D, Malcom S, Ischiropoulos H. Effects of peroxynitrite induced modifications to signal transdiction and protein degradation. FEBS Lett 1996; 385:63–66.
50. Kong S-K, Yim MB, Stadtman ER, Chock PB. Peroxynitrite disables the tyrosine phosphorylation regulatory mechanism: lymphocyte-specific tyrosine kinase fails to phosphorylate nitrated cdc2(6-20)NH_2 peptide. Proc Natl Acad Sci USA 1996; 93:3377–3382.
51. Davies KJA. Protein damage and degradation by oxygen radicals. J Biol Chem 1987; 262:9895–9901.
52. Davies KJA, Delsignore ME, Lin SW. Protein damage and degradation by oxygen radicals. J Biol Chem 1987; 262:9902–9907.
53. Ames BN, Shigenaga MK, Hagen TM. Oxidants, antioxidants and the degenerative diseases of aging. Proc Natl Acad Sci USA 1993; 90:7915–7922.
54. Moncada S, Palmer RMJ, Higgs E. Nitric oxide: physiology, pathophysiology, and pharmacology. Pharmacol Rev 1991 43:109–142.
55. Chao CC, Hu S, Molitor TW, Shaskan EG, Peterson PK. Activated microglia mediate neuronal cell injury via a nitric oxide mechanism. J Immunol 1992; 149:2736–2741.
56. Ischiropoulos H, Zhu L, Beckman JS. Peroxynitrite formation from macrophage-derived nitric oxide. Arch Biochem Biophys 1992; 298:446–451.
57. Carreras MC, Pargament GA, Catz DS, Poderoso JJ, Boveris A. Kinetics of nitric oxide and hydrogen peroxide production and formation of peroxynitrite during the respiratory burst of human neutrophils. FEBS Lett 1994; 341:65–68.
58. Kooy NW, Royall JA. Agonist-induced peroxynitrite production from endothelial cells. Arch Biochem Biophys 1994; 310:352–359.
59. Bonfoco E, Krainc D, Ankarcrona M, Nicotera P, Lipton SA. Apoptosis and necrosis: two distinct events induced, respectively, by mild and intense insults with N-methyl-D-aspartate or nitric oxide/superoxide in cortical cell cultures. Proc Natl Acad Sci USA 1995; 92:7162–7166.

60. Bolanos JP, Heals SJR, Land JM, Clark JB. Effect of peroxynitrite on the mitochondrial respiratory chain: differential susceptibility of neurones and astrocytes in primary culture. J Neurochem 1995; 64:1965–1972.
61. Estévez AG, Radi R, Barbeito L, Shin JT, Thompson JA, Beckman JS. Peroxynitrite-induced cytotoxicity in PC12 cells: evidence for an apoptotic mechanism differentially modulated by neurotrophic factors. J Neurosci 1995; 65: 1543–1550.
62. Lin K-T, Xue JY, Nomen M, Spur B, Wong PYK. Peroxynitrite-induced apoptosis in HL-60 cells. J Biol Chem 1995; 270:16487–16490.
63. Richter C. Pro-oxidants and mitochondrial Ca^{++}: their relationship to apoptosis and oncogenesis. FEBS Lett 1993; 325:104–107.
64. Ischiropoulos H. Exposure of endothelial cells to peroxynitrite inhibits tyrosine phosphorylation and induces apoptosis. Endothelium 1995; 3s:47–48.
65. Packer MA, Murphy MP. Peroxynitrite formed by simultaneous nitric oxide and superoxide generation induces cyclosporin-A-sensitive mitochondrial calcium efflux and depolarisation. Eur J Biochem 1995; 234:231–239.
66. Salman-Tabcheh S, Guerin MC, Torreilles J. Nitration of tyrosyl-residues from extra and intracellular proteins in human whole blood. Free Radic Biol Med 1995; 19:695–698.
67. Troy CM, Derossi D, Prochiantz A, Greene LA, Shelanski ML. Down regulation of Cu/Zn superoxide dismutase leads to cell death via the nitric oxide-peroxynitrite pathway. J Neurosci 1996; 16:253–261.
68. Anderson EW, Andelman RJ, Strauch JM. Effects of low-level carbon monoxide exposure on onset and duration of angina pectoris. Ann Intern Med 1973; 79:46–50.
69. Ginsberg MD, Myers RE, McDonagh BF. Experimental carbon monoxide encephalopathy in the primate. II. Clinical aspects, neuropathology, and physiologic correlation. Arch Neurol 1974; 30:209–216.
70. Cramlet SH, Erickson HH, Gorman HA. Ventricular function following acute carbon monoxide exposure. J Appl Physiol 1975; 39:482–486.
71. Garland H, Pearce J. Neurological complications of carbon monoxide poisoning. Q J Med 1967; 144:445–455.
72. Winter PM, Miller JN. Carbon monoxide poisoning. JAMA 1976; 236:1502–1505.
73. Choi S. Delayed neurologic sequelae in carbon monoxide intoxication. Arch Neurol 1983; 40:433–435.
74. Myers RAM, Bray P. Delayed treatment of serious decompression sickness. Ann Emerg Med 1985; 14:254–257.
75. Thom SR, Taber RL, Mendiguren II, Clark JM, Hardy KR, Fisher AB. Delayed neuropsychological sequelae following carbon monoxide poisoning and its prophylaxis by treatment with hyperbaric oxygen. Ann Emerg Med 1995; 25:474–480.
76. Ginsberg MD. Carbon monoxide intoxication: clinical features, neuropathology and mechanisms of injury. Clin Toxicol 1985; 23:281–288.
77. Brucher JM. Neuropathological problems posed by carbon monoxide poisoning and anoxia. Prog Brain Res 1967; 24:75–100.
78. Jain KK. Carbon monoxide poisoning. St. Louis: Warren H. Green, 1990, pp 91–92.

79. Lapresle J, Fardeau M. The central nervous system and carbon monoxide poisoning. In: Bour H, Ledingham IM, eds. Carbon Monoxide Poisoning. Amsterdam: Elsevier, 1967, pp 31–74.
80. Courville CB. The process of demyelination in the central nervous system. Demyelination as a delayed residual of carbon monoxide asphyxia. J Nerv Ment Dis 1957; 125: 504–546.
81. Meyer A. Experimentelle erfahrungen uber die kohlenoxydverguftung des zentralnervens system s. Zeitschrift fur die Ges. Neurol Psychiatr 1928; 112:187–212.
82. Putnam TJ, McKenna JB, Morrison LR. Studies in multiple sclerosis. JAMA 1991; 97:1591–1596.
83. Funata N, Okeda R, Takano T, Miyazaki Y, Higashino F, Yokoyama K, Manabe M. Electron microscopic observations of experimental carbon monoxide encephalopathy in the acute phase. Acta Pathol Jpn 1982; 32:219–229.
84. Maeda Y, Kawasaki Y, Jibiki I, Yamaguchi N, Matsuda H, Hisada K. Effect of therapy with oxygen under high pressure on regional cerebral blood flow in the interval form of carbon monoxide poisoning: observation from subtraction of technetium-99m HMPAOSPECT brain imaging. Eur Neurol 1991; 31:380–383.
85. Shimosegawa E, Hatazawa J, Nagata K, Okudera T, Inugami A, Ogawa T. Cerebral blood flow and glucose metabolism measurements in a patient surviving one year after carbon monoxide intoxication. J Nucl Med 1992; 33:1696–1698.
86. Ducasse JL, Celsis P, Marc-Vergnes JP. Non-comatose patients with acute carbon monoxide poisoning: hyperbaric or normobaric oxygenation. Undersea Hyperbaric Med 1995; 22:9–15.
87. DeReuck J, Decoo D, Lemahieu I, Strijckmans K, Boon P, Van Maele G, Buylaert W, Leys D, Petit H. A positron emission tomography study of patients with acute carbon monoxide poisoning treated by hyperbaric oxygen. J Neurol 1993; 240:430–434.
88. Sesay M, Bidabe AM, Guyot M, Bedry R, Caille JM, Maurette P. Regional cerebral blood flow measurements with xenon-CT in the prediction of delayed encephalopathy after carbon monoxide intoxication. Acta Neurol Scand 1996; 166:22–27.
89. Thom SR, Ohnishi ST, Ischiropoulos H. Nitric oxide released by platelets inhibits neutrophil β_2 integrin function following acute carbon monoxide poisoning. Toxicol Appl Pharmacol 1994; 128:105–110.
90. Zhang J, Piantadosi Ca. Mitochondrial oxidative stress after carbon monoxide hypoxia in the rat brain. J Clin Invest 1991; 90:1193-1199.
91. Thom SR. Carbon monoxide-mediated brain lipid peroxidation in the rat. J Appl Physiol 1990; 68:997–1003.
92. Thom SR. Dehydrogenase conversion to oxidase and lipid peroxidation in brain after carbon monoxide poisoning. J Appl Physiol 1992; 73:1581–1589.
93. Thom SR. Leukocytes in carbon monoxide-mediated brain oxidative injury. Toxicol Appl Pharmacol 1993; 123:234–247.
94. Thom SR, Ischiropoulos, H. Perivascular nitrotyrosine in rat aorta after carbon monoxide exposure. FASEB J 1996; 10:A271.
95. Luomanmaki K, Coburn RF. Effects of metabolism and distribution of carbon monoxide on blood and body stores. Am J Physiol 1969; 217:354–363.

96. Fenn WO. The burning of CO in the tissues. Ann NY Acad Sci 1970; 170:64–71.
97. Thom SR, Yu YA, Ischiropoulos H. Vascular endothelial cells generate peroxynitrite in response to carbon monoxide exposure. Chemical Res in Toxicol 1997; 10:1023–1031.
98. Tomaszewski C, Rudy J, Wathen J, Brent J, Rosenberg N, Kulig K. Prevention of neurologic sequelae from carbon monoxide by hyperbaric oxygen in rats. Ann Emerg Med 1992; 21:631–632.
99. Nabeshima T, Katoh A, Ishimaru H, Yoneda Y, Ogita K, Murase K, Ohtsuka H, Inari K, Fukuta T, Kameyama T. Carbon monoxide-induced delayed amnesia, delayed neuronal death and change in acetylcholine concentration in mice. J Pharmacol Exp Ther 1991; 256:378–384.
100. Ishimaru H, Nabeshima T, Katoh A, Suzuki H, Fukuta T, Kameyama T. Effects of successuve carbon monoxide exposures on delayed neuronal death in mice under the maintenance of normal body temperature. Biochem Biophys Res Commun 1991; 179:836–840.
101. Ishimaru H, Katoh A, Suzuki H, Fukuta T, Kameyama T, Nabeshima T. Effects of N-methyl-D-aspartate receptor antagonists on carbon monoxide-induced brain damage in mice. J Pharmacol Exp Ther 1992; 261:349–352.
102. Bianco F, Floris R. Transient disappearance of bilateral low-density lesions of the globus pallidus in carbon monoxide intoxication and MRI. J Neuroradiol 1988; 15:381–385.
103. Silverman CS, Brenner J, Murtagh FR. Hemorrhagic necrosis and vascular injury in carbon monoxide poisoning. MR demonstration. Am J Neuroradiol 1993; 14:168–170.
104. Song S-Y, Okeda R, Funata N, Higashino F. An experimental study of the pathogenesis of the selective lesion of the globus pallidus in acute carbon monoxide poisoning in cats. Acta Neuropathol (Berl) 1983; 61:232–238.
105. Meldrum BS. Excitatory amino acid and anoxic/ischemic brain damage. Trends Neurosci 1985; 8:47–48.
106. Raichle ME. The pathophysiology of brain ischemia. Ann Neurol 1983; 13:2–10.
107. Rothman SM, Olney JW. Glutamate and the pathophysiology of hypoxic-ischemic brain injury. Ann Neurol 1986; 19:105–111.
108. Choi DW. Calcium-mediated neurotoxicity: relationship to specific channel types and role in ischemic damage. Trends Neurosci 1988; 11:465–479.
109. Liu Y, Fechter LD. MK-801 protects against carbon monoxide-induced hearing loss. Toxicol Appl Pharmacol 1995; 132:196–202.
110. Mayer ML, Miller RJ. Excitatory amino acid receptors, second messengers and regulation of intracellular Ca^{2+} in mammalian neurons. Trends Neurosci 1990; 11: 254–260.
111. Lerner-Natoli M, Rondouin G, de Bock F, Bockaert J. Chronic NO synthase inhibition fails to protect hippocampal neurones against NMDA toxicity. Neuroreport 1992; 3:1109–1112.
112. Nowicki JP, Duval D, Poignet H, Scatton B. Nitric oxide mediates neuronal death after focal cerebral ischemia in the mouse. Eur J Pharmacol 1991; 204:339–342.
113. Pauwels PJ, Leysen JE. Blockade of nitric oxide formation does not prevent glutamate-induced neurotoxicity in neuronal cultures from rat hippocampus. Neurosci Lett 1992; 143:27–30.

114. Newby MB, Roberts RJ, Bhatnagar RK. Carbon monoxide and hypoxia-induced effects on catecholamines in the mature and developing rat brain. J Pharmacol Exp Ther 1978; 206:61–68.
115. Bindoli A, Rigobello MP, Deeble DJ. Biochemical and toxicological properties of the oxidation products of catecholamines. Free Radic Biol Med 1992; 13:391–405.
116. Piantadosi CA, Tatro L, Zhang J. Hydroxyl radical production in the brain after CO hypoxia in rats. Free Radic Biol Med 1995; 18:603–609.
117. Levitt P, Pintar JE, Breakfield XO. Immunocytochemical demonstration of monoamine oxidase B in brain astrocytes and serotonergic neurons. Proc Natl Acad Sci USA 1982; 79:6385–6389.
118. Reiderer P, Konradi C, Schay V, Kienzl E, Birkmayer G, Danielczyk W, Sofic E, Youdim MBH. Localization of MAO-A and MAO-B in human brain: a step in understanding the therapeutic action of L-deprenyl. Adv Neurol 1986; 45:111–118.
119. Hanbauer I, Wink D, Osawa Y, Edelman GM, Gally JA. Role of nitric oxide in NMDA-evoked release of [^{3}H]-dopamine from striatal slices. Neuroreport 1992; 3:409–412.
120. Montague PR, Gancayco CD, Winn MJ, Marchase RB, Friedlander MJ. Role of NO production in NMDA receptor-mediated neurotransmitter release in cerebral cortex. Science 1994; 263:973–977.
121. Merrill JE, Zimmerman RP. Natural and induced cytotoxicity of oligodendrocytes by microglia is inhibitable by TBFβ. Glia 1991; 4:327–331.
122. Verma A, Hirsch DJ, Glatt CE, Ronnett GV, Snyder SH. Carbon monoxide: a putative neural messenger. Science 1993; 259:381–384.
123. Zhuo M, Small SA, Kandel ER, Hawkins RD. Nitric oxide and carbon monoxide produce activity-dependent long-term synaptic enhancement in hippocampus. Science 1993; 260:1946–1950.
124. Brune B, Ullrich V. Inhibition of platelet aggregation by carbon monoxide is mediated by activation of guanylate cyclase. Mol Pharmacol 1987; 32:497–504.
125. Nathanson JA, Scavone C, Scanlon C, McKee M. The cellular Na^+ pump as a site of action for carbon monoxide and glutamate: a mechanism for long-term modulation of cellular activity. Neuron 1995; 14:781–794.
126. Ramos KS, Lin H, McGrath JJ. Modulation of cyclic guanosine monophosphate levels in cultured aortic smooth muscle cells by carbon monoxide. Biochem Pharmacol 1989; 38:1368–1370.
127. Keyse SM, Tyrrell RM. Heme oxygenase is the major 32-kDa stress protein induced in human skin fibroblasts by UVA radiation, hydrogen peroxide, and sodium arsenite. Proc Natl Acad Sci USA 1989; 86:99–103.
128. Llesuy SF, Tomaro ML. Heme oxygenase and oxidative stress. Evidence of involvement of bilirubin as physiological protector against oxidative damage. Biochim Biophys Acta 1994; 1223:9–14.

5

Structure-Activity Relationships in NOS Inhibitors

Roger A. Egolf
Pennsylvania State University, Fogelsville, Pennsylvania

Ned D. Heindel
Lehigh University, Bethlehem, Pennsylvania

I. INTRODUCTION

Research on biological NO is extremely active. In 1997 alone there were over 3000 papers published studying inhibitors of nitric oxide synthase (NOS) and about 100 of these dealt with new agent synthesis. To the medicinal chemist interested in structure-activity correlations, in chemical mechanisms of inhibition, and in models of drug-binding site interaction, the field of NOS inhibition stands now where the benzodiazepine development stood in the early 1970s.

Analog synthesis, the rational and serendipitous probing for active agents, and the search for selectivity in inhibition among nNOS, eNOS, and iNOS are definitely in the data collection stage. A clear picture of how the various inhibitors work will require more diversity in the candidate pharmaceutical test pool. It is in the nature of medicinal chemists not to rest until each inhibitor's action is understood at a molecular level and the various families or classes of inhibitors are justified by their molecular mechanisms. However, the subject of NOS inhibition has a long way to go before it reaches the current molecular-based state of understanding of the MAO inhibitors, the sulfas, or even the benzodiazepines.

From the clinical perspective it is clear that manipulation of NO levels ought to prove to be a tremendous therapeutic benefit. We already know a wide variety of biological functions in which NO appears to play a role, and it is also known that inhibition of NOS activity (at least in vitro) decreases available NO. This strong element of clinical promise will support and drive the chemistry of new candidate agent synthesis and testing.

This report is therefore just a snapshot—a status report of some of the major inhibitor classes and a view to their probable modes of action. Some molecular mechanisms for inhibition have been proposed and have received experimental support. Perhaps the most obvious one is active-site directed inhibition. The classic way to design an inhibitor is to take advantage of the specificity the enzyme (NOS) has for its substrate, in this case, L-arginine. In this fashion, scores of arginine analogs have been synthesized, screened, and found to be active antagonists. Most of these bear substituents such as benzyl, methyl, allyl, amino, nitro, or cyclopropyl on the terminal (or gamma) nitrogen of L-arginine. The N-allyl and N-methyl-L-arginines are interesting in that they appear to have both (a) a reversible, competitive and (b) an irreversible, enzyme-activated, mechanism-based inhibition (1).

Another major approach to inhibitor design is to target common regulatory systems, cofactors, or prosthetic functions such as the heme iron moiety of NOS. There is a wealth of prior art in other enzyme systems to select candidate inhibitors based on these principles. Five-membered heterocyclics and their benzo analogs appear likely to function by a competitive binding to an apical binding site on the NOS heme. Imidazole and N-phenylimidazole are claimed to inhibit cNOS by coordinating to the iron at the sixth binding site, thereby blocking oxygen's binding activation (2). There is a claim that 7-nitroindazole, an inhibitor of rat cNOS, and other isomeric nitroindazoles also act in this fashion (3).

Other NOS-inhibitory heterocyclics, less resembling imidazole and indazole, have unclear mechanisms of inhibition. The phenothiazine pharmaceutical chlorpromazine inhibits cNOS, and 3-amino-1,2,4-triazole and 3-amino-1,2,4-triazine both inhibited iNOS. One reviewer has suggested that the triazole and triazine may function by competing at the heme iron-binding site (4), but an alternative suggestion seems equally plausible (5). The imino tautomers of both of these amino-pendant heterocyclics resemble an N-gamma-amino-substituted guanidine. Additional support for the view that these heterocyclics may act as substrate mimics, was found when the isomeric 4-amino-1,2,4-triazole (not guanidine like) proved to be inactive (5).

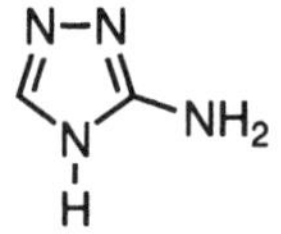

3-Amino-1,2,4-Triazole
Active

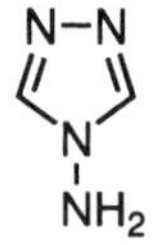

4-Amino-1,2,4-Triazole
Inactive

An aminoguanidine construct, an abbreviated version of an N-gamma-amino-L-arginine, has been synthesized and found to selectively inhibit iNOS (6). Moreover, Wolff has shown that this aminoguanidine is an active-site-directed, mechanism-based inhibitor of the NOS isoforms with a marked specificity for iNOS (7).

While the research objective of demonstrating selective inhibition of one isoform is readily understood in light of the clinical potential such an agent would possess, progress in the pursuit of that goal is most puzzling. Every reviewer takes pains to cite potency against a particular isoform (when observed), although understanding of such selectivity is almost never advanced (8). Furthermore, a recent structure-activity study of 2-iminopiperidines demonstrated that even the slightest molecular manipulation (e.g., the alteration in placement of a methyl at a position remote from the polar domain) can invert the selectivity ratios of iNOS over eNOS (9). Not only do we have the complication that minor alterations in the molecule alter selectivity but it also appears that multiple mechanisms of inhibition may exist for the same identical inhibitor (10). We clearly have a lot to learn about structural features that foster selectivity differences within a family of active analogs as well as about structures that can serve as novel lead inhibitors themselves.

If one focuses, however, on candidate inhibitors whose dominant mechanisms are interference with NOS regulatory pathways or with cofactors or prosthetic group function, one risks wider adverse pharmacology in vivo than if the drug were more specific for NOS. Since the various synthetic arginine mimics do demonstrate that one can develop agents with action against one particular isotype of NOS (4), albeit only by serendipity at present, the substrate analog approach holds most promise at this time.

Far more compounds have been prepared and tested as potential NOS inhibitors than have been studied mechanistically. At this time, no review of the medicinal chemistry can classify the structures (other than the most obviously architectural analogs of known inhibitors) according to mechanism of action, whether they be arginine substrate analogs, heme-iron directed inhibitors of oxygen binding, flavoprotein binders, or some other inhibitory type. The constraints of our knowledge base dictate that we classify most recently reported inhibitors by structural commonality.

Organization of NOS inhibitors by structure has been employed in several earlier reviews (11,12), and one of the best tabulations along structural lines has been Macdonald's (8). What follows is our amplification and updating of that categorization.

A crystal structure of iNOS is now available which establishes a juxtaposed hydrophobic oxygen-binding site (capable of being blocked by imidazoles) and a polar L-arginine binding site (capable of being blocked by aminoguanidine) (13). With an aid such as this, it ought to be possible to model some of the empirically-derived inhibitors, cited above, and anticipate not only where each may bind but also how that binding might be enhanced. Better NOS inhibitors by rational design are just around the corner.

II. STRUCTURAL CLASSES OF INHIBITORS

A. Amino Acids

1. Single Amino Acids

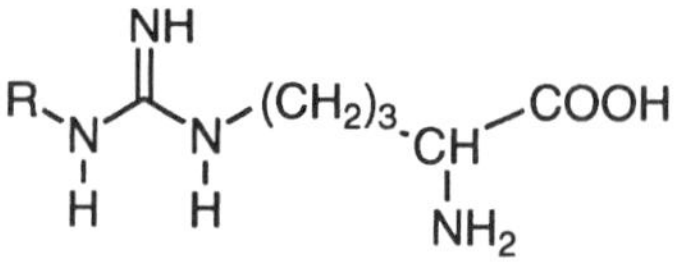

N$^{\omega}$-Substituted-L-Arginines

Of the several classes of inhibitors of NOS mentioned above, the earliest discovered were the substrate mimics, N$^{\omega}$-substituted L-arginines (12). N$^{\omega}$-methyl-L-arginine is slightly selective for nNOS and eNOS over iNOS, whereas N$^{\omega}$-ethyl-L-arginine is slightly selective for the iNOS form over the constitutive forms (14). The mechanism of inactivation of nNOS by N$^{\omega}$-allyl-L-arginine was described and N$^{\omega}$-propyl-L-arginine was shown to be a competitive inhibitor and inactivator of nNOS (15). The potency of inhibition of nNOS by N$^{\omega}$-propyl-L-arginine is 3158 times that of iNOS and 149 times that of eNOS (16). Unsaturation in the propyl side chain has a dramatic undesirable effect on selectivity. Both N$^{\omega}$-allyl-L-arginine and N$^{\omega}$-propargyl-L-arginine are weakly selective, the former having a selectivity of nNOS/iNOS of about 10 and nNOS/eNOS of 15, whereas the latter has a selectivity of less than 2. Obviously the geometry and size of the side chain are extremely important to the selectivity of inhibition.

2. Dipeptides

Along with single amino acids and their related analogs, dipeptides and dipeptide esters, all containing an arginine moiety, also function as inhibitors of NOS. A extensive family of N$^{\omega}$-nitroarginine (ArgNO$_2$)- and phenylalanine-containing dipeptides and dipeptide esters were found to be selective inhibitors of nNOS (17). All the compounds in this study were demonstrated to be competitive inhibitors of nNOS, iNOS, and eNOS, except those that contained the D-ArgNO$_2$ moiety. The latter were uncompetitive inhibitors of iNOS, but competitive in their inhibition of nNOS and eNOS. While N$^{\omega}$-nitro-L-arginine itself exhibited time-dependent inhibition of nNOS, several dipeptides (specifically 1, 2, 12, and 13 from Table 1) when tested specifically for such behavior displayed no similar time dependency against any of the isoforms. Clearly a different mode of action is implied for the dipeptides than for the monomeric nitropendant L-arginine. The order of the amino acids in the dipeptide is important to selectivity, which also depends on the chirality of the amino acids. Dipeptide benzyl esters favor iNOS over nNOS or eNOS inhibition; however, dipeptide methyl esters containing a D-amino acid exhibit an inhibitory preference for nNOS over iNOS and eNOS. The most impressive selectivities observed in this study were 1800- and 800-fold for compounds 12 and 13, respectively, in favor of nNOS over iNOS. Unfortunately, the selectivities of these compounds for nNOS over eNOS are only 2.5 and 5.3, respectively.

The promise of eventual pharmacologic specificity triggers, of course, the vigorous search for molecular selectivity of the candidate therapeutics within the various enzyme isoforms. In vitro bioassays used in such studies typically display experimental variability, frequently not quantified in the article, which makes perilous the rendering of precise statements of specificity when ratios of IC_{50}s among the isoforms are small. Table 1 summarizes the K_i and selectivity data from a study of inhibitory dipeptides and illustrates this problem since many ratios are under 10 (17).

B. Non-Amino Inhibitors

1. Isothioureas

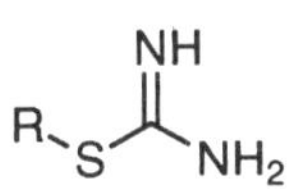

Isothioureas

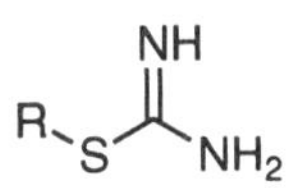

Isoselenoureas

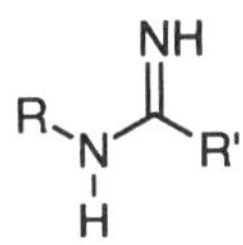

Amidines

Extending the structural definition of arginine mimics to bioisosteres of the guanidine end group has led to promising new families of non-amino acid inhibitors. N-Substituted phenyl S-alkyl isothioureas show selectivity for the inhibition of nNOS and bind to the enzyme competitively with L-arginine (18). The compounds with an S-ethyl group are most active, and the degree of selectivity favoring nNOS among the isoforms can be modified by placing a substituent group on the phenyl ring. The best selectivity toward nNOS was achieved with S-ethyl N-[4-(trifluoromethyl)phenyl]isothiourea, which exhibits 115-fold and 29-fold selectivity for nNOS versus iNOS and eNOS, respectively. 1-Amino-S-methylisothiourea has been identified as a mechanism-based inactivator selective for the nNOS isoform, is cell-impermeable, and inhibits the NOS enzyme irreversibly (19).

Interestingly, another study found that simple lower alkyl isothioureas were very potent inhibitors of all forms of NOS, with a slight selectivity toward the inducible form (20). In an effort to develop a tracer for probing iNOS levels in vivo utilizing positron emission tomography (PET), two positron-emitting iNOS selective inhibitors, S-[^{11}C]methylisothiourea and S-(2-[^{18}F]fluoroethyl)isothiourea, were prepared and determined to have a ninefold higher selectivity for iNOS compared to eNOS (21). Biodistribution in female Sprague-Dawley rats was examined and uptake was found to be significantly increased in the liver, kidney, lung, and heart

Table 1 K_i Values for Inhibition of NOS Isoforms by Nitroarginine-Containing Dipeptides and Dipeptide Esters

Inhibitors	nNOS (μM)	iNOS (μM)	eNOS (μM)	Selectivity		
				nNOS/iNOS	nNOS/eNOS	iNOS/eNOS
1. L-ArgNO$_2$-L-Phe	18	160	395	8.9	21.9	2.5
2. L-Phe-L-ArgNO$_2$	140	93	490	0.66	3.5	5.3
3. L-ArgNO$_2$-L-Phe-OMe	14	45	400	3.2	28.6	8.9
4. L-Phe-L-ArgNO$_2$-OMe	370	204	1350	0.55	3.6	6.6
5. L-ArgNO$_2$-L-Phe-OBn	55	18	125	0.33	2.3	6.9
6. L-Phe-L-ArgNO$_2$-OBn	110	45	250	0.41	2.3	5.6
7. L-ArgNO$_2$-D-Phe-OMe	92	100	525	1.1	5.7	5.3
8. D-ArgNO$_2$-L-Phe-OMe	290	6400[a]	1400	22.1	4.8	0.22
9. D-ArgNO$_2$-D-Phe-OMe	150	6500[a]	375	13.3	2.5	0.058
10. L-Phe-D-ArgNO$_2$-OMe	90	7500[a]	150	83.3	1.7	0.02
11. D-Phe-L-ArgNO$_2$-OMe	400	1200	8500	3	21.3	7.1
12. D-Phe-D-ArgNO$_2$-OMe	2	3600[a]	5	1800	2.5	0.0014
13. D-Phe-D-ArgNO$_2$	17	13600[a]	90	800	5.3	0.0066

Source: Ref. 17.
[a]Uncompetitive inhibition.

in lipopolysaccharide-pretreated rats compared to unpretreated controls, suggesting that these radiolabeled inhibitors may be useful for assessing induction of iNOS in vivo with PET.

2. Isoselenoureas

Aminoethylisoselenourea is a potent inhibitor of iNOS in homogenates of lung taken from endotoxin-treated rats, but has only a small inhibitory effect on rat eNOS activity in vivo, based on blood pressure measurements (22). Its aminopropyl analog has potent pressor effects similar to those of L-N-methylarginine. Both selenium analogs are unstable in aqueous solution at pH values above 6. Their disappearance from solution is accompanied by the appearance of a uncharacterized selenium-containing reductive species. These findings suggest that these compounds exert their inhibitory effects through intramolecular rearrangement to yield selenoalkylguanidines, which are the probable pharmacoactive agents responsible for the inhibition.

3. Amidines

Simple amidines may be viewed as nor-aza guanidines, a structural approximation of the metabolically reactive end of arginine. Thus it is not surprising to find some amidines which act as inhibitors of NOS which show little discrimination between the constitutive and inducible forms of the enzyme (23). However, N-[3-(aminoethyl)benzyl]acetamide has been shown to be a slow, tight-binding, and highly selective inhibitor of iNOS in vitro and in vivo (24).

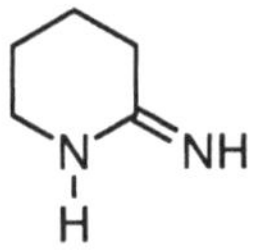

2-Iminopiperidine

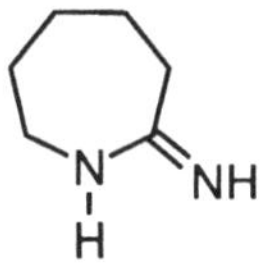

2-Iminohomopiperidine

Cyclic amidines with ring sizes ranging from 5 to 9 atoms are potent inhibitors of NOS, with 2-iminopiperidine and 2-iminohomopiperidine being the most active, with IC_{50} values of 1.0 and 2.0 μM, respectively, toward the human inducible isoform (25). This series of cyclic inhibitors has been further expanded to include analogs with heteroatoms in the 3 position of the 6-membered ring. This modification is tolerated for sulfur and oxygen, but nitrogen reduces their inhibitory potency. As mentioned earlier, analogs prepared to study the effect of substitution on the piperidine ring on inhibitory potency and selectivity among the isoforms of the enzyme, did show these properties to be highly dependent on the alkyl group substitution pattern around the piperidine ring (9). Methyl groups appended at the 4 or 6 positions afforded compounds which were 2- to 10-fold more potent than the parent 2-iminopiperidine at inhibiting all three isoforms. In contrast, analogs with methyl groups in the 3 or 5 positions were significantly less active than the parent compound. Compounds with methyl substitution on either amidine nitrogen showed a sharp drop in inhibitory potency. Larger substituents on the 4 position caused major decreases in potency against all isoforms; however, longer chain lengths in the substituents at the 6 position increased potency against the inducible isoform, along with an increase in selectivity. Bulky carbocyclic substituents at the 6-position caused a decrease in potency at inhibiting all the isoforms but were among the more selective examples.

Hydrochloride salts of pyrazole-1-carboxamidines inhibit all three isoforms of NOS (26). The parent compound of this series with no additional substituents on the pyazole ring inhibits all isoforms to a similar extent, with an IC_{50} of 0.2 μM. Its 3-methyl and 4-methyl analogs show reduced potencies but a preference for iNOS over the other isoforms. Inhibition of purified iNOS by these compounds can be reversed completely by excess L-arginine, while their inhibition of NO production by stimulated macrophages can be reversed by transfer to a drug-free medium, suggesting a competitive form of inhibition. Again, "selectivity" in these in vitro assays is modest, and structure activity statements about which substituent at which locus generates the most isoform-specific agent must be interpreted with caution.

4. Guanidines

The closest structural fragment of arginine is, of course, guanidine. Indeed, guanidines substituted with small alkyl groups are weak inhibitors of NOS, and thus define a minimum structure required for recognition at the binding site for amidines and guanidines of the NOS isoforms (8). 2-Ethylaminoguanidine displays a high isoform selectivity for iNOS compared with the other isoforms (19). Selenoalkylguanidines are also selective inhibitors of iNOS (22).

5. Heterocyclic and Other Compounds

As noted above, there is evidence that nucleophilic small-ring nitrogen heterocycles function as competitive NOS inhibitors by complexing at a heme iron apical valency. The closest histidine mimics, the azole antifungals itraconazole, ketoconazole, and miconazole, are known as weak inhibitors of iNOS in intact cells, with an $IC_{50} \geq 10$ μM.[27] At this level they are probably not practical drugs for iNOS inhibition, since these concentrations are at least 100 times greater than those used clinically to produce a 50% inhibition of yeast growth and ergosterol synthesis of Candida albicans after the same induction period.

As mentioned previously a number of less direct mechanisms appear to operate with other diverse structures now well known as NOS inhibitors. These include the calmodulin inhibition effects in cNOS, the diphenyleneiodonium and other flavoprotein inhibition effects in cNOS and iNOS, and other yet to be defined mechanisms (4,8). Examples are common of active compounds which at present defy detailed rational explanations for their mechanisms of inhibition.

Mention has already been made of the neuroleptic chlorpromazine, which has modest activity against rat nNOS (28). The oft-abused psychotomimetic phencyclidine has been shown to be a mechanism-based inhibitor of brain NOS (29). Furthermore, the anesthetic ketamine has been observed to be a weak inhibitor of both the iNOS activity and expression of the protein induced by lipopolysaccharide in rat alveolar macrophages, with an IC_{50} of slightly > 100 μM (30).

Such examples point again to the fact that structure-activity understanding in NOS inhibitors is clearly in the discovery phase. A healthy dose of serendipity and trial-and-error development must precede the implementation of a protocol for rational drug design.

REFERENCES

1. Olken NM, Marletta MA. N^G-Allyl and N^G-cyclopropyl-L-arginine: two novel inhibitors of macrophage nitric oxide synthase. J Med Chem 1992; 35:1137–1144.
2. Babbedge RC, Bland-Ward PA, Hart SL, Moore PK. Inhibition of rat cerebellar nitric oxide synthase by 7-nitroindazole and related substituted indazoles. Br J Pharmacol 1993; 110:225–228.
3. Wolff DJ, Grim BJ. The inhibition of the constitutive and inducible nitric oxide isoforms by indazole agents. Arch Biochem Biophys 1994; 311:300–306.
4. Fukuto JM, Komori Y. The enzymology and manipulation of nitric oxide synthase. In: Bristol JA, ed. Annu Rep Med Chem 1994; 29:83–92.
5. Buchmuller-Rouiller Y, Schneider P, Betz-Corradin S, Smith J, Mauel J. 3-Amino-1,2,4-triazole inhibits macrophage NO synthase. Biochem Biophys Res Commun 1992; 183:150–155.
6. Misko TP, Moore WM, Kasten TP, Nickols GA, Corbett JA, Tilton RG, McDaniel ML, Williamson JR, Currie MG. Selective inhibition of the inducible nitric oxide synthase by aminoguanidine. Eur J Pharmacol 1993; 233:119–125.
7. Wolff DJ, Lubeskie A. Aminoguanidine is an isoform-selective mechanism-based inhibitor of nitric oxide synthase. Arch Biochem Biophys 1995; 316:290–301.
8. Macdonald JE. Nitric oxide synthase inhibitors. In: Bristol JA, ed. Annu Rep Med Chem 1996; 31:221–230.
9. Webber RK, Metz S, Moore WM, Connor JR, Currie MG, Fok KF, Hagen, TJ, Hansen DW Jr, Jerome GM, Manning PT, Pitzele BS, Toth MV, Trivedi M, Zupec ME, Tjoeng FS. Substituted 2-iminopideridines as inhibitors of human nitric oxide synthase isoforms. J Med Chem 1998; 41:96–101.
10. Olken NM, Osawa Y, Marletta MA. Characterization of the inactivation of nitric oxide synthase by N-gamma-methyl-L-arginine. Biochemistry 1994; 33:14784–14791.

11. Kerwin JF Jr, Heller M. The argine-nitric oxide pathway: a target for new drugs. Med Res Rev 1994; 14:23–74.
12. Marletta MA. Approaches toward selective inhibition of nitric oxide synthase. J Med Chem 1994; 37:1899–1907.
13. Crane BR, Arvai AS, Ratan G, Wu C, Ghosh DK, Getzoff ED, Stuehr DJ, Tainer JA. The structure of nitric oxide synthase oxygenase domain and inhibitor complexes. Science 1997; 278:425–431.
14. Moore WM, Webber RK, Fok KF, Jerome GM, Kornmeier CM, Tjoeng FS, Currie MG. Inhbitors of human nitric oxide synthase isoforms with the carbamidine moiety as a common structural element. Bioorg Med Chem 1996; 4:1559–1564.
15. Zhang HQ, Dixon RP, Marletta MA, Nikolic D, Van Breemen R, Silverman RB. Mechanism of inactivation of neuronal nitric oxide synthase by N^{ω}-allyl-L-arginine. J Am Chem Soc 1997; 119:10888–10902.
16. Zhang HQ, Fast W, Marletta MA, Martasek P, Silverman RB. Potent and selective inhibition of neuronal nitric oxide synthase by N^{ω}-propyl-L-arginine. J Med Chem 1997; 40:3869–3870.
17. Silverman RB, Huang H, Marletta MA, Martasek P. Selective inhibition of neuronal nitric oxide synthase by N--nitroarginine- and phenylalanine-containing dipeptides and dipeptide esters. J Med Chem 1997; 40:2813–2817.
18. Shearer BG, Lee S, Oplinger JA, Frick LW, Garvey EP, Furfine ES. Substituted N-phenylisothioureas: potent inhibitors of human nitric oxide synthase with neuronal isoform selectivity. J Med Chem 1997; 40:1901–1905.
19. Wolff DJ, Gauld DS, Neulander MJ, Southan G. Inactivation of nitric oxide synthase by substituted aminoguanidines and aminoisothioureas. J Pharmacol Exp Ther 1997; 283:265–273.
20. Garvey EP, Oplinger JA, Tanoury GJ, Sherman PA, Fowler M, Marshall S, Harmon MF, Paith JE, Furfine ES. Potent and selective inhibition of human nitric oxide synthases. J Biol Chem 1994; 269:26669–26676.
21. Zhang J, McCarthy TJ, Moore WM, Currie MG, Welch MJ. Synthesis of evaluation of two positron-labeled inhibitors, S-[^{11}C]methylisothiourea and S-2-[^{18}F]fluoroethylisothiourea, as potential positron emission tomography tracers. J Med Chem 1996; 39:5110–5118.
22. Southan GJ, Salzman AL, Szabo C. Potent inhibition of the inducible isoform of nitric oxide synthase by aminoethylisoselenourea and related compounds. Life Sci 1996; 58:1139–1148.
23. Tjoeng FS, Fok KF, Webber RK. PCT Pat. Appl 1995: WO95/11014.
24. Garvey EP, Oplinger JA, Furfine ES, Kiff RJ, Laszlo F, Whittle BJR, Knowles RG. 1400W is a slow, tight binding, and highly selective inhibitor of inducible nitric-oxide synthase in vitro and in vivo. J Biol Chem 1997; 272:4959–4963.
25. Moore WM, Webber RK, Fok KF, Jerome GM, Connor JR, Manning PT, Wyatt PS, Misko TP, Tjoeng FS, Currie MG. 2-Iminopiperidine and other 2-iminoazaheterocycles as potent inhibitors of human nitric oxide synthase isoforms. J Med Chem 1996; 39:669–672.
26. Southan GJ, Gauld D, Lubeskie A, Zingarelli B, Cuzzocrea S, Salzman AL, Szabo C, Wolff DJ. Inhibition of nitric oxide synthase with pyrazole-1-carboxamidine and related compounds. Biochem Pharmacol 1997; 54:L409–417.

27. Vermuyten K, Laurijssens L, Vanden Bossche H. Azole antifungals: weak inhibitors of inducible nitric oxide synthase in mouse and human cells. Mycoses 1997; 40:119–125.
28. Palacios M, Rojas GL, Delgado R, Knowles R, Moncada S. Chlorpromazine inhibits both the constitutive nitric oxide synthase and the induction of nitric oxide synthase after LPS challenge. Biochem Biophys Res Commun 1993; 196:280–286.
29. Osawa Y, Davila JC. Phencyclidine, a psychotomimetic agent and drug of abuse, is a suicide inhibitor of brain nitric oxide synthase. Biochem Biophys Res Commun 1993; 194:1435–1439.
30. Li CY, Chou TC, Wong CS, Ho ST, Wu CC, Yen MH, Ding YA. Ketamine inhibits nitric oxide synthase in lipopolysaccharide-treated rat alveolar macrophages. Can J Anaesth 1997; 44:989–995.

6

Homeostatic and Pathological Actions of Nitric Oxide in the Kidney

Lori A. Hoerrner and Aaron J. Kowalski
Rutgers University, Piscataway, New Jersey

Jonathan R. Diamond
Pennsylvania State University College of Medicine, Hershey, Pennsylvania

Patricia D. Wilson
Mount Sinai School of Medicine, New York, New York

David T. Denhardt
Rutgers University, Piscataway, New Jersey

I. INTRODUCTION

Nitric oxide is an important signaling and effector molecule that has been implicated in many biological processes. Its remarkable nature (small size, membrane permeability, short half-life, biological reactivity) has prompted many studies into its role in the physiology and pathophysiology of the kidney (reviews: 1–4). NO is involved in the maintenance of basal vascular tone, in renal and glomerular hemodynamics, and in the tubuloglomerular feedback response. It inhibits contraction and proliferation of glomerular mesangial cells. In this chapter we review the impact of NO on the renin-angiotensin system and its role in the cellular response to transient hypoxia. We also briefly discuss hypertension and the involvement of osteopontin in renal homeostasis and pathology.

Abbreviations: ACE, angiotensin converting enzyme; angII, angiotension II; IFN, interferon; L-NAME, N^{G}-nitro-L-arginine methyl ester; 7-NI, 7-nitroindazole; NOS, nitric oxide synthase (NOS1 = neural NOS; NOS2 = inducible macrophage NOS; NOS3 = endothelial NOS); LPS, lipopolysaccharide; OPN, osteopontin; PTE, proximal tubule epithelial cells; RT-PCR, reverse transcriptase-polymerase chain reaction; TG, tubuloglomerular.

II. KIDNEY STRUCTURE AND FUNCTION

The human kidneys are paired organs composed of approximately 1 million nephrons each. Idealized nephrons are illustrated in Figure 1 (5). Each nephron, which is composed of a tubule and a vascular element called the glomerulus, is bound together by blood vessels, nerves, lymphatics, and small amounts of connective tissue. The tubule is composed of a single layer of epithelial cells that differ significantly in structure and function along the length of the tubule. It originates as the Bowman's capsule, which on one side is associated with the glomerulus (the renal corpuscle in Fig. 1) and on the other side opens into the highly coiled

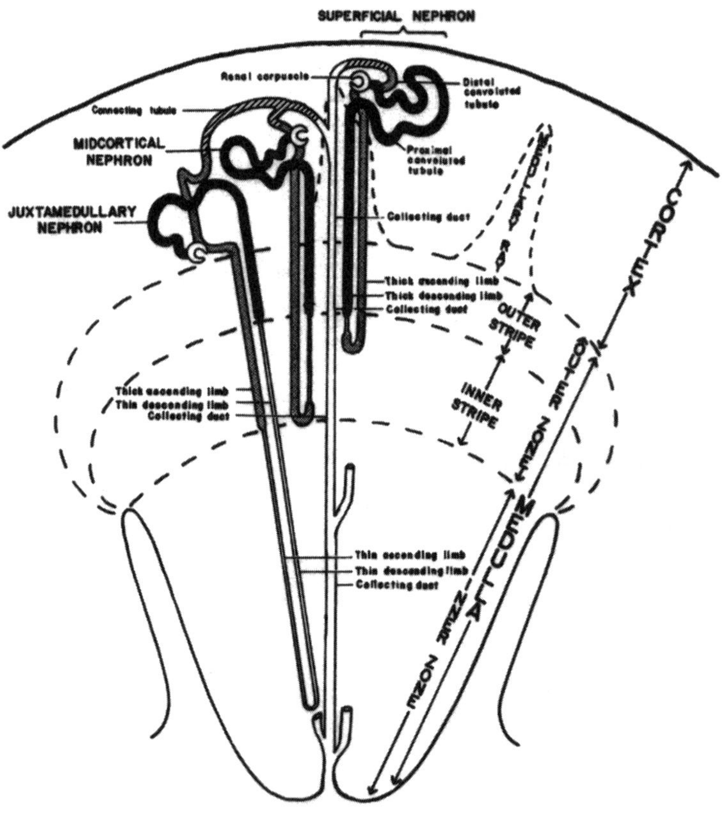

Figure 1 Cartoon illustrating nephron organization in the kidney. The upper end of each nephron (the glomerulus and proximal convoluted segment) lies in the cortex while the loop of Henle and the collecting ducts are located in the medulla. (From Ref. 5.)

proximal convoluted tubule, which is structurally and functionally divided into the initial S1 portion, the intermediate S2 portion, and the straighter S3 portion, or pars recta. The next segment of the tubule is a hairpinlike structure called Henle's loop, which varies in length depending upon location of the glomerulus within the cortex. Henle's loop is divided into a descending portion with a thin epithelial lining and an ascending limb that consists of an initial thin section followed by a section with a thicker epithelium.

From Henle's loop, after transversing back past the glomerulus via the juxtaglomerular apparatus, the tubule once again coils and becomes the distal convoluted tubule. Connection with the highly branched collecting tubule is made via a distinctive connecting tubule. The collecting ducts form a large portion of the inner kidney, the renal medulla, and they drain into the renal pelvis. The outer portion of the kidney, the renal cortex, contains all the glomeruli. Blood enters the kidney via the renal artery, which progressively divides into smaller, arcuate arteries between the cortex and medulla, giving off interlobular branches that further subdivide into lateral branches, the afferent arterioles, each of which terminates in a capillary tuft, the glomerulus. Filtration of fluid occurs from the capillaries through a thin layer of tissue into the Bowman's capsule. The glomerular capillaries recombine to form another set of arterioles, the efferent arterioles through which the blood leaves the glomerulus. The efferent arterioles from the cortical glomeruli first form a peritubular mesh surrounding the cortical tubules and then empty into the interlobular veins. In contrast, the efferent arterioles from the juxtamedullary nephrons feed into the medulla via descending and ascending vasta recta (reviews: 6,7).

A. Renal Distribution of NOS Isoforms

NO is formed from the terminal guanidino nitrogen of L-arginine by NO synthase (NOS). This enzyme exists in at least three isoforms: the constitutive NOS1 ("neural" or nNOS) and NOS3 ("endothelial" or eNOS) forms, and the inducible NOS2 (iNOS) form. Several laboratories have localized the major NOS isoforms within the kidney using immunohistochemistry or in situ hybridization. Immunohistochemical studies by Bachmann et al. (8) placed NOS1 in the glomeruli and vasculature as well as the macula densa, collecting duct, and the inner medullary limb. NOS3 protein was detected in the endothelium of the glomerular capillaries, the afferent and efferent arterioles, the intrarenal arteries, and the medullary vasta recta (Fig. 2). eNOS mRNA was identified in glomeruli, preglomerular vasculature, proximal tubules, thick ascending limbs and collecting ducts, renal vortex, and medulla of the rat (9).

Mohaupt et al. identified two structurally distinct iNOS isoforms that occurred *constitutively* in the rat kidney in vascular smooth muscle at the juxtaglomerular apparatus and in the tubule epithelium in various segments (10). Singh and

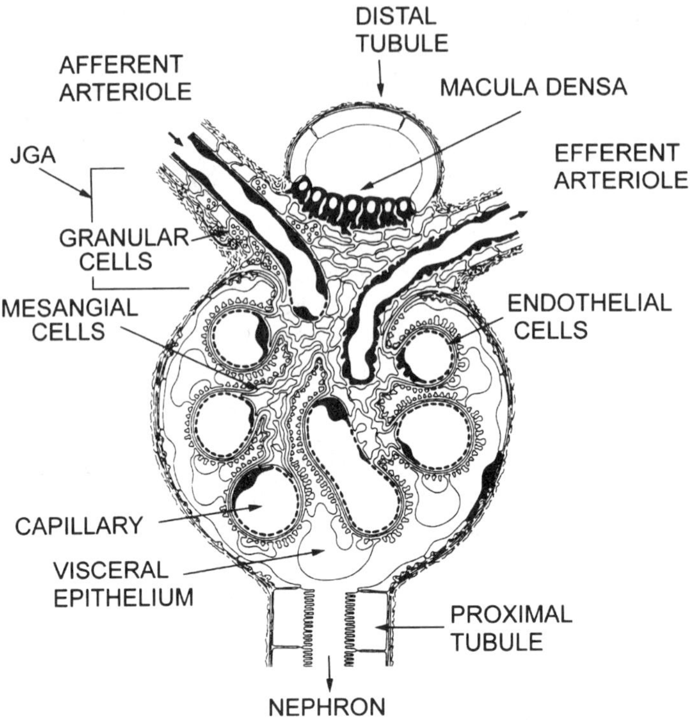

Figure 2 Schematic longitudinal section through the glomerulus and juxtaglomerular apparatus. NOS1 is indicated by solid black; it is found in the macula densa cells, endothelium of efferent arterioles, and single cells of the parietal epithelium of Bowman's capsule. The presence of NOS3 in other vascular endothelia is indicated by a gray shading. (From Ref. 8.)

colleagues (11) reported the presence of an unusual NOS in rat kidney. This enzyme activity did not require calcium or calmodulin and was found at elevated levels during pregnancy. The enzyme activity was capable of producing nitrite and nitrate as measured by high-performance capillary electrophoresis, but was resistant to inhibition by arginine analogs. Biochemical studies have shown that the renal medulla has a greater capacity to generate NO than the renal cortex (12).

B. Regulation of NOS Activity and Actions of NO

The so-called constitutive isoforms NOS1 and NOS3 are dependent upon calcium and calmodulin. Both may be activated by increases in intracellular calcium triggered by exogenous or endogenous signals. Calcium binds to calmodulin and promotes its binding to NOS(1 or 3), resulting in NOS(1 or 3) activation. The inducible isoform, NOS2, contains tightly bound calmodulin and, unlike NOS1 and NOS3, does not appear to be regulated by fluctuations in calcium ion levels, though it does require Ca^{2+} for activity. NOS2 is more active in the presence of exogenous Ca^{2+} than in its absence (13). It is transcriptionally induced by various agents in a cell lineage-dependent manner. For example, endotoxin (LPS, lipopolysaccharide), which acts via the transcription factor NF-κB, and interferon-γ, which requires interferon regulatory factor-1 and can synergize with LPS to augment NOS2 induction in some cell types (14). As discussed elsewhere in this volume, NO has a multitude of effects both on the cell that produces it and on nearby cells into which it diffuses. For example: It can elevate cGMP levels by activating guanylate cyclase; it can inactivate a variety of proteins, particularly those with iron-sulfur centers; and it can react with superoxide to produce the highly reactive peroxynitrite.

III. NO AND ANGIOTENSIN II

The regulation of glomerular microcirculation by NO is complex. A key feature of renal hemodynamics is that renal perfusion pressure is one of major determinants that govern the amount of glomerular filtrate produced. Acute systemic inhibition of all NOS isoforms with L-arginine analogs produces a dose-dependent increase in arterial blood pressure and renal vasculature resistance that is reversible with excess L-arginine. The increase in renal vascular resistance due to inhibition of NO synthesis reduces renal plasma flow and the glomerular filtration rate (3).

The modulation of afferent arteriolar tone by a number of vasoactive substances, including NO, is an area of intense experimental and clinical investigation. The afferent arteriole is a major vascular segment controlling glomerular hemodynamics, and NO (produced by endothelial and macula dense cells) modulates its activity. The vasodilating activity of NO is in opposition to angiotensin II (angII)-induced vasoconstriction. In addition to direct interactions between NO and vasoconstrictors in control of vascular tone, NO also controls glomerular hemodynamics via the tubuloglomerular feedback system (15). NO has been implicated in the regulation of renin synthesis and medullary blood flow. McKee et al. (12) suggest that NO produced by the action of angII on renal tubules acts directly on tubular cells to inhibit sodium reabsorption in proximal tubules and collecting ducts by blocking active Na^+ transport.

A. The Renin-Angiotensin System

The renin-angiotensin system is a critical regulator, via angII, of renal hemodynamics. The vascular pole of the glomerulus forms a specialized structure, called the juxtaglomerular apparatus (JGA in Fig. 2), at the point of interaction with the epithelial cells of the ascending Henle's loop segment of the nephron. Specialized afferent arteriolar cells within the juxtaglomerular apparatus synthesize the enzyme renin, which is then stored in cytoplasmic granules and secreted into the blood when required. Renin cleaves the decapeptide angI from the large plasma protein angiotensinogen. AngI then undergoes further enzymatic cleavage by angiotensin-converting enzyme (ACE) to form the highly bioactive peptide hormone angII. Figure 3 diagrams these relationships. AngII is a potent vasoconstrictor, as well as stimulator of aldosterone secretion (16).

The rate-limiting factor in angiotensin formation is the concentration of plasma renin. The angiotensinogen concentration in the plasma is maintained by controlled release from the liver, where it is synthesized and stored. Control of renin secretion by the kidneys serves as a major regulator of the renin-angiotensin system. Renin is released from stores in the juxtaglomerular cells of the afferent

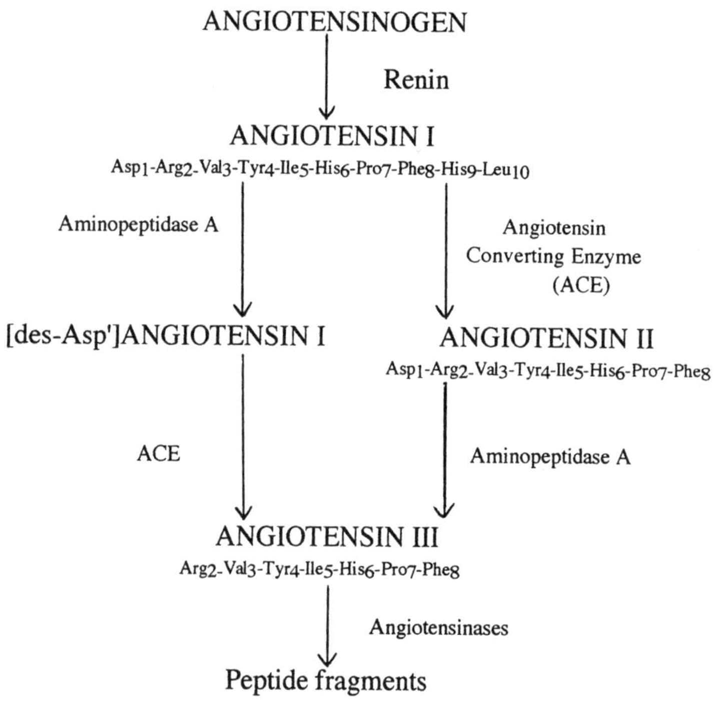

Figure 3 Renin-angiotensin cascade. The cleavage of angI from angiotensinogen is renin-dependent. The major mechanism of the conversion of angI to angII is through the angiotensin-converting enzyme (ACE). A variety of peptidases further cleave angII into biologically active fragments.

arteriole of the kidney in response to a number of signals. The primary stimuli for renin secretion are reductions in arterial pressure or Na^+ levels. Thus, stimulation of renin release occurs during acute hemorrhage, administration of diuretics that stimulate water and salt secretion, low salt intake, or acute renal artery clamping (17,18).

The proximal tubular epithelium is another source of the components of the renin-angiotensin system. Angiotensinogen is synthesized by proximal tubular cells (19,20). Tank et al. (21) have demonstrated that following uninephrectomy in the rat, proximal tubular renin mRNA is markedly increased without any changes in circulating renin levels or renin expression in juxtaglomerular cells. ACE is found in greatest concentrations in the proximal tubules with smaller amounts in arterioles and glomeruli (19). On a weight basis there is five- to sixfold more ACE found in the kidney than in the lung. The endogenous production of physiological amounts of angII has been detected in the proximal tubule using micropuncture techniques. Thus, the proximal tubule has all the components of a local angII-generating system (22).

B. Control of Renin Secretion by NO

How NO controls renin release and/or synthesis is not fully understood. The epithelial cells of the macula densa (see Fig. 2) are believed to control renin secretion. They appear to sense the sodium chloride load of the distal tubule and to regulate the rate of secretion of renin by the juxtaglomerular cells inversely to the concentration of salt in the distal tubular fluid (23). NOS1 has been found in the renal cortex, predominantly in the macula densa (8). NO derived from the macula densa provides feedback regulation of renin biosynthesis in response to altered sodium levels. Inhibition of endogenous NO synthesis is the technique most frequently used to study the role of NO. The majority of NOS inhibitors have been the substituted L-arginine analogs, which, however, are relatively nonspecific and block all NOS isoforms when given in high doses. These nonspecific NOS inhibitors can also produce systemic effects, such as an increase in renal vascular resistance and mean arterial pressure.

Studies in rats by Beierwaltes (24) have shown that NOS1 is a critical component in the stimulation of renin production induced by dietary sodium restriction. The stimulation of renin activity in response to low NaCl levels was reversed in rats chronically treated with 7-nitroindazole (7-NI). 7-Nitroindazole is a relatively specific inhibitor of NOS1 at or below 10 μM levels; above 100 μM levels it may inhibit other isoforms [25]. NOS1 inhibition by 7-NI did not affect blood pressure, renal blood flow, or renin activity in control rats on a normal diet, indicating that NOS2 and NOS3 levels were likely not significantly inhibited. The stimulation of renin activity in response to low NaCl levels was reversed in rats chronically treated (20 mg/kg body weight) with 7-NI but not in rats acutely treated (50 mg/kg body weight). Macula densa stimulation of renin secretion in response to a low-sodium diet, clamping of the renal artery, or treatment with an angII antagonist or with furosemide was investigated by Schricker et al. (26). (Furosemide stimulates renin secretion by acting as a diuretic on macula densa cells, inhibiting salt transport.) Noncompetitive RT-PCR experiments indicated that only NOS1 mRNA levels changed concordantly with the levels of renin. Further analysis revealed that the renal level of NOS1 mRNA was increased by about 50% after a low-sodium diet or hypoperfusion of the kidney.

The effect of inhibiting NOS on the renin secretory response to furosemide was investigated in rabbits by Reid and Chou (27). Inhibition of NOS activity with the non-isoform-selective, but specific, inhibitor N^G-nitro-L-arginine methyl ester (L-NAME) increased mean arterial pressure, decreased heart rate, and markedly suppressed the renin response to furosemide. (The transient increase in blood pressure following furosemide treatment would be expected to decrease, rather than increase, renin activity in the kidney.) These results are consistent with other studies implicating NO in the modulation of renin secretion in the macula densa. The involvement of prostaglandins and NO in the macula densa-dependent stimulation of renin in vivo was examined in rats chronically infused with furosemide (26). Activation of renin gene expression by the furosemide block of Na^+-K^+-$2Cl^-$ cotransport in the macula densa was dependent on both NO and prostaglandin formation, whereas stimulation of renin secretion required only NO formation.

Studies by Tharaux, Chatziantoniou, Ardaillou, and colleagues on freshly isolated rat afferent arterioles have shown that endogenous NO is necessary for renin formation (28,29). Renin synthesis was stimulated by ACE inhibition (ramapril) or by furosemide treatment, with or without inhibition of endogenous NO synthesis. Renin secretion, total renin content, and renin mRNA levels were measured. In both cases inhibition of NO synthesis by L-NAME impaired renin synthesis. The inhibitory effect of L-NAME on the activation of renin secretion was abolished when afferent arterioles were treated with a Ca^{2+} channel blocker. Inhibition of NO synthesis appeared to impair renin secretion and synthesis through changes in the cytosolic Ca^{2+} concentration. The adenylyl cyclase-mediated mechanism regulating renin release was also impaired when NO synthesis was inhibited. Stimulation of adenylyl cyclase by forskolin failed to increase renin secretion despite increased cAMP levels.

Harding et al. (30) investigated the basal renal renin content of mice deficient in NOS1 ($NOS1^{-/-}$). In mice maintained on a normal diet, renal renin content was reduced in the NOS1 knockout mice compared with the control strain. The macula densa-mediated changes in renin were studied in mice maintained on a low-sodium diet. Neither selective inhibition of NOS1 by 7-NI in control mice nor ablation of NOS1 activity (i.e., in the $NOS1^{-/-}$ mice) affected the ability of the animal to increase renal renin output in response to low sodium.

C. Synopsis

Regulation of the renin-angiotensin system by NO is exquisitely sensitive to NO levels and dependent upon the target tissue. NO is able both to stimulate and to inhibit renin release. Studies of the effect of NO on renin secretion revealed that an increase in NO in the vicinity of the juxtaglomerular apparatus could cause either an inhibition or a stimulation of renin secretion. Under conditions of low afferent arteriolar shear stress and low NaCl concentration, it seemed that NO produced in the macula densa by NOS1 stimulated renin secretion, whereas under conditions of vascular shear stress and elevated NaCl, NO, apparently produced by the endothelial cells of the afferent arteriole, inhibited renin secretion (31).

IV. ROLE OF NO IN RENAL PHYSIOLOGY

The glomerular filtration rate of a single nephron is controlled by the tone of the preglomerular (afferent) and postglomerular (efferent) arterioles, and angII may exert direct effects on either (32). In the kidneys of conscious dogs, NO modulates the preglomerular response to angII, reducing its ability to inhibit glomerular filtration (33). Thus NO can provide protection against severe reductions in glomerular filtration under conditions of elevated renal angII levels.

In the rat, intravenous infusion of the relatively nonselective NOS inhibitor N^G-monomethyl-L-arginine (L-NMMA) increased efferent arteriolar resistance more than afferent arteriolar resistance (34). This suggested that NO affected the rat postglomerular vessels. In another study, NO modulated the angII-induced vasoconstriction in isolated rabbit afferent arterioles but not efferent arterioles (35). Juncos and colleagues (36) have shown in isolated microperfused rabbit glomerular afferent and efferent arterioles that inhibition of NO synthesis enhanced angII action in the afferent arterioles but not in the efferent arterioles. Flow through the lumen of the microperfused afferent arteriole stimulated the endothelium to produce NO and prostaglandins, which in turn counteracted the vasoconstricting action of angII.

Schnackenberg et al. (37) reported that in dogs NO could protect pre- and postglomerular vessels from angII. The effect of intrarenal infusions of angII in the presence and absence of NO was examined in the kidney stop-flow model. During vehicle treatment, angII infusion preferentially increased *postglomerular* resistance. However, when NO synthesis was blocked, angII dramatically increased *preglomerular* resistance. NO may protect preglomerular vessels and to a lesser extent postglomerular vessels from angII-induced renal vasoconstriction.

Angiotensinogen gene-knockout ($Atg^{-/-}$) mice exhibit chronic hypertension and an increase in renin gene expression and renin protein (38). In studies by

Kihara et al. (39) macula densa-derived NO was implicated in the increased expression of renin in the kidneys of $Atg^{-/-}$ mice. Semiquantitative RT-PCR revealed an increase in NOS1 mRNA abundance in the renal cortical tissue of $Atg^{-/-}$ mice compared to $Atg^{+/+}$ mice. In addition, selective inhibition of NOS1 with 7-NI significantly decreased the level of renal renin mRNA in the $Atg^{-/-}$ mice. These results suggest that increased NOS1 activity in the macula densa stimulates renin overproduction in the $Atg^{-/-}$ mice. Roles for NO and angII in the control of glomerular hemodynamics are evident, but the mechanisms underlying their actions require further study.

Targeted disruption of each of the three NOS isoforms has been accomplished in mice (reviewed by D. Laskin elsewhere in this volume), and studies of renal function in these mice are under way (e.g., 30). NOS1-deficient mice are less sensitive to cerebral ischemia, and the males exhibit higher levels of aggressive behavior (40,41). Mice lacking the gene for NOS2 exhibit greater susceptibility to infection than wild-type mice (42). NOS3-deficient animals are hypertensive and more sensitive to cerebral ischemia (43,44). Analysis of renal function in these NOS-deficient animals will help untangle the critical functions of these enzymes in the kidney.

AngII is not the only active peptide derived from angiotensinogen (16) (see Fig. 3). Several cleavage products of angII also possess bioactivities, some of which are listed in Table 1. AngIII, which is formed from angII by deletion of

Table 1 Summary of Active Fragments of Angiotensin II (16,45)

		Receptors	Biological effects
Angiotensin II[a]	AngII(1–8)	AT1	Vasoconstriction
		AT2	Aldosterone secretion
			Increased salt appetite
Angiotensin III	AngII(2–8)	AT1	Vasoconstriction
		AT2	Aldosterone secretion
			Increased salt appetite
			Vasopressin secretion
Angiotensin IV	AngII(3–8)	AT4	Increased intracellular Ca^{2+}
			Vasoconstriction
			Vasodilation
	AngII(1–7)	Specific	Increased glomerular filtration
		receptor	Increased urinary output
		AT1	Increased sodium excretion rate
	AngII(3–7)		Facilitates recognition memory in brain

[a]Several of the degradation products of angII exhibit biological activity. These peptides are formed by the action of a variety of peptidases and either counterbalance the effects of angII or are responsible for effects previously attributed to angII.

the N-terminal Asp residue by aminopeptidase A, possesses many of the properties of angII and shares its receptors. AngIV, which is formed by the actions of aminopeptidase la and N, and has a deletion of the two N-terminal amino acids, possesses its own distinct receptors and has effects that are directly opposed to those of angII-renal vasodilation, for example. AngII(1-7) is obtained by deletion of the C-terminal amino acid via the action of prolylendopeptidases and carboxypeptidases. It has been shown to be biologically active and capable of binding to specific receptors, distinct from the AT1 and AT2 receptors, that stimulate prostaglandin E_2 and prostaglandin I_2 synthesis in porcine endothelial cell. AT1 and AT2 are the classic angII receptors. AngII(1-7) exhibits renal effects resulting from angII(1-7)-dependent formation of NO and vasodilatory prostaglandins, and it can control the electrolyte balance by increasing glomerular filtration, urinary output, and sodium excretion (45).

The impact of NO and prostaglandins on the renal vascular response to angIV was investigated by Yoshida and colleagues (46). Their studies in the spontaneously hypertensive and normotensive rat revealed that intrarenal infusion of angIV induced a sustained vasoconstriction of lesser magnitude than the initial rapid transient constriction. A stable vasoconstriction was observed with angII infusion. Vasoconstriction induced by either angIV or angII was enhanced by L-NAME but not affected by indomethacin. The combination of these inhibitors enhanced the angII-induced vasoconstriction in normotensive, but not in the spontaneously hypertensive rat. These results suggest that NO and prostaglandins modulate the renal vascular response to angIV and to angII in different ways.

A. Effect of Acute and Chronic Inhibition of NO Synthesis

In several species, including humans, acute systemic inhibition of NO synthesis by L-arginine analogs produces a dose-dependent increase in arterial blood pressure and renal vascular resistance that is reversible with excess L-arginine. The response to systemic NOS inhibition closely resembles the response to angII infusion, suggesting that angII participates in NOS inhibition-induced vasoconstriction (3). The intravenous administration of L-arginine, a precursor of NO, decreases blood pressure in humans by its direct vasodilatory effects. L-arginine or the salt vehicle were administered to 10 healthy male subjects. L-arginine was found to reduce blood pressure and renal vascular resistance. It was also found to reduce ACE and plasma angII levels (47). By contrast, in the anesthetized rat, infusion of L-arginine caused diuresis and natriuresis, possibly via the formation of NO in the kidney; endogenous angII did not appear to be to be involved in the L-arginine-induced renal effects (48). During chronic inhibition of NO synthesis, angII participates in maintaining hypertension. The ability of nitric oxide to modulate the chronic renal and arterial responses to angII was investigated in dogs by Schnackenberger et al. (49). They found that chronic inhibition of NO synthesis within the kidney prolonged the long-term renal and arterial pressure responses to angII.

B. Control of Blood Flow by NO

Vascular tone is reduced by NO. In anesthetized rats, NO has been found to modulate the renal vasoconstriction due to angII or norepinephrine (33). The antagonism between vasoconstrictors and NO is not due to a constrictor-induced production of NO, because exogenous NO and endogenous NO are equally effective (50). In rats, Ikenaga et al. (51) monitored the arteriolar lumen diameter response to exogenous angII before and during treatment with the NOS inhibitor N(ω) nitro-L-arginine (L-NNA). They found that L-NNA potentiated the juxtamedullary afferent and efferent arteriolar vasoconstrictor responses to angII, suggesting that basal NO production curtails the vasoconstrictor action of angII. Thorup et al. (52) showed that synthesis of NO by isolated renal resistance arteries was enhanced by angII.

C. Effect of Glucose on NO Synthesis

Abnormal glomerular hemodynamics is one of the consequences of the early stages of diabetes. A reduction in preglomerular vascular resistance, regulated by the afferent arterioles, is suspected to be responsible for the increase in glomerular perfusion in response to elevated glucose (53). The glucose-induced vasodilation may be mediated by insulinlike growth factor I, which has been shown to stimulate NO release from the vascular endothelium (54). The direct effects of high glucose levels on the afferent arteriole are not fully understood. Arima et al. (55) examined the action of high glucose on the vascular reactivity of rabbit afferent arterioles. In the in vitro microperfused rabbit afferent arterioles, high glucose caused constriction rather than dilation. Exposure of the afferent arterioles to high glucose concentrations also augmented the action of angII, to an extent similar to that observed during NO synthesis inhibition. The effects of high glucose were abolished by pretreatment with L-arginine. Vascular abnormalities occur in diabetes and may be due to a glucose-induced inhibition of NO synthesis in the afferent arterioles.

D. NO in the Developing Kidney

How NO and angII contribute to the regulation of the development of renal function is poorly understood. It is known that plasma renin activity and circulating angII concentrations are elevated in the newborn, and that plasma renin activity decreases with advancing age (56). NO is an important intrarenal vasodilator in the developing animal. Inhibition of basal NO activity by intrarenal infusion of L-NAME produced alterations in renal hemodynamics in both the developing piglet and the adult pig kidney (57). Greater changes in renal blood flow and renal vascular resistance were observed in the developing animal. Intrarenal infusion of the angII AT1 receptor antagonist ATX(A-81988) produced greater increases in renal blood flow in the developing piglet than in the adult pig (58). Preinfusion of the AT1 receptor antagonist attenuated the renal blood flow and renal vascular responses to L-NAME in both age groups. In the developing animal NO is an important intrarenal vasodilator; it counteracts vasoconstricting activity of angII.

E. Shear Stress and NO

Nonchemical factors can prolong the vasorelaxation mechanisms attributed to NO. Flow-dependent dilation occurs as a result of shear stresses imposed on vascular beds by blood flow. Shear stress enhances production of NO by vascular endothelial cells (59–61). The mechanisms by which shear stress stimulates NO production by vascular endothelial cells are complex. Vessel wall shear stress during physiological states varies at different segments of the renal vasculature. However, for any given vessel, shear stress increases directly with blood flow and inversely with vascular radius. Shear stress-induced release of NO from endothelial cells is dependent upon calcium and L-arginine. In a study by Deng and colleagues of anesthetized rats, short-term infusions of angII increased renal vascular resistance and thereby endothelial shear stress and NO release (62). In contrast, prolonged stimulation by angII infusion did not increase NO release. Shear stress-induced increases in renal NO generation were lost during prolonged angII infusion. These results suggest an uncoupling of renal NO release from shear stress during long-term angII infusion.

F. Role of NO in Tubuloglomerular Feedback

In addition to interactions between NO and vasoconstrictors in the control of tone, NO also controls glomerular hemodynamics via the tubuloglomerular (TG) feedback system. The TG feedback mechanism regulates blood flow and glomerular filtration of a single nephron by changing the afferent arteriolar resistance in response to the salt concentration in the macula densa. NO produced in the juxtaglomerular apparatus apparently regulates the TG feedback response. Local NO opposes angII in modulating TG feedback. The inhibition of endogenous NO production results in significant renal hemodynamic changes, an event similar to that observed after angII administration (4,63,64). Kawata and colleagues have studied the angiotensin receptor antagonist-2-ethoxy-1-[[2′-(1H-tetrazol-5-

YL)biphenyl-4-yl]methyl]-1H-benzimidazole-7-carboxylic acid (CV11974) during NOS blockade in rats (65). CV11974 attenuated the TG feedback.

Augmentation of TG feedback resulting from inhibition of endogenous NO production depended in part on intrinsic angII activity. In a study of peritubular capillary infusion in rats, Braam and Koomans found that the influence of NO on the preglomerular vasculature and the TG feedback system varied with angII activity (66). NO counteracted the TG feedback-dependent and -independent actions of angII on the afferent arteriole. This is because angII and NO interact at the level of the afferent arteriole, and under conditions of high angII, NO synthesis inhibition can cause afferent arteriole vasoconstriction that does not require the actions of the TG feedback system. Thorup and Persson examined the effects of both NOS1 and NOS2 on glomerular capillary pressure and TG feedback in rats (67). The proximal stop-flow pressure technique was used to study changes in glomerular capillary pressure obtained by activation of TG feedback. Through the use of the selective NOS1 inhibitor 7-NI, it appeared that NOS1 in the macula densa regulated TG feedback and glomerular capillary pressure. The NOS2-selective inhibitor aminoguanidine revealed that NOS2-derived NO had little influence on the regulation of glomerular capillary pressure under normal conditions. The presence of NOS2 in both glomerular and tubular cells in the normal kidney indicates that this isoform may be of importance under pathological conditions, such as diabetes, where NO synthesis is increased.

G. NO and Hypertension

End-stage renal disease resulting from hypertension continues to increase in the aging human population. Chronic inhibition of NO synthesis by L-NAME in the spontaneously hypertensive rat induced a sustained increase in blood pressure and amplified the pressor response to infused angII (68). The increased blood pressure and enhanced pressor responsiveness that occurred with chronic inhibition of NO synthesis to angII were partly the result of increased vascular angII receptor expression. The role of NO in the development of naturally occurring end-stage renal disease in aged (73- and 85-week-old) and young (20-week-old) spontaneously hypertensive rats was investigated by Frolich (69). He found that the renal and intrarenal hemodynamic pathology could be reversed with ACE inhibition therapy. The implications and possible mechanisms involved in the development of both naturally occurring and NO-exacerbated renal disease are multifactorial, involving the endothelial NO-generating system and its interaction with angII.

Recent studies have provided strong evidence that a deficit in NO production can contribute to hypertension. For instance, Mattson and Bellehumeur (70) found that when NOS1 was inhibited by infusing an antisense oligodeoxynucleotide into the renal medulla, there was a decrease in NOS activity accompanied by an increase in blood pressure in rats maintained on a high-salt diet. Salt-induced hypertension can be alleviated by the administration of L-arginine, which increases NO production (71,72). The afferent arteriole of patients with essential hypertension exhibit increased vasoconstriction relative to normotensive controls, and there is defective intrarenal synthesis of NO (73).

H. Synopsis

NO is a controlling factor in the maintenance of renal hemodynamics, glomerular filtration, and blood pressure. It appears that NO originating from various cellular sources in diverse cellular environments can have different effects on renin activation. NO has a dual role in renin control. Depending upon the site of NO production and its target cells, NO can stimulate or inhibit renin production and activity. NO applied directly to granular cells inhibits renin secretion; however, stimulation of NO production by the macula densa leads to an increase in renin secretion. NO generated by the macula densa and the afferent arterioles controls glomerular hemodynamics, glomerular capillary blood pressure, and glomerular plasma flow. There is good evidence that NO regulates renal hemodynamics by modulating the renal vasoconstricting activity of angII. Chronic administration of NOS inhibitors produces a sustained hypertension and glomerular damage. Despite these findings, inhibition of angII had little effect on the renal vasoconstrictor response in chronic hypertension induced by inhibition of NO synthesis in rats. NO deficiency may also be a factor in other forms of hypertension, such as salt-dependent hypertension. Although both NO and angII play key roles in the development of hypertension, this is nevertheless a multifactorial process and many mechanisms remain to be elucidated.

V. ISCHEMIA AND HYPOXIA

Ischemia results from a reduced supply of blood to an organ or tissue, while hypoxia is defined as a deficiency of oxygen at the tissue level. Ischemic hypoxia occurs when a tissue or organ, the kidney, for example, suffers a reduction in its blood supply and consequently oxygen and substrate availability. In the hemodynamics of the kidney, a major feature is that renal perfusion pressure is the primary force that determines the amount of glomerular filtrate produced. Renal blood flow is directed mostly to the cortex to optimize glomerular filtration and the reabsorption of solute. By contrast, blood flow to the renal medulla is low, to preserve osmotic gradients and enhance urinary concentration (74). The outer medulla is a target for tubular hypoxic injury because of this regional hypoxia and the presence of tubules vulnerable to hypoxia (the straight segment of the proximal tubule, S_3, and the thick ascending limbs of Henle) (75). These nephron segments are vulnerable because of their high ATP requirements for active solute transport. The inner medulla is far less vulnerable because of lower metabolic demands. The S_3 proximal tubules located in the glomerulus-deficient medullary rays of the cortex (see Fig. 1) are also very susceptible to hypoxic injury since this region of the cortex is the most removed from the arterial circulation.

A. Experimental Models of Ischemia

Renal ischemia induced by total cessation of renal blood flow is the most extensively studied experimental animal model of acute tubule necrosis. In the clamped renal artery animal model, and in humans with acute tubular necrosis, the proximal straight tubule is the segment most affected by ischemia (76). In vitro studies using primary cultures of defined segmental epithelia confirmed the hypersensitivity of human and rabbit proximal tubules to hypoxic injury (77). Studies of oxygen deprivation injury in isolated perfused kidneys have yielded conflicting results. Brezis and colleagues found that the isolated kidney perfused by a conventional erythrocyte-free buffer was characterized by necrosis of the medullary thick ascending limb cells and not the proximal tubule cells (78). By contrast, Endre and colleagues found that in isolated kidneys perfused by buffer containing erythrocytes, the proximal tubule segment not the thick limb cells was injured in response to ischemia and hypoxia (79). Addition of erythrocytes to the perfusate seemed to facilitate delivery of oxygen to the medullary thick ascending limbs in the outer medulla.

B. NO in Models of Ischemia-Induced Acute Renal Failure

NO, a vasodilator, has been implicated in renal ischemia. Studies of postischemic renal failure using antisense oligodeoxynucleotides has revealed functional roles for specific NOS isoforms. Goligorsky and colleagues used a phosphorothiolated oligodeoxynucleotide complementary to the translation initiation region of NOS2 to suppress the synthesis of NO by this enzyme. Production of NO by renal tubular epithelial cells (BSC-1) exposed to LPS or oxidant stress (H_2O_2) could be specifically inhibited by the antisense oligodeoxynucleotide (80). More dramatically, treatment of rats with the oligodeoxynucleotide prevented renal dysfunction after an ischemic insult (81). This specific inhibition of NOS2 synthesis provides strong evidence that it is NO produced by NOS2 that is responsible for the cytotoxicity. In contrast, the inhibition of all the NOS isoforms by the nonselective inhibitor L-NAME was ineffective and resulted in the deterioration of renal function compared to nontreated animals. This provides evidence for the functional disparity of the various NOS isoforms found in the kidney and suggests that the consequences of inhibiting all isoforms of NOS could dominate over the possible benefits of inhibiting NOS2 alone

Supporting these data are the findings in rats by Conger and associates that in the norepinephrine-induced model of acute renal failure, infusion of L-NAME resulted in a significantly greater renal vasoconstrictor response than in normal or sham acute renal failure kidneys (82). They also found that early in the course of norepinephrine-induced acute renal failure, NOS3 activity was not reduced but was probably maximal, as this activity could not be further stimulated by endothelium-dependent dilators. Lack of response to endothelium-dependent vasodilators has been considered support of reduced NO activity in models of ischemic injury. This suggests that NOS activity in the postischemic renal vasculature could be at maximal levels in response to ischemia-induced renal vasoconstrictor activity.

The severity of ischemia-induced acute renal failure in rats worsens during acute systemic blockade of NO production. The inhibition of NOS with L-NAME in combination with renal ischemia resulted in a 70% reduction in cardiac output and substantial mortality (83). It was also accompanied by an increase in total pulmonary resistance and systemic vascular resistance, emphasizing the importance of NO as a vasodilator. An investigation of renal ischemia in rabbits revealed that the enhancement of NO synthesis caused by systemic L-arginine infusion could best preserve glomerular and tubular function in the ischemic reperfused kidney when given before the ischemic insult (84).

C. Ischemia in the Diabetic Kidney

Kidneys of diabetic rats exposed to ischemia have been found to develop a more severe ischemic acute renal failure than those of their nondiabetic controls (85). Goor et al. (86) recently examined the time-dependent production of NO in the model of ischemic acute renal failure in streptozotocin-induced diabetes mellitus rats. They found that the plasma and urine nitrite and nitrate levels were significantly higher in the untreated diabetic rats than in the untreated normal rats before ischemia. Acute renal failure in nondiabetic rats resulted in increased plasma and urinary NO_2^- and NO_3^- levels. In the diabetic rat kidney, plasma NO appeared unchanged, and, contrary to the response seen in nondiabetic acute renal failure rats, urinary NO_2^- and NO_3^- were decreased. The higher basal level of NO in the diabetic rats offered no protection against ischemia, suggesting possibly an impaired response to NO. The decrease in NO production in the diabetic rat after ischemia suggests an impaired production of NO. The vulnerability of the diabetic kidney seems to be associated with low NO activity and poor renal function. These findings are supported by similar studies in the streptozotocin-induced diabetes mellitus rat (87).

D. Ischemia-Reperfusion and Inflammation

During the reperfusion period following ischemic injury an inflammatory and cytotoxic cascade is activated. In a model of polymorphonuclear neutrophil (PMN)-mediated injury to the isolated perfused rat kidney, the involvement of NO was examined (88). NO appeared to be both beneficial and harmful, with data suggesting that the effect of NO on ischemia-reperfusion depended on the presence or absence of circulating neutrophils. Linas and colleagues (88) found that NO augmented ischemic injury in the absence of PMNs. However, NO donors accelerated the recovery of renal function in the presence of neutrophils. NO is an important regulator of neutrophil adhesion and function, playing a major role in neutrophil-mediated acute renal failure.

α-Melanocyte-stimulating hormone is a neuropeptide with broad anti-inflammatory properties. It inhibits neutrophil migration and the production of both neutrophil chemokines and NO. The precise mechanism of the α-melanocyte-stimulating hormone anti-inflammatory action is unknown (89). Chiao et al. (90) reported that α-melanocyte-stimulating hormone inhibited renal injury in a model of bilateral renal ischemia in both the mouse and the rat. In this model of renal ischemia, α-melanocyte-stimulating hormone significantly reduced ischemia-induced renal damage by its inhibition of inflammatory and cytotoxic pathways activated during reperfusion. α-Melanocyte-stimulating hormone was also effective at inhibiting further renal damage even when administered in the reperfusion period.

E. Ischemic Injury in Transplantation

During transplantation, donor kidneys are subjected to varying degrees of ischemia. Renal damage occurs from prolonged lack of oxygen or from reactive oxygen species generated after reperfusion. This injury contributes to transplant tubular necrosis and delayed graft function. Nitric oxide may play a role in both recovery from ischemia and promotion of rejection of kidney transplants. NO produced by NOS2 has been implicated as a mediator in allograft rejection (91,92). Direct measurement of NOS activity in ischemic rat renal tissue by Shoskes et al. (93) revealed an increase during the first 6 hours following reperfusion. However, after 24 hours, NOS activity decreased in the ischemic kidney and did not recover for 21 days. The reduced NOS levels could increase intrarenal vasoconstriction and be damaging to recovery from ischemia. The decrease in NOS activity was partly prevented by preoperative treatment with oral L-arginine.

F. Regulation of NOS Synthesis and Activity

NOS2 expression and activity are regulated by inflammatory cytokines. Several cytokine-response elements and a hypoxia-responsive enhancer element have been identified in the promoter of the NOS2 gene (94–97). The first evidence that NOS2 was a hypoxia-inducible gene came from studies by Melillo and colleagues in murine macrophages (96). The NOS2 promoter in the murine macrophage cell line ANA-1 contains a sequence homologous to a hypoxia-responsive enhancer, and NOS2-hypoxia-responsive enhancer activity was induced by hypoxia. Hypoxia in combination with interferon-γ (IFNγ) activated the NOS2 promoter in transient transfection assays, inducing NOS2 transcription and mRNA expression. The hypoxia-inducible factor (HIF-1) studied by Semenza et al. (98) is a basic helix-loop-helix protein that activates the transcription of hypoxia-inducible genes, including NOS2. Hypoxia response elements consist of a HIF-1 binding site that contains the core sequence 5′-CGTG-3′. The 5′-flanking region of the NOS3 gene does not contain an HIF-1 consensus sequence (99).

The effect of hypoxia on NOS3 synthesis and activity is less clear. The constitutive isoforms of NOS (NOS1 and NOS3) isoforms are regulated by Ca^{2+} via calmodulin. Thus Ca^{2+} influx into neural or endothelial cells prompted by acetylcholine, bradykinin, or shear stress stimulated NO biosynthesis (100). However, not much is known about the mechanisms controlling NOS3 expression in renal vascular or glomerular endothelium. Arnet and colleagues found that bovine aortic endothelial cells increased NOS3 protein and mRNA levels in response to hypoxia, but with no reported changes in mRNA stability (101). The total activity of NOS3 was unchanged after 24 hours of hypoxia, although the amount of enzyme doubled. It is not known why the activity did not increase along with the amount of enzyme. In contrast, studies of endothelial cells from various vascular beds (umbilical vein, aortic endothelial, pulmonary artery) exposed to hypoxia revealed that hypoxia suppressed both the transcriptional rate of the NOS3 gene and the stability of its mRNA (102). Differences in the response of NOS3 to hypoxia may be due to experimental differences or the use of endothelial cells of different lineages.

G. NO and Hypoxia/Reoxygenation Injury in Proximal Tubules

Hypoxia is an important regulator of blood vessel tone. Acute hypoxia causes pulmonary vasoconstriction, and chronic hypoxia causes smooth muscle cell replication and extracellular matrix deposition. The cellular responses to hypoxia involved complex cell-cell interactions mediated by the release of growth factors, cytokines, and biological messengers. It is thought that hypoxia-induced vasoconstriction is due to the inhibition of NO synthesis.

Renal proximal tubules are capable of producing NO by both NOS2 and NOS3 (10,103). Yu, Schrier, and colleagues investigated the involvement of NO in hypoxia/reoxygenation injury in isolated rat proximal tubule suspension (104). In this in vitro model, suspensions of freshly isolated rat proximal tubules were used to study tubular epithelial cells in the absence of other sources of NO, such as endothelial cells and neutrophils. L-arginine addition to the isolated rat proximal tubules enhanced rat tubular hypoxia/reoxygenation injury, while L-NAME almost completely prevented injury. In a follow-up study of Yaqoob et al., hypoxia-stimulated NO generation was continuously monitored in a suspension of rat proximal tubules using an NO-selective electrode (105). Hypoxia-stimulated NO generation was characterized by an initial rise and a subsequent sustained increase. This increase preceded cell membrane damage. L-NAME reduced the levels of NO with parallel protection against hypoxia-induced membrane damage in a dose-dependent manner. The hypoxia-induced elevation in NO and the associated membrane injury were both markedly prevented by low pH. This study established that during hypoxia NO was generated in the rat proximal tubules. It further suggested that the protective effect of low pH against hypoxic rat tubular injury was associated with an inhibition of NO production.

H. Synopsis

Renal ischemia is a common cause of renal failure. The contribution NO makes to renal failure has been examined in many experimental models of acute renal failure. NO has been shown both to enhance and to diminish the extent of ishemic/hypoxic injury. The results obtained with antisense oligomers to NOS2 highlight this disparity. The in vivo targeting of NOS2 with oligodeoxynucleotides protected rat kidneys against ischemia. By contrast, the nonselective inhibitor L-NAME was ineffective and led to the deterioration of renal function. The different functional roles of the NOS isoforms in renal disease remain to be elucidated. Unfortunately, the L-arginine analogs most widely used to study the effects of chronic NOS inhibition are relatively nonspecific and block all NOS isoforms when administered in high doses. Investigation of the functional role of the different NOS isoforms in renal disease awaits highly selective inhibitors for each isoform.

VI. ON THE FUNCTION OF OSTEOPONTIN

A potential regulator of NO synthesis in the kidney is osteopontin (OPN). OPN is a secreted acidic glycosylated phosphoprotein that contains the integrin-binding GRGDS motif (106). It may serve an important protective function, possibly by attracting macrophages to sites of injury (107) and/or ameliorating NO-induced cellular injury by virtue of its ability to inhibit induction of NOS2 by LPS and IFNγ (108).

A. Localization and Regulation of OPN Expression

To begin to understand the function of OPN in the kidney, the sites of its synthesis, the regulation of its expression, and the distribution of the protein must all be considered. A number of studies have been undertaken to localize OPN expression in the kidney in both normal and pathological settings. Nomura et al. (109) and Ullrich et al. (110) detected OPN mRNA in the epithelium of both proximal and distal tubules of the kidneys of newborn mice and rats, respectively. Brown et al. (111) found OPN protein in the epithelial cells of the distal tubules, the collecting ducts, and the renal pelvis of normal adult human kidneys. OPN mRNA was most abundant in the epithelium of the thin descending limbs of Henle's loop, presumably the major site of synthesis, and consistent with the presence of the protein in the tubule fluid. OPN is abundant in urine, where it has been called "uropontin" (112). Basal levels of OPN may function to inhibit kidney stone formation.

Lopez et al. (113) detected OPN expression, both protein and message, in the distal tubules and along with the thick ascending limbs of the loop of Henle in normal mouse kidneys. Curiously, there was considerable variation in expression among the tubules. Of particular interest was the observation that OPN was overexpressed in the parietal epithelium of glomeruli undergoing sclerosis. This is consistent with other observations that OPN is highly expressed in some pathological conditions (107). Kleinman et al. (114) detected OPN mRNA and protein primarily in cells of the descending thin limb of the loop of Henle and in the papillary surface epithelium in the rat kidney. Some of the differences in the localization of OPN reported by different groups may reflect species diversity, the existence of undetected pathologies, variable ischemia during tissue preparation, differences in fixation and staining protocols, or the reaction of anti-OPN antibodies with epitopes on other proteins.

OPN expression is regulated by a variety of growth factors and cytokines in a cell type-dependent manner. OPN mRNA and protein synthesis in cultured rat mesangial cells is enhanced by fetal calf serum and TGFα, but not by PDGF, IL-1β, or TGFβ (115). In a rat renal epithelial cell line (NRK52E), Malyankar et al. (116) found that TGFβ1 and EGF, but not IGF-1, FGF, angII, or PDGF BB, strongly induced OPN expression (mRNA and protein). Aspects of the regulation of OPN expression by angII, PDGF, and TGF-β have been discussed (117). Endothelin-1, whose expression is increased during ischemia, has been shown to stimulate OPN expression in a rat mesangial cell line (118).

Singh et al. (119) showed that the synthetic glucocorticoid dexamethasone caused a 10-30-fold increase in OPN mRNA levels in heart muscle in vivo and in myocytes and microvascular endothelial cells in vitro, while suppressing induction of NOS2. Confirming the report of Hwang et al. (108) that OPN activated a signaling pathway that suppressed induced NOS2 transcription, they reported that a 20-mer peptide identical to the sequence bracketing and containing the GRGDS sequence in rat OPN (thus interferring with its interaction with the $\alpha v\beta_3$ integrin) was able to decrease NOS2 mRNA and protein levels. Further, transfection with a vector expressing an antisense OPN RNA decreased OPN mRNA levels while increasing NOS2 mRNA abundance. It appears that OPN may be a means through which glucocorticoids suppress NOS2 in cardiac myocytes and microvascular endothelial cells.

B. OPN Expression in Response to Injury

OPN is expressed at increased levels at sites of injury or inflammation. Rats with interstitial inflammation and fibrosis induced by protein-overload proteinuria exhibit increased OPN expression by cortical tubular cells (120). AngII infusion in rats leads to hypertension and focal tubulointerstitital cell injury, which is associated with an increased OPN expression in the tubular epithelium and Bowman's capsule cells (121). The increase in OPN expression preceded the morphological changes. Padanilam et al. measured an 18-fold increase in OPN messenger RNA compared with controls in rats with acute ischemic renal injury induced by clamping both renal arteries for 60 min (122). Elevated OPN expression was localized to cells of the distal tubule and the thick ascending limb of the loop of Henle. The authors proposed that OPN may reduce the extent of injury.

Enhanced OPN levels have been observed in a number of pathological settings. In cyclosporine-induced nephropathy in rats, OPN expression is enhanced in the tubular epithelial cells (123). Macrophages appeared to accumulate at the sites of OPN expression, and there was a good correlation between the extent of OPN expression and the severity of interstitial fibrosis. Antibodies to OPN reduced renal injury in a rat model of crescentic glomerulonephritis, possibly by inhibiting infiltration by macrophages and T-cells (124). Release of NO and various inflammatory mediators by the infiltrating cells is believed to be a major cause of tissue injury.

Quantification of OPN expression induced by puromycin aminonucleoside, likely involving angII, in rat cortical tubules uncovered a close correlation between OPN protein levels, urinary protein excretion, proliferating cell nuclear antigen levels, and the intraglomerular monocyte count (117). Although OPN is primarily localized to, and secreted from, the luminal (apical) aspect of the cortical tubular epithelial cells, it is not excluded that some may be released from the interstitial (distal) aspect of injured tubular cells, where it could attract monocytes and macrophages.

Experimental hydronephrosis induced by clamping the ureter in the rat results in a rapid (4 h) induction of OPN expression in the obstructed, but not the unobstructed, kidney (125). The T-cell-mediated tubulointerstitial nephritis that develops in kd/kd mice is accompanied by increased expression of CD44 and its ligands hyaluronic acid and OPN, all three of which colocalize in areas of nephron injury (126).

C. Relation to NO Synthesis

What is the relation between OPN and NO synthesis in the kidney? A number of studies (e.g., 119) have provided evidence that OPN may be a regulator of NOS2 gene expression. In primary cultures of proximal tubule epithelial cells derived from mouse kidneys, OPN can suppress nitric oxide synthesis induced with LPS and γ-IFN (108). The increase in NOS2 mRNA and protein in response to LPS and IFN-γ exposure was significantly inhibited by OPN at pM concentrations. Further studies, utilizing primary cultures of proximal tubule epithelia derived from human kidneys at various ages, showed a positive correlation between age and expression of both the constitutive brain NOS isoform (NOS1) and OPN mRNA levels (127). Cultures of PTE cells derived from older kidneys were found to have higher basal levels of OPN and NOS1 mRNAs than those derived from younger kidneys. When subjected to hypoxia-reoxygenation, OPN levels could be augmented only in cells derived from the younger kidneys, particularly during reoxygenation. The authors suggested that this may due to a diminished ability to respond to hypoxia-reoxygenation in older cells or that these cells may be fully expressing OPN and cannot further upregulate OPN expression during hypoxia-reoxygenation. In contrast, while the levels of NOS1 mRNA were increased in cells derived from both younger and older kidneys during hypoxia, there was a decrease during reoxygenation, perhaps due to the increased OPN expression.

What, then, may OPN be doing in the kidney, particular vis-à-vis the action of NO? In pathological situations OPN may act as a chemotactic agent for monocytes and macrophages (121,128). OPN may also protect cells or tissues from oxidant injury, such as that encountered in ischemic episodes. Hwang and colleagues described a simultaneous increase in OPN mRNA and decrease in NOS2 mRNA levels in human proximal tubule epithelial cells during reoxygenation after a period of hypoxia (127). They hypothesized that OPN may protect cells during stress by downregulating NO production. Other observations support this hypothesis. For example, less NO is secreted by macrophages obtained from older mice than younger mice when stimulated with LPS and γ-IFN, possibly due to the observed increased OPN expression in the older mice; also, activated macrophages from older mice were less cytotoxic (129). OPN inhibited NO production and cytotoxicity in this model. Rollo et al. observed that recombinant OPN was able to inhibit NO production by RAW264.7 cells, a macrophage line, stimulated with LPS and γ-IFN (130) These researchers reported that OPN inhibited the cytolytic activity of activated macrophages toward the NO-sensitive P815 mouse mastocytoma cells. Magil et al. have discussed the unresolved question of whether increased cortical tubular OPN expression is harmful (because it attracts macrophages, which may cause injury) or has a protective effect (by reducing NO production) (117).

The increase in the synthesis of OPN during stress may protect cells by inhibiting NO-induced apoptosis. NO has been shown to induce apoptosis in glomerular mesangial cells, epithelial cells, and endothelial cells (131). Rogers et al. showed that OPN is expressed in developing nephrons and the ureteric bud of rat embryos (132). They found that anti-OPN, as well as anti-$\alpha_v\beta_3$ integrin, prevented normal tubulogenesis in culture, and that increased apoptosis was observed in blastemal cells when cultured with anti-OPN antibody. The authors suggested that OPN, through the $\alpha_v\beta_3$ receptor, may inhibit apoptosis during nephron formation. This may also be the case in situations such as those described above (hypoxia, ischemia, etc.). OPN upregulation may act to downregulate NOS2 expression and prevent NO-induced apoptosis.

The signal transduction pathways that OPN activates remain to be elucidated. It is well known that OPN binds α_v integrins, particularly the $\alpha_v\beta_3$ integrin (133). Also, Weber et al. have reported that OPN is a ligand for the CD44 receptor (134). There are several types of posttranslational modifications that OPN may be subject to, dependent upon the circumstances. OPN may be phosphorylated, sulfated, glycosylated, and/or sialylated. Shanmugam, Mukherjee, and associates observed that sialylation of OPN controlled binding to the $\alpha_v\beta_3$ receptors on a transformed Rat-1 cell line (tsB77) (135). It may be that differential posttranslational modification of epitopes on the protein in various settings may direct interactions with different receptors, which in turn could modulate distinct signaling cascades.

D. Synopsis

It is evident that OPN has the potential both to exacerbate and to ameliorate injury, particularly injury resulting from NO synthesis, in the kidney. OPN may exacerbate injury by stimulating excessive influx of monocytes/macrophages. Activated macrophages release various cytokines and inflammatory mediators that may cause tissue injury as well as fending off pathogens. OPN may ameliorate injury by reducing the amount of NO produced, by proximal tubule cells, for example, in response to inducing stimuli such as LPS and IFNγ. Complicating a full understanding of the role of OPN in kidney homeostasis and pathology is the possibility that the consequences of elevated OPN expression may vary with the age of the organism. Although the above discussion has focused on what OPN might do within kidney tissue, there is also its role in the tubule fluid, which likely includes maintaining the solubility of calcium salts.

ACKNOWLEDGEMENTS

We thank Dr. Michael Goligorsky for comments on the manuscript. J.R.D., P.D.W., and D.T.D. gratefully acknowledge research support from the National Institutes of Health.

REFERENCES

1. Kone BC. Nitric oxide in renal health and disease. Am J Kidney Dis 1997; 30:311–333.
2. Goligorsky MS, Gross SS. Nitric Oxide and the Kidney: Physiology and Pathology. New York: Chapman & Hall, 1997.
3. Bachmann S, Mundel P. Nitric oxide synthesis in the kidney: synthesis, localization and function. Am J Kidney Dis 1994; 24:112–129.
4. Raij L, Baylis C. Glomerular actions of nitric oxide. Kidney Int 1995; 48:20–32.
5. Hebert SC, Kriz W. Structural-functional relationships in the kidney. In: Schrier RW, Gottschalk CW, eds. Diseases of the Kidney. Boston: Little, Brown and Co., 1997:1–5.
6. Wardener HF, ed. The Kidneys. 5th ed. New York: Churchill Livingston, 1985.
7. Heptinstall RH. Pathology of the Kidney. Vol. 1. 4th ed. Boston: Little Brown & Co., 1992.
8. Bachmann S, Bosse HM, Mundel P. Topography of nitric oxide synthesis by localizing constitutive NO synthases in the mammalian kidney. Am J Physiol (Renal Fluid Electrolyte Physiol 37) 1995; 268:F885–F898.
9. Ujiie K, Yuen J, Hogarth L, Danziger R, Star RA. Localization and regulation of endothelial NO synthase mRNA expression in rat kidney. Am J Physiol (Renal Fluid Electrolyte Physiol 36) 1994; 267:F296–F302.
10. Mohaupt MG, Elzie JL, Ahn KY, Clapp WL, Wilcox CS, Kone BC. Differential expression and induction of mRNAs encoding two inducible nitric oxide synthases in rat kidney. Kidney Int 1994; 46:653–665.
11. Singh R, Pervin S, Rogers NE, Ignarro LJ, Chaudhuri, G. Evidence for the presence of an unusual nitric oxide and citrulline-producing enzyme in rat kidney. Biochem Biophys Res Commun 1997; 232:672–677.
12. McKee M, Scavone C, Nathanson JA. Nitric oxide, cGMP, and hormone regulation of active sodium transport. Proc Natl Acad Sci USA 1994; 91:12056–12060.
13. Venema RC, Sayegh HS, Kent JD, Harrison DG. Identification, characterization

and comparison of the calmodulin binding domains of the endothelial and inducible nitric oxide synthases. J Biol Chem 1996; 271:6435–6440.
14. Nathan C, Xie QW. Nitric oxide synthases: roles, tolls and controls. Cell 1994; 78:915–918.
15. Kone BC, Baylis C. Biosynthesis and homeostatic roles of nitric oxide in the normal kidney. Am J Physiol (Renal Physiol 4) 1997; 272:F561–F578.
16. Gunning M, Ingelfinger J, King A, Brenner BM. Vasoactive peptides and the kidney. In: Brenner BM, ed. Brenner and Rector's The Kidney. Philadelphia: W.B. Saunders, 1996:627–635.
17. Morgan T, Brunner HR. The renin-angiotensin system and the heart: beyond 2000. Heart (suppl 3) 1996; 76:98–103.
18. Rosendorff C. The renin-angiotensin system and vascular hypertrophy. J Am Coll Cardiol 1996; 28:803–812.
19. Ingelfinger JR, Dzau VJ. Molecular biology of renal injury: emphasis on the role of the renin angiotensin system. J Am Soc Nephrol 1991; 2:S9-S20.
20. Burns KD, Homma T, Harris RC. The intrarenal renin-angiotensin system. Semin Nephrol 1993; 13:13–30.
21. Tank JL, Moe OR, Star RA, Henrich WL. Differential regulation of rat glomerular and proximal tubular renin mRNA following uninephrectomy. Am J Physiol 1996; 270:F776–F783.
22. Seikaly MG, Arant BS, Seney FD. Endogenous angiotensin concentrations in specific intrarenal fluid compartments of the rat. J Clin Invest 1990; 86:1352–1357.
23. Schricker K, Hamann M, Kurtz A. Nitric oxide and prostaglandins are involved in the macula densa control of the renin system. Am J Physiol (Renal Fluid Electrolyte Physiol 38) 1995; 269:F825–F830.
24. Beierwaltes WH. Macula densa stimulation of renin is reversed by selective-inhibition of neuronal nitric oxide synthase. Am J Physiol (Regul Integrative Comp Physiol 41) 1997; 272:R1359–R1364.
25. Wolff DJ, Lubeskie A, Umansky S. The inhibition of the constitutive bovine endothelial nitric oxide synthase by imidazole and indazole agents. Arch Biochem Biophys 1994; 314:360–366.
26. Schricker K, Potzl B, Hamann M, Kurtz A. Coordinate changes of renin and brain-type nitric oxide synthase messenger RNA levels in rat kidneys. Pflugers Arch Eur J Physiol 1996; 432:394–400.
27. Reid IA, Chou L. Effect of blockade of nitric oxide synthesis on the renin secretory response to furosemide in conscious rabbits. Clin Sci 1995; 88: 657–663.
28. Tharaux PL, Dussaule JC, Pauti MD, Vassitch Y, Ardaillou R, Chatziantoniou C. Activation of renin synthesis is dependent on intact nitric oxide production. Kidney Int 1997; 51:1780–1787.
29. Chatziantoniou C, Pauti MD, Pinet F, Promeneur D, Dussaule JC. Ardaillou R. Regulation of renin release is impaired after nitric oxide inhibition. Kidney Int 1996; 49:626–633.
30. Harding P, Sigmon DH, Alfie ME, Huang PL, Fishman MC, Beierwaltes WH, Carretero OA. Cyclooxygenase-2 mediated increased renal renin content induced by low sodium diet. Hypertension 1997; 29:297–302.

31. He XR, Greenberg SG, Briggs JP, Schnerman JB. Effect of nitric oxide on renin secretion. II. Studies in the perfused juxtaglomerular apparatus. Am J Physiol (Renal Fluid Electrolyte Physiol 37) 1995: 268:F953–F959.
32. Navar LG, Rosivall L. Contribution of the renin-angiotensin system to control of intrarenal hemodynamics. Kidney Int 1984; 25:857–868.
33. Alberola AM, Salazar FJ, Nakamura T, Granger JP. Interaction between angiotensin II and nitric oxide in control of renal hemodynamics in conscious dogs. Am J Physiol 1994; 267:R1472–R1478.
34. Zatz R, De Nucci G. Effects of acute nitric oxide inhibition on rat glomerular microcirculation. Am J Physiol (Renal Fluid Electrolyte Physiol 30) 1991; 261:F360–F363.
35. Ito S, Arima S, Ren YL, Juncos LA, Carretero OA. Endothelium-derived relaxing factor/nitric oxide modulates angiotensin II action in the isolated microperfused rabbit afferent, but not efferent arteriole. J Clin Invest 1993; 91:2012–2019.
36. Juncos LA, Ren RL, Arima SJ, Gavin J, Carretero OA, Ito S. Angiotensin II action in isolated microperfused rabbit afferent arterioles is modulated by flow. Kidney Int 1996; 49:374–381.
37. Schnackenberg CG, Wilkens FC, Granger JP. Role of nitric oxide in modulating the vasoconstrictor actions of angiotensin II in preglomerular and postglomerular vessels in dogs. Hypertension 1995; 26:1024–1029.
38. Niimura F, Lobosky PA, Kakuchi J, Okubo S, Yoshida H, Oikawa T, Ichiki T, Naftilan AJ, Fogo A, Inagami T, Hogan BLM, Ichikawa I. Gene targeting in mice reveals a requirement for angiotensin in the development and maintenance of kidney morphology and growth factor regulation. J Clin Invest 1995; 96:2947–2954.
39. Kihara M, Umemura S, Kadota T, Yabana M, Tamura K, Nyuui N, Ogawa N, Murakami K, Fukamizu A, Ishii, M. The neuronal isoform of constitutive nitric oxide synthase is up-regulated in the macula densa of angiotensinogen gene-knock out mice. Lab Invest 1997; 76:285–294.
40. Huang PL, Dawson TM, Bredt DS, Snyder SH, Fishman MC. Targeted disruption of the neuronal nitric oxide synthase gene. Cell 1993; 75:1273–1286.
41. Nelson RJ, Demas GE, Huang PL, Fishman MC, Dawson VL, Dawson TM, Snyder SH. Behavioural abnormalities in male mice lacking neuronal nitric oxide synthase. Nature 1995; 378:383–386.
42. MacMicking JD, Nathan C, Hom G, Chartrain N, Fletcher DS, Trumbauer M, Stevens K, Xie QW, Sokol K, Hutchinson N, Chen H, Mudgett JS. Altered responses to bacterial infection and endotoxic shock in mice lacking inducible nitric oxide synthase. Cell 1995; 81:641–650.
43. Huang, PL, Huang Z, Mashimo H, Bloch KD, Moskowitz MA, Bevan JA, Fishman MC. Hypertension in mice lacking the gene for endothelial nitric oxide synthase. Nature 1995; 377:239–242.
44. Huang ZH, Huang PL, Ma JY, Meng W, Ayata C, Fishman MC, Moskowitz MA. Enlarged infarcts in endothelial nitric oxide synthase knockout mice are attenuated by nitro-L-arginine. J Cereb Blood Flow Metab 1996; 16:981–987.
45. Ardiallou R. Active fragments of angiotensin II. Enzymatic pathways of synthesis and biological effects. Curr Opin Nephrol Hypertens 1997; 6:28–34.

46. Yoshida M, Kikukawa M, Hisa H, Satoh S. Modulation by nitric oxide and prostaglandin of the renal vascular response to angiotensin II (3-8). Br J Pharm 1996; 117:885–890.
47. Higashi Y, Oshima T, Ono N, Higara H, Yoshimura M, Watanabe M, Matsuura H, Kambe M, Kajiyama G. Intravenous administration of L-arginine inhibits an angiotensin-converting enzyme in humans. J Clin Endocrinol Metab 1995; 80:2198–2202.
48. Urabe K, Matsumura Y, Nonaka H, Nishiura M, Maeda K, Morimoto S. Role of endogenous angiotensin II in renal hemodynamic and excretory responses to L-arginine infusion. Biol Pharm Bull 1996; 19:1145–1153.
49. Schnackenberg CG, Tucker B, Pigg K, Granger L. Role of nitric oxide in modulating the chronic renal and arterial responses to angiotensin II. Am J Hypertens 1997; 10:226–229.
50. Parekh N, Dobrowolski L, Zou AP, Steinhausen M. Nitric oxide modulates angiotensin II-dependent and norepinephrine-dependent vasoconstriction in rat kidney. Am J Physiol (Regul Integrative Comp Physiol 39) 1996;270:R630–R635.
51. Ikenaga H, Fallet RW, Carmines PK. Basal nitric oxide production curtails arteriolar vasoconstrictor responses to angiotensin II in rat kidney. Am J Physiol (Renal Fluid Electrolyte Physiol 40) 1996; 271:F365–F373.
52. Thorup C, Kornfeld M, Winaver JM, Goligorsky MS, Moore LC. Angiotensin-II stimulates nitric oxide release in isolated perfused renal resistance arteries. Pflügers Arch Eur J Physiol 1998; 435:432–434.
53. Zatz R, Dunn BR, Meyer TW, Anderson S, Rennke HG, Brenner BM. Prevention of diabetic glomerulopathy by pharmacological amelioration of glomerular capillary hypertension. J Clin Invest 1986; 77:1925–1930.
54. Tsukahara H, Gordienko DV, Tonshoff B, Telato MC, Goligorsky MS. Direct demonstration of insulin-like growth factor I-induced nitric oxide production by endothelial cells. Kidney Int 1994; 45:598–604.
55. Arima S, Ito S, Omata K, Takeuchi K, Abe K. High glucose augments angiotensin II action by inhibiting NO synthesis in in vitro microperfused rabbit afferent arterioles. Kidney Int 1995; 48:683–689.
56. Osborn JL, Hook JB, Bailie MD. Regulation of plasma renin in developing piglets. Dev Pharm Ther 1980; 1:217–228.
57. Solhaug MJ, Wallace MR, Granger JP. Endothelium-derived nitric oxide modulates renal hemodynamics in the developing piglet. Pediatr Res 1993; 34:750–754.
58. Solhaug MJ, Wallace MR, Granger JP. Nitric oxide and angiotensin II regulation of renal hemodynamics in the developing piglet. Pediatr Res 1996; 39:527–533.
59. Rubanyi GM, Romero JC, Vanhoutte PM. Flow-induced release of endothelium-derived relaxing factor. Am J Physiol (Heart Circ Physiol 19) 1986; 250:H1145–H1149.
60. Buga GM, Gould ME, Fukoto JM, Ignarro LJ. Shear stress-induced release of nitric oxide from endothelial cells grown on beads. Hypertension 1991; 17:187–193.
61. Noris M, Morigi M, Donadelli R, Aiello S, Foppolo M, Todeschini M, Orisio S, Remuzzi G, Remuzzi A. Nitric oxide synthesis by cultured endothelial cells is modulated by flow conditions. Circ Res 1995; 76:536–543.

62. Deng XL, Welch WJ, Wilcox CS. Role of nitric oxide in short term and prolonged effects of angiotensin II on renal hemodynamics. Hypertension 1996; 5:1173–1179.
63. Ito S, Ren Y. Evidence for the role of nitric oxide in macula densa control of glomerular hemodynamics. J Clin Invest 1993; 92:1093–1098.
64. Thorup C, Persson AEG. Inhibition of locally produced nitric oxide resets tubuloglomerular feedback mechanism. Am J Physiol (Renal Fluid Electrolyte Physiol 36) 1994: 267:F606–F611.
65. Kawata T, Hashimoto S, Koike T. The angiotensin receptor antagonist 2-ethoxy-1-[[2′-(1H-tetrazol-5-YL)biphenyl-4-YL]methyl]-1H-benzimidazole-7-carboxylic acid (CV11974) attenuates the TGF response during NOS blockade in rats. J Pharm Exp Ther 1996; 277:572–577.
66. Braam B, Koomans HA. Nitric oxide antagonizes the actions of angiotensin II to enhance tubuloglomerular feedback responsiveness. Kidney Int 1995; 48:1406–1411.
67. Thorup C, Persson AEG. Macula densa-derived nitric oxide in regulation of glomerular capillary pressure. Kidney Int 1996; 49:430–436.
68. Yang Y, Macdonald GJ, Dagan KA. A study of angiotensin II receptors after chronic inhibition of nitric oxide synthase in the spontaneously hypertensive rat. Clin Exp Pharm Physiol 1996; 23:441–443.
69. Frohlich ED. Influence of nitric oxide and angiotensin II on renal involvement in hypertension. Hypertension 1997; 29:188–193.
70. Mattson DL, Bellehumeur TG. Neural nitric oxide synthase in the renal medulla and blood pressure. Hypertension 1996; 28:297–303.
71. Hu L, Manning RD Jr. Role of nitric oxide in regulation of long-term pressure-natriuresis relationship in Dahl rats. Am J Physiol 1995; 268:H2375–H2383.
72. Laurant P, Demolombe B, Berthelot A. Dietary L-arginine attenuates blood pressure in mineralcorticoid-salt hypertensive rats. Clin Exp Hypertens 1995; 17:1009–1024.
73. Gómez-Alamillo C, Sanchez-Casajús A, Sierra M, Huarte E, Díez J. Vasoconstriction of the afferent arteriole and defective renal synthesis of nitric oxide in essential hypertension. Kidney Int 1996; 49(suppl 55):S129–S131.
74. Brezis M, Rosen S. Hypoxia of the renal medulla—its implications for disease. N Engl J Med 1995; 332:647–654.
75. Shanley PF, Rosen MD, Brezis M, Silva P, Epstein FH, Rosen S. Topography of focal proximal tubular necrosis after ischemia with reflow in the rat kidney. Am J Pathol 1986; 122:462–468.
76. Brady HR, Brenner BM, Lieberthal W. Acute renal failure. In: Brenner, BM ed. Brenner and Rector's The Kidney. Philadelphia: W. B. Saunders, 1996:1207–1221.
77. Wilson PD, Schrier RW. Nephron segment and calcium as determinants of anoxic cell death. Kidney Int 1986; 29:1172–1179.
78. Brezis M, Rosen S, Silva P, Epstein FH. Selective vulnerability of the medullary ascending limb to anoxia in the isolated perfused rat kidney. J Clin Invest 1984; 73:182–190.
79. Endre ZH, Ratcliffe PJ, Tange JD, Ferguson DJP, Radda GK, Ledingham JGG.

Erythrocytes alter the pattern of renal hypoxic injury: predominance of proximal tubular injury with moderate hypoxia. Clin Sci 1989; 76:19–29.
80. Peresleni T, Noiri E, Bahou WF, Goligorsky MS. Antisense oligodeoxynucleotides to inducible NO synthase rescue epithelial cells from oxidative stress injury. Am J Physiol 1996; 270:F971–F977.
81. Noiri E, Peresleni T, Miller F, Goligorsky MS. In vivo targeting of inducible NO synthase with oligodeoxynucleotides protects rat kidney against ischemia. J Clin Invest 1996; 97:2377–2383.
82. Conger J, Robinette J, Villar A, Raji L, Schultz P. Increased nitric oxide synthase activity despite lack of response to endothelium dependent vasodilators in postischemic acute renal failure in rats. J Clin Invest 1995; 96:631–638.
83. Atanasova I, Burke TJ, McMurty IF, Schrier RW. Nitric oxide synthase inhibition and acute renal ischemia: effect on systemic hemodynamics and mortality. Renal Failure 1995; 17:389–403.
84. Dagher F, Pollina RM, Rogers DM, Gennaro M, Ascer E. The value and limitations of L-arginine infusion on glomerular and tubular function in the ischemic/reperfused kidney. J Vasc Surg 1995; 21:453–459.
85. Wald H, Markowitz H, Zevin S, Popovtzer MM. Opposite effects of diabetes on nephrotic and ischemic acute tubular necrosis. Proc Soc Exp Biol 1990; 195:51–56.
86. Goor Y, Peer G, Iaina A, Blum M, Wollman Y, Chernihovsky T, Siverberg C, Cabili S. Nitric oxide in ischaemic acute renal failure of streptozotocin diabetic rats. Diabetologia 1996; 39:1036–1040.
87. Bank A, Aynedjian HS. The role of EDRF (NO) in diabetic renal hyperfiltration. Kidney Int 1993; 43:1306–1312.
88. Linas S, Whittenburg D, Repine JE. Nitric oxide prevents neutrophil-mediated acute renal failure. Am J Physiol (Renal Physiol) 1996; 272:F48–F54.
89. Star RA, Rajora N, Huang J, Stock RC, Catania A, Lipton JM. Evidence of autocrine modulation of macrophage nitric oxide synthase by α-melanocyte-stimulating hormone. Proc Natl Acad Sci USA 1995; 92:8016–8020.
90. Chiao H, Yukimasa K, McLeroy P, Craig L, Housini I, Star RA. α-Melanocyte-stimulating hormone protects against renal injury after ischemia in mice and rats. J Clin Invest 1997; 99:1165–1172.
91. Langrehr JM, White DA, Hoffman RA, Simmons RL. Macrophages produce nitric oxide at allograft sites. Ann Surg 1993; 218:159–166.
92. Russell ME, Wallace AF, Wyner LR, Newell JB, Karnovsky MJ. Upregulation and modulation of inducible nitric oxide synthase in rat cardiac allografts with chronic rejection and transplant arteriosclerosis. Circulation 1995; 92:457–464.
93. Shoskes DA, Xi YN, Gonzalez-Cadavid NF. Nitric oxide synthase activity in renal ischemia-reperfusion injury in the rat: implications for renal transplantation. Transplantation 1997; 63:495–500.
94. Alley EW, Murphy WJ, Russell SW. A classical enhancer element responsive to both lipopolysaccharide and interferon-γ augments induction of the iNOS gene in mouse macrophages. Gene 1995; 158:247–251.
95. Martin E, Nathan C, Xie QW. Role of interferon regulatory factor 1 in induction of nitric oxide synthase. J Exp Med 1994; 180:977–984.

96. Melillo G, Musso T, Sica A, Taylor LS, Cox GW, Varesio L. A hypoxia-responsive element mediates a novel pathway of activation of the inducible nitric oxide synthase promoter. J Exp Med 1995; 182:1683–1693.
97. Devera ME, Shapiro RA, Nussler AK, Mudgett JS, Simmons RL, Morris SM, Billiar TR, Geller DA. Transcriptional regulation of human inducible nitric oxide synthase (NOS2) gene by cytokines: initial analysis of the human NOS2 promoter. Proc Natl Acad Sci USA 1996; 93:1054–1059.
98. Semenza GL, Agani F, Booth G, Forsythe J, Iyer N, Jiang BH, Leung S, Roe R, Wiener C, Yu A. Structural and functional analysis of hypoxia-inducible factor-1. Kidney Intl 1997; 51:553–555.
99. Robinson, LJ, Weremowicz S, Morton CC, Michel T. Isolation and chromosomal localization of the human endothelial nitric oxide synthase (NOS3) gene. Genomics 1994; 19:350–357.
100. Lowenstein CJ, Snyder SH. NO, a novel biologic messenger. Cell 1992; 70:705–707.
101. Arnet UA, McMillan A, Dinerman JL, Ballermann B, Lowenstein CJ. Regulation of endothelial nitric oxide synthase during hypoxia. J Biol Chem 1996; 271:15069–15073.
102. Kourembanas S, Morita T, Liu YX, Christou H. Mechanisms by which oxygen regulates gene expression and cell-cell interaction in the vasculature. Kidney Int 1997; 51:438–443.
103. Markewitz BA, Michael JR, Kohan DE. Cytokine-induced expression of a nitric oxide synthase in rat renal tubule cells. J Clin Invest 1993; 91:2138–2143.
104. Yu L, Gengaro PE, Niederberger M, Burke TJ, Schrier RW. Nitric oxide: a mediator in rat tubule hypoxia/reoxygenation injury. Proc Natl Acad Sci USA 1994; 91:1691–1695.
105. Yaqoob M, Edelstein CL, Wieder ED, Alkhunaizi AM, Gengaro PE, Nemenoff RA, Schrier RW. Nitric oxide kinetics during hypoxia in proximal tubules—effects of acidosis and glycine. Kidney Int 1996; 49:1314–1319.
106. Denhardt DT, Guo X. Osteopontin: a protein with diverse functions. FASEB J 1993; 7:1475–1482.
107. Giachelli CM, Schwartz SM, Liaw L. Molecular and cellular biology of osteopontin. Trends Cardiovasc Med 1995; 5:88–95.
108. Hwang SM, Lopez CA, Heck DE, Gardner CR, Laskin DL, Laskin JD, Denhardt DT. Osteopontin inhibits induction of nitric oxide synthase gene expression by inflammatory mediators in mouse kidney epithelial cells. J Biol Chem 1994; 261:711–715.
109. Nomura S, Wills AJ, Edwards DR, Heath JK, Hogan BLM. Developmental expression of 2ar (osteopontin) and SPARC (osteonectin) RNA as revealed by in situ hybridization. J Cell Biol 1988; 106:441–450.
110. Ullrich O, Mann K, Haase W, Koch-Brandt C. Biosynthesis and secretion of an osteopontin-related 20-kDa polypeptide in the Madin-Darby canine kidney cell line. J Biol Chem 1991; 266:3518–3525.
111. Brown LF, Berse B, Van de Water L, Papadopoulos-Seriou A, Perruzzi CA, Manseau EJ, Dvorak HF, Senger DR. Expression and distribution of osteopontin

in human tissues: widespread association with luminal epithelial surfaces. Mol Biol Cell 1992; 3:1169–80.

112. Shiraga H, Min W, VanDusen WJ, Clayman MD, Miner D, Terrell CH, Sherbotie JR, Foreman JW, Przysiecki C, Neilson EG, Hoyer JR. Inhibition of calcium oxalate crystal growth in vitro by uroprontin: another member of the aspartic acid-rich protein superfamily. Proc Natl Acad Sci USA 1992; 89:426–430.
113. Lopez CA, Hoyer JR, Wilson PD, Waterhouse P, Denhardt DT. Heterogeneity of osteopontin expression among nephrons in mouse kidneys and enhanced expression in sclerotic glomeruli. Lab Invest 1993; 69:355–363.
114. Kleinman JG, Beshensky A, Worcester EM, Brown D. Expression of osteopontin, a urinary inhibitor of stone mineral crystal growth in rat kidney. Kidney Int 1995; 47:1585–1596.
115. Nagasaki T, Ishimura E, Shioi A, Jono S, Inaba M, Nishizawa Y, Morii H, Otani S. Osteopontin gene expression and protein synthesis in cultured rat mesangial cells. Biochem Biophys Res Commun 1997; 233:81–85.
116. Malyankar UM, Almeida M, Johnson RJ, Pichler RH, Giachelli CM. Osteopontin regulation in cultured rat renal epithelial cells. Kidney Int 1997; 51:1766–1773.
117. Magil AB, Pichler RH, Johnson RJ. Osteopontin in chronic puromycin aminonucleoside nephrosis. J Am Soc Nephrol 1997; 8:1383–1390.
118. Nambi P, Gellai M, Wu HL, Prabhakar U. Upregulation of osteopontin in ischemia-induced renal failure in rats: a role for ET-1? Biochem Biophys Res Commun 1997; 241:212–214
119. Singh K, Ballignad JL, Fischer TA, Smith TW, Kelly RA. Glucocorticoids increase osteopontin expression in cardiac myocytes and microvascular endothelial cells. Role in regulation of inducible nitric oxide synthase. J Biol Chem 1995; 270:28471–28478.
120. Eddy AA, Giachelli CM. Renal expression of genes that promote interstitial inflammation and fibrosis in rats with protein-overload proteinuria. Kidney Int 1995; 47:1546–1557.
121. Giachelli CA, Pichler R, Lombardi D, Denhardt DT, Alpers CE, Schwartz SM, Johnson RM. Osteopontin expression in angiotensin II-induced tubulointerstitial nephritis. Kidney Int 1994; 45:515–524.
122. Padanilam BJ, Martin DR, Hammerman MR. Insulin-like growth factor I-enhanced renal expression of osteopontin after acute ischemic injury in rats. Endocrinology 1995; 137:2133–2140.
123. Pichler RH, Franceschini N, Young BA, Hugo C, Andoh TF, Burdmann EA, Shankland SJ, Alpers CE, Bennett WM, Couser WG. Pathogenesis of cyclosporine nephropathy: roles of angiotensin II and osteopontin. J Am Soc Nephrol 1995; 6:1186–1196.
124. Yu XQ, Nikolic-Paterson DJ, Mu W, Giachelli CM, Atkins RC, Johnson RJ, Lan HY. A functional role for osteopontin in experimental crescentic glomerulonephritis in the rat. Proc Assoc Am Physicians 1998; 110:50–64.
125. Diamond JR, Kees-Folts D, Ricardo SD, Pruznak A, Eufemio M. Early and persistent up-regulated expression of renal cortical osteopontin in experimental hydronephrosis. Am J Pathol 1995; 146:1455–1466.

126. Sibalic V, Fan X, Loffing J, Wuthrich RP. Upregulated renal tubular CD44, hyaluronan, and osteopontin in kd/kd mice with interstitial nephritis. Nephrol Dial Transplant 1997; 7:1344–1353.
127. Hwang SM, Wilson PD, Laskin JD, Denhardt DT. Age and development-related changes in osteopontin and nitric oxide synthase mRNA levels in human kidney proximal tubule epithelial cells: contrasting responses to hypoxia and reoxygenation. J Cell Physiol 1994; 160:61–68.
128. Singh RP, Patarca R, Schwartz J, Singh P, Cantor H. Definition of a specific interaction between the early T lymphocyte activation 1 (Eta-1) protein and murine macrophages in vitro and its effect upon macrophages in vivo. J Exp Med 1990; 171:1931–1942.
129. Rollo EE, Denhardt DT. Differential effects of osteopontin on the cytotoxic activity of macrophages from young and old mice. Immunology 1996; 88:642–647.
130. Rollo EE, Laskin DL, Denhardt DT. Osteopontin inhibits nitric oxide production and cytotoxicity by activated RAW264.7 macrophages. J Leukoc Biol 1996; 60:397–404.
131. Muhl H, Sandau K, Brune B, Briner VA, Pfeilschifter J. Nitric oxide donors induce apoptosis in glomerular mesangial cells, epithelial cells and endothelial cells. Eur J Pharmacol 1996; 317:137–149.
132. Rogers SA, Padanilam BJ, Hruska KA, Giachelli CM, Hammerman MR. Metanephric osteopontin regulates nephrogenesis in vitro. Am J Physiol 1994; (4 Pt 2):F469–F476.
133. Hu DD, Lin ECK, Kovach NL, Hoyer JR, Smith JW. A biochemical characterization of the binding of osteopontin to integrins $\alpha_v\beta_1$ and $\alpha_v\beta_5$. J Biol Chem 195; 270:26232–26238.
134. Weber GF, Ashkar S, Glimcher MJ, Cantor H. Receptor-ligand interaction between CD44 and osteopontin (Eta-1). Science 1996; 71:509–512.
135. Shanmugam V, Chackalaparampil I, Kundu GC, Mukherjee AB, Mukherjee BB. Altered sialylation of osteopontin prevents its receptor-mediated binding on the surface of oncogenically transformed tsB77 cells. Biochemistry 1997; 36:5729–5738.

7

Roles of Protein Kinase C in the Synthesis and Cellular Action of Nitric Oxide

Susan A. Rotenberg
Queens College, City University of New York, Flushing, New York

I. INTRODUCTION

As a naturally occurring signaling molecule, nitric oxide is known to be an active component of the inflammatory response, neurotransmission, and vasodilatation. To study each of these phenomena, investigators have adopted certain mode-systems in which to examine the genesis and cellular effects of NO. Thus, macrophages (anti-inflammatory effects), cardiac myocytes (contractile events), endothelial cells (vasoconstriction), and neurons (anoxic cell death) have provided the experimental settings for many of the NO-related studies published to date. Of present interest is the finding that in each of these systems, protein kinase C (PKC) has been implicated in either the regulation of NO synthase (NOS) activity or its expression. Although other protein kinases have been associated with the regulation of NOS, much more information about PKC is available with regard to its regulation, subcellular localization of its related isoforms, experimental methods for studying its cellular activity, and its role as a component of signaling pathways that govern the expression of NOS. This chapter first reviews aspects of the PKC literature that focus on the function and regulation of this enzyme in a cellular context with emphasis on the existence and functional specificity of its related isoforms. Following this general overview, discussion addresses the complex interplay among PKC, NOS, and NO. A model summarizing the reciprocal interactions occurring between PKC and NOS is presented. It is proposed that while PKC is instrumental in directing the expression of NOS, the product NO may itself be involved in the regulation of PKC, as reflected by the sensitivity of PKC to NO and other cellular oxidants.

II. MECHANISMS THAT CONTROL PKC ACTIVITY IN VITRO AND IN CELLS

A. Structure-Function Relationships of PKC

Protein kinase C (PKC) is a Ca^{2+}- and phospholipid-dependent protein kinase that phosphorylates cellular target proteins on serine/threonine residues. Upon activation by agonist-induced elevations in Ca^{2+} and diacylglycerol levels, PKC associates more tightly with the membrane, where it phosphorylates target proteins and thus participates in transmitting intracellular messages that culminate in altered transcriptional and translational events. Ultimately, the actions of this enzyme are expressed in the global cellular changes that underlie the phenomena of cell growth and development.

It is well known that PKC exists as a family of related isoforms (1) that are divided into three major categories: (a) the conventional isoforms (cPKCs) PKC-α, βI, βII, and γ, which are stimulated by Ca^{2+}, diacylglycerol, phorbol ester, and phospholipid such as phosphatidylserine (1); (b) the novel PKC isoforms (nPKCs) PKC-δ, ε, η, and θ, which are not affected by the presence of Ca^{2+} but are stimulated by diacylglycerol, phorbol ester, and phospholipid (2–6); and (c) the atypical isoforms (aPKCs) PKC-ζ and -ι are stimulated by additional lipid cofactors such as phosphatidylinositol but are not affected by Ca^{2+}, diacylglycerol, or phorbol ester (7–9). Another isoform identified in mouse, PKC-λ, is generally considered equivalent to the human PKC-ι since they share a 98% sequence homology (10). Additional members of the PKC family that do not fit readily into the established categories are PKCμ (11) and PKD (12).

Figure 1 depicts the structures of each category of PKC isoform and highlights the structural features that characterize their functional similarities and dissimilarities. All PKC isoforms are monomeric proteins that consist of a regulatory domain, located at the amino terminal half of the protein, and a catalytic domain, located at the carboxyl terminal half of the protein. The two domains are separated by a short segment termed the hinge region which, as the name implies, is believed to impart flexibility to the molecule. Activating cofactors (Ca^{2+} plus PS, phorbol ester plus PS, or diacylglycerol plus PS) bind to sites located at the regulatory domain (13,14), while substrates (ATP, peptides, and protein targets) bind to the catalytic domain (15).

1. Pseudosubstrate Domain

The inactive form of PKC is generally envisioned as a closed structure in which a segment located at the N-terminus, i.e., the pseudosubstrate sequence, is tucked into the substrate binding site in the catalytic domain. Upon stimulation of the enzyme by activating cofactors, the pseudosubstrate is released from the cata-

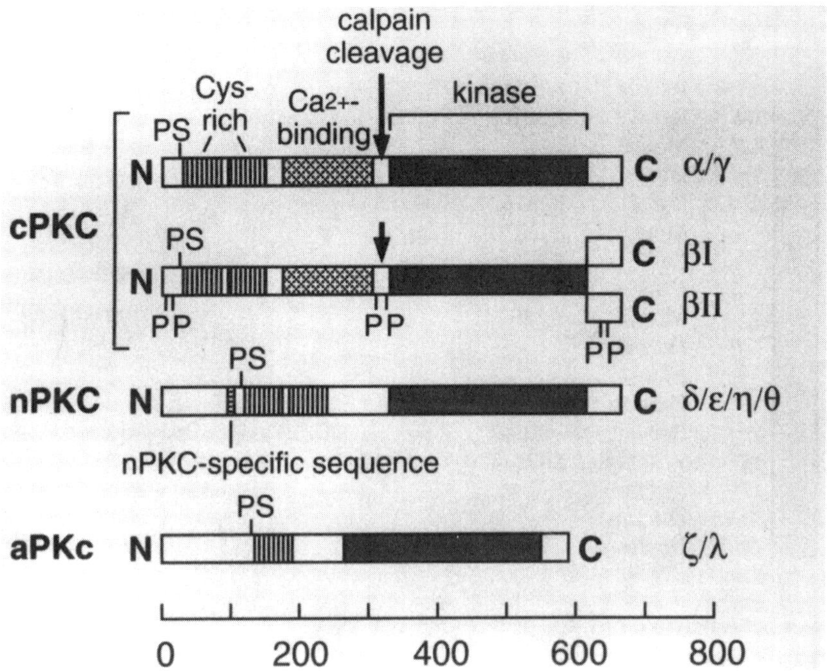

Figure 1 Structural features of PKC isoforms. Segments of identical coloration denote conserved regions; white segments denote variable sequences. The length in amino acids is given by the scale at the bottom. Details are discussed in the text. Abbreviations: PS, pseudosubstrate sequence; P, autophosphorylation site. (Reprinted (From Ref. 105).)

lytic domain, thereby allowing the entry and subsequent phosphorylation of substrate proteins. This model was formulated on the basis of three fundamental observations: that a synthetic peptide based on the sequence of the pseudosubstrate segment (residues 19–36) produced potent inhibition of the enzyme, while a modified peptide bearing a phosphorylatable serine in place of alanine at position 25 (Ala^{25}), served as a potent substrate peptide (16); that an antibody directed against the pseudosubstrate sequence caused PKC activation (17); and that site-specific mutagenesis of Ala^{25} to a glutamate was sufficient to produce the activated enzyme (18).

2. Phorbol Ester-Binding Domain

Located C-terminal to the pseudosubstrate segment is the high-affinity binding domain for tumor-promoting phorbol esters, the most potent of which is phorbol

12-O-tetradecanoyl 13-acetate (TPA). (An older nomenclature identifies this compound as phorbol 12-myristate 13-acetate [PMA].) A phorbol ester such as TPA is isolated from the plant *Croton tiglium*, and was originally used as an experimental tool to study tumor promotion on mouse skin (19). In 1982, Nishizuka and his colleagues discovered that PKC is potently activated by phorbol esters both in vitro and in platelets (20). In this landmark publication, it was reported that in platelets, nanomolar concentrations of TPA dramatically activated PKC in the absence of diacylglycerol, the endogenous PKC activator normally produced by the phosphatidylinositol hydrolysis stimulated by extracellular growth factors. Thus, by directly activating PKC in cells, tumor-promoting phorbol esters acted by circumventing the normal agonist-induced pathway of PKC activation.

The phorbol ester-binding domain has been localized to a segment in the C1 region (Fig. 1) that is characterized by two cysteine-rich domains, CYS1 and CYS2 (21,22), each of which contains two zinc ions coordinated by cysteine and histidine residues (23,24). Each cysteine-rich domain is capable of binding a molecule of phorbol ester with nanomolar specificity, as determined by Scatchard analysis with ^{3}H-phorbol 12,13-dibutyrate (^{3}H-PDBu) (22). It should be noted that these cysteine-rich, zinc-bound projections are structurally dissimilar to transcriptional regulator proteins that contain Zn fingers. The structure of each cysteine-rich unit was studied by atomic absorption (24) and by x-ray absorption fine structure (EXAFS) (23). The CYS2 phorbol ester-binding domain of PKC-γ was defined as a 43-amino acid peptide in the C1 region (residues 102–144) (25), and the x-ray crystal structure of the CYS2-binding domain for PKC-δ was recently elucidated (26).

The significance of the zinc atoms to the phorbol ester-binding function of the cysteine-rich domains was documented by studies of ^{3}H-PDBu binding to defined segments of the C1 region expressed in E. coli as glutathione-S-transferase or TrpE fusion mutant proteins and analyzed for zinc content either by use of ^{65}Zn during biosynthesis (27) or by atomic absorption of the fusion protein (25). Recent evidence has suggested that the CYS1 and CYS2 phorbol ester-binding domains apparently function with different affinities for diacylglycerol and phorbol ester (28). Interestingly, it is CYS2 and not CYS1 that is reportedly the predominant binding domain of phorbol ester to PKC-δ, having been implicated as the determining factor of phorbol ester-induced translocation of this isoform into membranes of NIH 3T3 cells (29). Remaining unresolved, however, are the anomalous findings with the atypical isoform PKC-ζ. This isoform possesses only a single cysteine-rich Zn complex and is neither activated nor bound by phorbol ester (7). Curiously, neither function can be restored by inserting a phorbol ester-binding sequence into PKC-ζ using mutagenesis techniques (30), implying that additional structural information is required.

3. Ca^{2+}-Binding Domain

A comparison of primary structures for the Ca^{2+}-dependent (cPKCs) and Ca^{2+}-independent isoforms (nPKCs) revealed an initial clue as to the location of the Ca^{2+}-binding domain. With the cloning of the first Ca^{2+}-independent isoform, PKC-ε, it became clear that this isoform lacked the entire C2 region (Fig. 1). This finding indirectly implicated the C2 region as the calcium-binding domain. Structural evidence has emerged to support this idea by using the x-ray crystal structure of a highly homologous protein segment as a three-dimensional scaffolding. For this purpose, a recent study employed the Ca^{2+}-binding domain of synaptotagmin I, which has a high degree of sequence homology with the C2 region of cPKC and other Ca^{2+}-binding proteins (31) to model the analogous region in PKC via computer-assisted methods that substituted individual amino acids of the PKC C2 primary sequence into the three-dimensional structure of the synaptotagmin I Ca^{2+}-binding domain (31). Experimental evidence demonstrated by nuclear magnetic resonance that the C2 region of PKC-β alone is sufficient to bind phospholipids in a Ca^{2+}-dependent manner and with micromolar specificity (32), similar to the conditions required for Ca^{2+}- and phospholipid-dependent activation of PKC. This finding challenged the results of another laboratory which claimed that, while complexed with PS or phorbol ester, the C1 region may be the actual site of Ca^{2+} binding (14) and that the C2 region participates by conferring Ca^{2+} specificity (33).

4. Catalytic Domain

While the regulatory domain houses the structural features unique to PKC, the catalytic domain possesses the structural hallmarks of a typical protein kinase such as an ATP-binding site (34). The substrate-binding domain, however, is as yet undefined and probably involves the participation of noncontiguous portions of the catalytic domain that includes a critical cysteine residue (35). Deletion analysis of the C-terminus of bovine PKC-α revealed that truncation of the terminal 10-amino acids resulted in an inactive enzyme (36). For the purpose of studying the catalytic domain fragment independently of the regulatory domain, the enzyme can be subjected to limited proteolysis with either trypsin (34) or the endogenous protease calpain (37), which causes specific cleavage at the hinge region and releases the catalytic fragment (Fig. 1).

B. Experimental Methods for Studying Intracellular PKC Activity Translocation/Downregulation by Phorbol Ester

As the primary target of a phorbol ester agent, PKC becomes tightly associated with the membrane within minutes, an event termed translocation since it entails a decrease of PKC activity in the soluble fraction with concomitant elevation of PKC activity in the particulate fraction. Upon prolonged treatment of cells with phorbol ester (18 to 24 h), membrane-associated PKC undergoes degradation by cellular proteases (such as calpain), a phenomenon termed downregulation. This event occurs independently of PKC catalytic function since kinase-defective mutants of PKC-α and PKC-γ can be downregulated (38,39).

The apparently exclusive target specificity of TPA for cellular PKC led investigators for many years to use either short-term (1 h) treatment (to activate PKC specifically), or prolonged exposure to TPA (to degrade PKC protein specifically) as routine methods to study the involvement of PKC in signaling pathways activated by a specific extracellular stimulus, or in the transcriptional activation of TPA-responsive genes (40). In view of the identification of additional cellular targets of phorbol esters (27,41) and the existence of phorbol ester-unresponsive isoforms (PKC-ζ and -ι), the traditional use of this method is subject to misinterpretation.

1. PKC Inhibitors

A common strategy for studying the involvement of PKC in signaling pathways is to employ small molecule inhibitors that are targeted to the enzyme. The central problem of such studies lies in the assumption that the inhibitor is exclusively specific for PKC and therefore does not influence the activity of other cellular components. In most if not all cases, this assumption has proved incorrect (reviewed in 42). As the late Prof. Efraim Racker frequently admonished: the specificity of an inhibitor decreases with the length of time that it is studied (!).

Most commercially available inhibitors that were routinely used with cell systems in the past few years (e.g., H-7, calphostin C, and staurosporine [42]) are now generally considered to be nonselective for PKC. Those inhibitors that are considered to be both potent and apparently specific for PKC include the newer staurosporine derivatives such as the indolecarbazoles (43) and the bis-indolylmaleimides (44–47). Both classes of compounds have nanomolar specificity and have been shown to exhibit high selectivity in vitro for PKC with respect to other protein kinases (42). One recently identified bis-indolylmaleimide (LY333531) exhibits a specificity of 5 nM for the PKC-β isoforms, which represented a 50-fold or greater selectivity for the β-isoforms relative to other PKC isoforms (48).

Novel approaches to the design of potent and specific PKC inhibitors may improve the utility of small molecule inhibitors as specific probes of intracellular PKC-related events. For example, the "bisubstrate approach" (49) entails a combination built into the same inhibitor molecule of both an ATP mimic (isoquinoline-5-sulfonamide) and a substrate mimic (a substrate peptide). Although the most potent combination proved to have nanomolar specificity for both PKC (Ki = 100 nM) and the cAMP-dependent protein kinase (Ki = 4 nM), the approach nonetheless offers an interesting model on which future combinations can be based. An intriguing application of this approach is that the most potent bisubstrate combination can yield information about the approximate distance between the ATP- and substrate-binding sites in the enzyme. In this study the spacer distance between the two substrate moieties was calculated as 16.3 Å.

Another approach under study is the use of the PKC inhibitor dequalinium in combination with long-wave UV light (366 nm), which induces the inhibitor to covalently modify its sites of binding on PKC to produce irreversible inactivation (50; unpublished studies). Photolinking of dequalinium to a target molecule has been used successfully with PKC both in vitro and in cells, and was originally described in work with the F1 ATPase (51). Low micromolar concentrations of dequalinium can reversibly inhibit PKC (52) as well as other potential cellular targets such as a calmodulin-dependent phosphodiesterase (53), K^+ channels (54), and the mitochondrial F1 ATPase (51). Recent studies with B16 cells, however, have shown that intracellular PKC activity can be inhibited with nanomolar concentrations of dequalinium when cells are concomitantly irradiated with UV light (S. A. Rotenberg, unpublished results). In vitro studies have further shown that, at concentrations up to 800 μM, dequalinium has no effect on cAMP-dependent protein kinase (PKA), calmodulin-dependent myosin light-chain kinase, or the tyrosine protein kinase pp60src (S. A. Rotenberg, unpublished results).

2. Phosphorylation of PKC in Cells

There is considerable evidence to support the idea that intracellular PKC is a phosphoprotein. Borner et al. [55] reported that in breast cancer cells PKC is initially synthesized as an unphosphorylated, inactive enzyme that subsequently undergoes multiple phosphorylations that mature the enzyme to full activity and responsiveness to lipid-mediated activation.

Several years ago it was proposed that stimulation of a PKC-directed signaling pathway engenders phosphorylation of PKC and consequently activates the enzyme (56). Since then several studies have supported this idea. For example, PKC-ε and -δ, both phosphoproteins in quiescent Swiss 3T3 cells, are each further phosphorylated in response to the mitogen bombesin (57). In another study, which employed NIH-3T3 fibroblasts cell lines each transfected individually with α, δ, ε, η, or ζ PKC isoforms, it was found that only PKC-δ underwent phosphorylation on tyrosine residues when cells were treated with TPA. When tested in vitro with a tyrosine protein kinase (e.g., pp60src), PKC-δ was again phosphorylated and concomitantly activated (58). In epidermal keratinocytes (59), however, the activation of the EGF receptor or a member of the Src kinase family produced phosphorylation of PKC-δ, which consequently caused inhibition of the enzyme.

These phosphorylation events are believed to be carried out by other cellular protein kinases (transphosphorylation) at PKC sites (60) that are distinct from the sites of autophosphorylation (61,62) (Fig. 1). Potential sites of transphosphorylation have been explored by employing site-specific mutagenesis techniques. In conjunction with an earlier study implicating the three threonine residues at positions 494, 495, and 497 in PKC-α (63), it was further proposed that the threonine residue at an analogous position of PKC-βII (T500) occurs at the "activation loop," which serves as the entrance to the catalytic site (64). Mutagenesis of the cDNA of PKC-βII at T500 to a glutamate residue (in order to simulate the negative charge of a phosphate group) with subsequent expression in COS cells was sufficient to produce a catalytically active enzyme, while substitution of a valine residue at T500 produced an inactive enzyme. A recent report further identified a cooperative interaction between T497 and T638 of PKC-α (65). Although the occurrence of phosphorylation of T638 is not required for catalytic activity, it appears to modulate the sensitivity of the enzyme to inactivation by phosphatases. Thus, multisite phosphorylation events regulate the duration of PKC activation.

C. PKC Isoform Diversity and Subcellular Localization

In any given cell type, the expressed PKC activity is usually defined by two or more isoforms which can be localized to the particulate, soluble, or nuclear fractions. Interestingly, PKC has been proposed to possess subcellular localization domains and proteolytic recognition sites (66). Evidence indicates, however, that both the expression profile of PKC isoforms and their subcellular distribution are highly characteristic of the cell type being studied. Furthermore, a specific PKC isoform can have a different response to a stimulus in different cell types (67). Because of this so-called context dependence of the functions of PKC isoforms, it is difficult at present to discern general principles of behavior for a given isoform that are independent of cell type. Nonetheless, an important realization from the sampling of studies presented in Table 1 is that, for a given cell type, functional differences do indeed exist, thus opening up possibilities for isoform-selective substrate targets, specific responses to cell stimuli, and sequence-based targeting of isoforms to specific subcellular sites (see below).

D. Adaptor Molecules and the Physical Organization of Signaling Molecules

An emerging theme in the structural biology of cellular metabolism focuses on the physical compartmentation of enzymes that participate in signaling pathways

Table 1 Differential Responses of PKC Isoforms to Physiological Stimuli

Cell type	Stimulus	PKC Isoform					Ref.
		α	β	δ	ε	ζ	
Swiss 3T3 cells	Bombesin	Increased expression after 30 h treatment	NE	Rapid mem assoc & downreg	Rapid mem assoc & downreg	No response	147
	PDGF	No response	NE	Rapid mem assoc & downreg	Rapid mem assoc & downreg	No response	
IIC9 Fibroblasts	α-Thrombin	Rapid mem assoc (transient); rapid nuclear trans-location	NE	NE	Rapid nonnuclear mem assoc (sustained)	No response	148, 149
	PDGF	No response	NE	NE	Slow mem assoc	No response	
Polymorphonuclear leukocytes	fMLP – Ca^{2+}		Transient mem assoc			Transient mem assoc	150
	fMLP + Ca^{2+}		Rapid mem assoc, sustained (βI, βII)			Rapid mem assoc, sustained	
	TPA – Ca^{2+}		Mem assoc (βII fast, βI slow)			No response	
GH4C1 pituitary cells	TRH (acute)	Rapid mem assoc	Rapid mem assoc	Rapid mem assoc	Rapid mem assoc	ND	151
	TRH (6 h)	No response	No response	No response	Downreg	ND	

NE, not expressed in these cells; ND, not determined; PDGF, platelet-derived growth factor; TPA, phorbol 12-O-tetradecanoyl 13-acetate; TRH, thyrotropin-releasing hormone; fMLP, formyl-methionyl-leucyl-phenylalanine.
Source: Ref. 2.

(68). Although compartmentation generally connotes localization to specific organelles, a more subtle form of compartmentation is gaining common acceptance. Given that a signaling pathway involves the participation of some enzymes but not others, it is likely that a specific and highly conserved recognition sequence in participating enzymes instructs their physical assembly (or disassembly) during periods of cell growth or quiescence, and their dynamic reorganization during periods of metabolic transition. This principle of organization via sequence "modules" first gained attention with the discovery of adaptor proteins containing SH2 consensus sequences (Src homology-2 domains), which orchestrate interactions among signaling enzymes that possess SH2 recognition sequences and have undergone phosphorylation on tyrosine residues as a consequence of an activated receptor tyrosine protein kinase (69). Since then additional sequence modules have been described: the SH3 module (recognizes proline-rich sequences of cytoskeletal phosphotyrosine proteins) (70,71); the PH module (pleckstrin homology-binding) (72,73); and the PTB (phosphotyrosine-binding) module (74).

Consistent with the idea that signaling molecules are physically organized, several classes of proteins that bind PKC have been identified and are distinguished by their subcellular localization (68). The protein-protein interactions described below represent emerging areas of research in the organization of signaling pathways.

1. Located at the cytoskeleton and membrane, PKC substrate/binding proteins (S-BPs) (75,76) are proposed to bind PKC in the presence of PS; the pseudosubstrate domain was also found to be a major determinant for PS-dependent binding of PKC to S-BPs (77). Interestingly, phosphorylation of these substrate proteins brings about redistribution of PKC in the cell (75). The substrate/binding proteins have been identified as providing structural links between cytoskeleton and membrane (e.g., vinculin, talin), or interacting with the actin cytoskeleton (e.g., adducins, annexins).

Also localized to the membrane is AKAP79, an adaptor protein that binds PKC, cAMP-dependent protein kinase (PKA) and the calmodulin-dependent protein phosphatase-2B (PP-2B) (78).

Although speculative at this time, the pleckstrin homology (PH) domains of two other kinases (the Burton tyrosine kinase and RAC kinase) may be the basis for their observed binding interaction to PKC at the membrane. This event may directly cause enhanced PKC activity, as suggested by the interaction between PKC and the PH domain of the protein stratifin (79).

2. Receptors for activated C-kinase (RACKs) are found in the cytoplasm or the detergent-insoluble fraction, and bind to PKC independently of PS at a site in the C2 domain (80,81). The PKC-RACK interaction was shown to be critical to insulin-induced PKC translocation in Xenopus oocytes (82).

3. PICKs, or "proteins that interact with C-kinase," were identified by two-hybrid screening (83), and at least one of these proteins, PICK-1, is a peri-

nuclear PKC substrate that, regardless of its state of phosphorylation, selectively binds the catalytic fragment of PKC-α but to neither the native enzyme nor the regulatory domain (84). It is possible that this specificity identifies the binding of only activated PKC molecules in which the target site is presumably fully accessible, while in the inactive holoenzyme the target site is concealed. Indeed, in an earlier study (85), deletion of the regulatory domain of PKC-α such that only the hinge and catalytic domain remained, produced a form of the enzyme that preferentially interacted with a receptor in the nuclear envelope.

III. REGULATION OF PKC ACTIVITY BY CELLULAR OXIDANTS: REVERSIBLE AND IRREVERSIBLE INHIBITION

It is useful to consider that the cell's capacity to carry out reduction and oxidation ("redox") reactions is a critical determinant of molecular signaling events. The possibility that the activated forms of NO (86) and reactive oxygen species (ROS) are poised in counterpoint to the cell's available reducing power, adds an important new dimension to our understanding of the molecular design of signal-transducing processes and their responsiveness to the extracellular environment.

By definition, a redox-sensitive cellular component undergoes a change in activity in the presence of reactive oxygen, NO, or reactive NO products. Interestingly, a given redox sensor may exhibit different sensitivities to ROS, NO, and peroxynitrite (a highly reactive oxidant that forms from NO and superoxide [87]), such as observed with the inactivation of aconitase by peroxynitrite but not by NO (88,89). The cellular targets of NO and its reactive products generally consist of proteins bearing metal-reactive sites or sulfhydryls, including iron-sulfur complexes (90). There is a single reported case of an enzyme (ribonucleotide reductase) with a tyrosyl radical intermediate that may act to scavenge NO radicals (91). It is noteworthy that peroxynitrite was reported to be a more reactive oxidant of protein thiol groups than was H_2O_2, which is the experimental model for ROS (92). Prominent target molecules identified to date have been located in several subcellular compartments: nuclear transcriptional regulating proteins (NF-κB, AP-1), and an array of cytosolic and membrane-associated enzymes, including PKC (93,94).

Each redox target site may contribute an exquisitely sensitive mechanism for responding to subtle changes in the balance between NO/ROS and glutathione, a major source of cellular reducing power, with concomitant adjustments made to cellular metabolism. One example of the reciprocity that exists among redox components is given by the increase in NOS activity when cellular glutathione levels are depleted (95). Another example entails the interrelationships that exist between iron and NO: iron can regulate NOS levels by controlling its transcription (96), and NO can chelate iron, which alters the binding of the iron-regulatory protein (a master regulator) to iron-responsive elements (97). Thus, iron and NO are intricately involved in regulating the cellular levels of one another.

Redox control of an enzyme is usually less amenable to detection and contrasts sharply with enzymatic control by covalent modification, which can be detected experimentally by the presence or absence of the modification. A redox-sensitive component is by its very nature susceptible to the facile loss and gain of electrons or protons, making technically difficult its detection and isolation in buffers that usually include thiol reagents such as dithiolthreitol or 2-mercaptoethanol. Proteins that are sensitive to a change in the redox balance could reflect the imbalance by having free sulfhydryls on neighboring cysteine residues convert to disulfide bonds (94).

In view of the observation that phorbol esters not only induce the translocation of PKC into membranes but also cause the formation of reactive oxygen species (98), PKC offers a prime target for oxidative effects on its activity. Oxidative inactivation was initially speculated to occur to PKC in phorbol ester-treated cells (99). Later studies with C6 glioma or B16 melanoma cells treated with H_2O_2 showed a time-dependent loss of Ca^{2+}/PS-dependent activity that correlated with the transient formation of an oxidatively modified form of PKC exhibiting elevated activity in the absence of activators (100). The authors speculated that the enzyme undergoes an initial reversible activation under mildly oxidizing conditions but can be subsequently inactivated by further oxidation. These oxidative effects involved both the catalytic domain and the phorbol ester-binding domain. Recent studies with human neutrophils indicated that Src-related tyrosine protein kinases can also be activated by reactive oxygen species (101).

Direct evidence of an essential cysteine at the PKC active site in the catalytic domain was given by evidence of inhibition at 10 μM by a modified peptide analog (RKRCLRRL), which bears a cysteine residue instead of the phosphorylatable threonine residue of the actual substrate peptide (35). Inhibition was demonstrated for the catalytic domain fragment as well as for the intact enzyme (α, β, and γ), and was not reversed by dilution. Inhibition was, however, reversed by the presence of dithiothreitol. Protection against inhibition was shown by substrate peptide and MgATP, suggesting that the peptide is covalently modifying an active site cysteine residue. Using a biotinylated analog of RKRCLRRL, the authors demonstrated with the catalytic domain fragment that both inactivation and covalent labeling of an active site cysteine were coincident (102).

In view of the existence of at least one critical thiol at the active site, it is of interest to consider the finding that PKC undergoes reversible inactivation in vitro by micromolar concentrations of NO (IC_{50} = 75 μM) using the NO-generating agent S-nitrosocysteine (103). Similar results were obtained with the catalytic fragment of PKC (IC_{50} = 50 μM). At higher concentrations of this agent (>100 μM), phorbol ester binding was only partially lost. Inactivation of both the intact enzyme and catalytic fragment implied the nitrosylation by NO of an essential cysteine residue(s) known to exist at the active site in the catalytic domain. The loss of phorbol ester binding is consistent with interaction by NO with free sulfhydryls that may occur within the cysteine-rich, zinc finger motif that defines the phorbol ester-binding site.

Although the primary sequence of the catalytic domain of Ca^{2+}-dependent PKC isoforms predicts six cysteine residues (34), their spatial positions with respect to one another will not be clarified until the three-dimensional structure of PKC has been elucidated. Nonetheless, Gopalakrishna et al. (103) speculated that NO interacts with vicinal thiols, which then undergo oxidation to form a disulfide bridge in the enzyme to produce inactivation that can be reversed by thiol reagents. In support of this model, they observed that the loss of enzyme activity was blocked by the NO scavenger oxyhemoglobin and reversed by the addition of dithiothreitol. In addition, NO gas as well as other NO-generating systems (S-nitroso-N-acetylpenicillamine and sodium nitroprusside) produced similar effects on phosphotransferase and phorbol ester binding. Although relatively high concentrations of NO were required in vitro for this reversible inhibition, the authors suggested that this may reflect the high local concentrations of NO that accumulate in cells.

The effect of NO on PKC activity was observed in intact cells such as B16 melanoma and the macrophage cell line IC-21 (103). In the latter cells, NO levels were elevated by induction with lipopolysaccharide and the Ca^{2+} ionophore A23187. For the time interval 3 to 12 h postinduction, PKC was isolated in buffer containing 1 mM dithiothreitol, and both phosphotransferase activity and phorbol ester binding were assayed. The results showed that the two parameters of PKC function underwent similar rates of inhibition, whose reversibility was a function of the length of exposure to NO. For short periods of time (5 min), a reversible modification of PKC at vicinal thiols occurred, while a chronic exposure (3 to 12 h) to NO produced irreversible inactivation. As is common during inflammation, chronic exposure to NO also correlated with cytotoxicity. It is important to note that no effect on cAMP-dependent protein kinase (PKA) in these samples was detected. Addition to the cells of the NOS inhibitor N^G-monomethyl-L-arginine was shown to prevent both the synthesis of NO and PKC inactivation. The authors suggested that the reversible short-term exposure effects on PKC by NO may act to desensitize the cell temporarily to intracellular signal events, thus producing a cytoprotective effect that can be repaired quickly by the presence of intracellular thiols.

When considered within the context of the role of PKC in inducing NOS expression, particularly in macrophages and cardiac myocytes, this short-term exposure to NO could act as a feedback mechanism for controlling cellular NO levels by suppressing the synthesis of new NOS. That PKA is not similarly affected even by long-term exposure to NO is an interesting observation in view of the fact that PKA also participates in inducing NOS expression. It is worth noting that while PKA and PKC share a high degree of sequence homology in their catalytic domains (38%) (104), PKA has only two cysteines in this domain, while PKC has six cysteines (105). It is therefore possible that the number of cysteine residues, and thus the likelihood of disulfide bond formation at the active site, disposes PKC to temporary inhibition by intracellular NO levels.

IV. PKC AND THE REGULATION AND SYNTHESIS OF INDUCIBLE NOS

PKC is known to influence the levels of nitric oxide in the cell. How PKC might bring about this effect could be accomplished by either direct phosphorylation of nitric oxide synthase (NOS), or as an early and controllable component of a signaling pathway that involves the activation of mitogen-activated protein (MAP) kinases (ERK1 and ERK2), and that culminates in the nucleus in a higher level of expression of iNOS, the inducible form of the enzyme. Phorbol esters (106), cytokines (107,108), lipopolysaccharide, and IFNγ (109) are all widely recognized activators of PKC-directed signaling pathways that induce iNOS expression in several cell types. Although there is an extensive literature on this subject, the following references serve to illustrate the general nature of iNOS induction via PKC-directed pathways. Some of the various cell types that have been studied include hepatocytes (109,110), astrocytes (111), lung fibroblasts (112), macrophages (113), renal cells (114,115), endothelial cells (116), and cardiac myocytes (117–119).

Other extracellular agonists such as IL-13 (120), TGFβ1 (121), PDGF (122), and anti-inflammatory steroids such as glucocorticoids, dexamethasone, and progesterone, suppress the induction of iNOS by cytokines (110,123–126). Additional mechanisms of iNOS induction include: NF-κB activation (127) which can be modulated by phorbol ester (128) and by antioxidants (129); and a non-PKC-directed pathway utilized by IFNγ that involves the participation of Jak1 and Jak2 tyrosine protein kinases (130) and activation of the Jak/STAT1α complex (131). Further discussion of these non-PKC-directed pathways is beyond the scope of this review, but they are included nonetheless in the summary of iNOS induction pathways depicted in Figure 2.

The following discussion highlights some of the model systems used to study the role of PKC in iNOS expression. Wherever possible, emphasis is given to the identification of PKC isoforms and/or their subcellular distribution, since this characteristic may be a complicating factor in the regulation of iNOS expression.

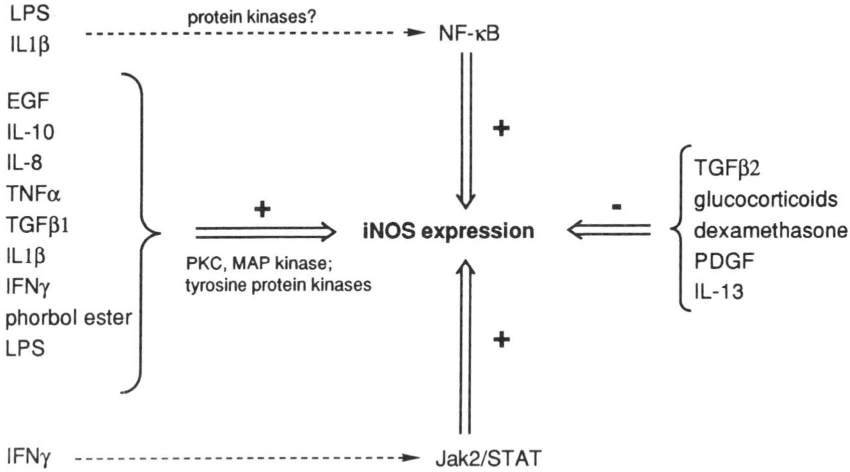

Figure 2 Signaling pathways that alter iNOS expression. Details and references are given in the text.

A. PKC-Directed Signaling Pathways that Control iNOS Expression

1. Macrophages

Because of their role in the anti-inflammatory response, macrophages offer a model system for the study of NOS expression and NO cellular activity (113). In murine macrophages (J774 and RAW 264.7), either lipopolysaccharide (LPS) or IFNγ was observed to stimulate both NO production and PKC activity, directly implicating PKC in the pathway (132). Consistent with this finding, treatment with the bis-indoleylmaleimide PKC inhibitor Ro31-8220 significantly reduced IFNγ-stimulated NO production (133). When cells were subjected to prolonged pretreatment with phorbol ester (which downregulates PKC), IFNγ-stimulated NO production was significantly reduced (133) or abolished (132).

In a similar study with J774 cells which express PKC-βII, ε, and ζ, PKC was selectively downregulated by prolonged treatment with phorbol ester. Of particular interest was the finding that these PKC isoforms underwent different rates of downregulation (or almost no effect of all, as with PKC-ζ) (134). In response to LPS, these TPA-treated cells exhibited significantly reduced NO production, thus implicating specific PKC isoforms (PKC-βII and/or -ε) in pathways governing LPS-induced NO production.

While the above-described experiments involved the downregulation of PKC, other studies tested the effect of PKC activation, which is achieved by short-term treatment of cells with phorbol ester. In these studies, short-term treatment of macrophages with phorbol ester had no effect on NO synthesis, implying that PKC activation alone was insufficient for iNOS induction. Furthermore it could be demonstrated that while IFNγ alone stimulated NO production to some degree, both IFNγ and phorbol ester in combination induced significant NO synthesis. This observation has been interpreted as an indication that PKC and cytokines (such as IFNγ and LPS) act in synergy with one another to induce NOS expression in murine macrophages (109,135). The proposed synergy may involve a collaboration between distinct pathways that operate at both the transcriptional and posttranscriptional levels. In this regard, IFNγ was observed to activate the NOS gene promoter (109), while phorbol ester treatment was seen to engender posttranscriptional stabilization of IFNγ-induced iNOS mRNA (135). In experiments in which PKC was downregulated, this posttranscriptional stabilization was apparently lost.

2. Cardiac Myocytes and Endothelial Cells

A recent study with adult rat ventricular myocytes and cardiac microvascular endothelial cells (CMEC) examined the importance of PKC, MAP kinases (ERK1/ERK2), and farnesyl transferase in mediating signals originating from either IFNγ or IL-1β and culminating in the induction of iNOS mRNA (136). By use of inhibitors of PKC, ERK1/ERK2, and farnesyl transferase, each enzyme was found to be essential for iNOS mRNA induction in myocytes treated with either IFNγ or IL-1β. By contrast, CMECs were responsive to IL-1β with regard to iNOS mRNA induction but were insensitive to IFNγ. The apparent regiospecificity of IFNγ for myocytes and not CMEC suggests that only certain areas or cells in heart tissue will respond under conditions of IFNγ release. When IL-1β is present, as during anti-inflammatory activity, there is apparently a more generalized response that could mobilize rapid production and release of NO.

In further consideration of the different transcriptional responses given by CMEC and cardiac myocytes to IFNγ, it is possible that each cell has a different PKC isoform expression profile and/or subcellular distribution, particularly in the nucleus. As determined by Western blot analysis, adult rat ventricular cardiac myocytes were found to consist of PKC-α, -δ, -ε, and ζ, of which -δ and -ε were expressed in the nucleus (137). In the same study, it was shown by runoff assays that transcriptional stimulation of an opioid peptide gene occurred in nuclei isolated from phorbol ester-treated myocytes, as well as in nuclei isolated from control cells and subsequently treated with phorbol ester. Although NOS was not a focus of this study, nuclear PKC activity was nonetheless implicated as an important factor for regulating transcription in cardiac myocytes. Whether the insensitivity to IFNγ exhibited by CMEC can be attributed to a different nuclear PKC isoform expression profile provides an avenue for further investigation.

3. Ischemic Neural Cell Death

Following anoxia, NO release is a causal factor in neuronal cell death. In a study conducted with primary cultured hippocampal neurons, the actions of specific protein kinases (PKC, PKA, CaM-II kinase, and cGMP-dependent kinase) were examined as possible modulators of NO release and the attendant neuronal toxicity (138). Compounds known to inhibit PKC were protective of these cells. Prolonged treatment of neurons with phorbol ester sufficient to downregulate PKC increased cell survival fourfold during anoxia and elevated NO levels. Interestingly, this effect was also evident during anoxic conditions in which NOS was inhibited by N-ω-methyl-L-arginine, suggesting that PKC participates in anoxic cell death by mechanisms in addition to NOS induction. In this regard, it was reported that in fetal brain, ischemia triggers the redistribution of PKC isoforms (139).

The role of other protein kinases in neuronal cell death was explored by testing specific inhibitors targeted to these enzymes. Inhibition of cGMP-dependent protein kinase was protective of cells during exposure to NO, while inhibition of CaM-II was not. By contrast, activation rather than inhibition of PKA enhanced neuronal survival in the presence of NO, revealing the possibility that PKC- and PKA-activated pathways carry out opposing functions during the process of ischemic cell death.

4. NO and NOS in Brain Tissue

Okada recently reported that NOS activity in rat cerebellar slices is modulated by PKC (140), and exhibited a different sensitivity to PKC as compared with in vitro assays (141). Biochemical studies with neurogranin, a calmodulin-binding protein and substrate of PKC in rat brain, showed that NO oxidatively modifies this protein and renders it both a poorer substrate for PKC and a weaker calmodulin-binding protein (142).

B. Regulation of NOS by Phosphorylation

The significance of phosphorylation and other modes of covalent modification as mechanisms for controlling NOS activity continues to be examined despite earlier inconclusive studies (143). There is evidence of extensive phosphorylation of NOS, specifically on serine residues, when immunoprecipitated from extracts of cells treated with phorbol ester. Furthermore, PKC, PKA, and calmodulin-activated kinase II (CaM-II) are all known to phosphorylate NOS in vitro. In addition, phosphorylation of NOS by a tyrosine protein kinase was recently reported as a posttranslational regulatory event (144) that apparently produces NOS activation (131). Phosphorylation of NOS has also been attributed to subcellular translocation of NOS in endothelial cells (145).

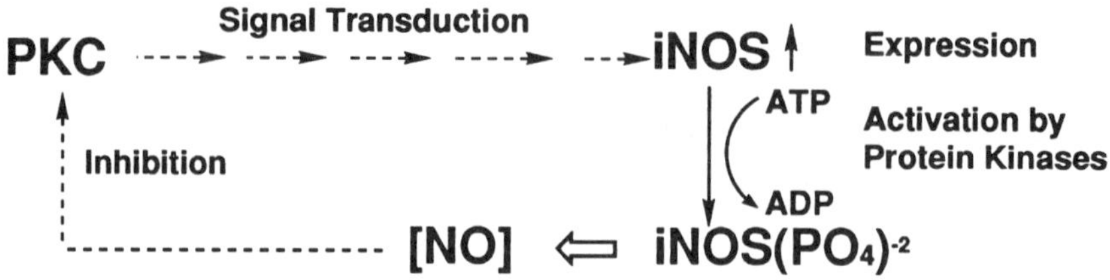

Figure 3 Reciprocal control of PKC and NOS activities.

V. CLOSING REMARKS

Over the past 5 years, studies of PKC have evolved from in vitro characterization to exploring the behavior of this enzyme in a cellular context. To carry out such "in vivo" enzymology, researchers have begun to harness genetic methods (84), immunocytochemistry (146), and confocal fluorescent microscopy of tissue samples. Such approaches are warranted in view of the diverse functions and subcellular locations that enable PKC to participate in various signaling pathways, such as NOS induction. Underscoring the complex behavior of PKC in these pathways are the multiplicity of its isoforms and their characteristic responses; the subcellular distribution of these isoforms, perhaps as instructed by specific adaptor molecules; and the redox-sensing properties of PKC in general such that it can respond to NO and reactive oxygen species.

The activity of PKC and NOS are related in a manner that suggests classical feedback inhibition. A conceptual model (Fig. 3) seeking to unify the available evidence depicts PKC as a key element in a signaling pathway that induces NOS expression with the consequence that NO levels are elevated. As the end product of the pathway, NO acts as an inhibitory signal to PKC, producing reversible or irreversible inhibition of the enzyme. The additional possibility that direct phosphorylation of NOS by PKC can modulate both activity and subcellular localization of NOS and thereby alter the effective levels of NO directly in specific subcellular compartments (145), further characterizes the complex interplay between these prominent signaling entities.

REFERENCES

1. Parker PJ. Protein kinase C: a structurally related family of enzymes. In: Lester DS, Epand RM, eds. Protein Kinase C. Current Concepts and Future Perspectives. Chichester, U.K.: Ellis Horwood Limited, 1992:3–24.
2. Schaap D, Parker PJ. Expression, purification and characterization of protein kinase C-ε. J Biol Chem 1990; 265:7310–7307.

3. Osada S-i, Mizuno K, Saido TC, Akita Y, Suzuki K, Kuroki T, Ohno S. A phorbol ester receptor/protein kinase, nPKCη, a new member of the protein kinase C family predominantly expressed in lung and skin. J Biol Chem 1990; 265:22434–22440.
4. Mischak H, Bodenteich A, Kolch W, Goodnight J, Hofer F, Mushinski JF. Mouse protein kinase C-δ, the major isoform expressed in mouse hemopoietic cells: sequence of the cDNA, expression patterns, and characterization of the protein. Biochemistry 1991; 30:7925–7931.
5. Bacher N, Zisman Y, Berent E, Livneh E. Isolation and characterization of PKC-L, a new member of the protein kinase C-related gene family specifically expressed in lung, skin, and heart. Mol Cell Biol 1991; 11:126–133.
6. Ogita K, Miyamoto S-i, Yamaguchi K, Koide H, Fujisawa N, Kikkawa U, Sahara S, Fukami Y, Nishizuka Y. Isolation and characterization of δ-subspecies of protein kinase C from rat brain. Proc Natl Acad Sci USA 1992; 89:1592–1596.
7. Nakanishi H, Exton JH. Purification and characterization of the ζ isoform of protein kinase C from bovine kidney. J Biol Chem 1992; 267:16347–16354.
8. Kazanietz MG, Areces LB, Bahador A, Mischak H, Goodnight J, Mushinski JF. Characterization of ligand and substrate specificity for the calcium-dependent and calcium-independent protein kinase isoenzymes. Mol Pharmacol 1993; 44:298–307.
9. Selbie LA, Schmitz-Peiffer C, Sheng Y, Biden TJ. Molecular cloning and characterization of PKCι, an atypical isoform of protein kinase C derived from insulin-secreting cells. J Biol Chem 1993; 268:24296–24302.
10. Akimoto K, Mizuno K, Osada S, Hirai S, Tanuma S, Suzuki K, Ohno S. A new member of the third class in the protein kinase C family, PKCλ, expressed dominantly in an undifferentiated mouse embryonal carcinoma cell line and also in many tissues and cells. J Biol Chem 1994; 269:12677–12683.
11. Johannes F-J, Prestle J, Eis S, Oberhagemann P, Pfizenmaier K. PKCμ is a novel, atypical member of the protein kinase C family. J Biol Chem 1994; 269:6140–6148.
12. Valverde AM, Sinnett-Smith J, Lint JV, Rozengurt E. Molecular cloning and characterization of protein kinase D: a target for diacylglycerol and phorbol esters with a distinctive catalytic domain. Proc Natl Acad Sci USA 1994; 91:8572–8576.
13. Burns DJ, Bell RM. Lipid regulation of protein kinase C. In: Lester DS, Epand RM, eds. Protein Kinase C. Current Concepts and Future Perspectives. Chicester, U.K.: Ellis Horwood Limited, 1992:25–40.
14. Luo J-H, Weinstein IB. Calcium-dependent activation of protein kinase C. J Biol Chem 1993; 268:23580–23584.
15. Newton AC. Protein kinase C: structure, function, and regulation. J Biol Chem 1995; 270:28495–28498.
16. House C, Kemp BE. Protein kinase C contains a pseudosubstrate prototope in its regulatory domain. Science 1987; 238:1726–1728.
17. Makowske M, Rosen OM. Complete activation of protein kinase C by an anti-peptide antibody directed against the pseudosubstrate prototope. J Biol Chem 1989; 264:16155–16159.
18. Pears CJ, Kour G, House C, Kemp BE, Parker PJ. Mutagenesis of the pseudo-

substrate site of protein kinase C leads to activation. Eur J Biochem 1990; 194:89–94.
19. Hecker E. Structure-activity relationships in diterpene esters irritant and co-carcinogenic to mouse skin. In: Slaga TJ, Sivak A, Boutwell RK, eds. Carcinogenesis: A Comprehensive Survey. Mechanisms of Tumor Promotion and Cocarcinogenesis. Vol. 2. New York: Raven, 1978:11–48.
20. Castagna M, Takai Y, Kaibuchi K, Sano K, Kikkawa U, Nishizuka Y. Direct activation of calcium-activated, phospholipid-dependent protein kinase by tumor-promoting phorbol esters. J Biol Chem 1982; 257:7847–7851.
21. Ono Y, Fujii T, Igarashi K, Takayoshi K, Tanaka C, Ushio K, Nishizuka Y. Phorbol ester binding to protein kinase C requires a cysteine-rich zinc-finger-like sequence. Proc Natl Acad Sci USA 1989; 86:4868–4871.
22. Burns DJ, Bell RM. Protein kinase C contains two phorbol ester binding domains. J Biol Chem 1991; 266:18330–18338.
23. Hubbard SR, Bishop WR, Kirschmeier P, George SJ, Cramer SP, Hendrickson WA. Identification and characterization of zinc binding sites in protein kinase C. Science 1991; 254:1776–1778.
24. Quest AFG, Bloomenthal J, Bardes ESG, Bell RM. The regulatory domain of protein kinase C coordinates four atoms of zinc. J Biol Chem 1992; 267:10193–10197.
25. Quest AFG, Bardes ESG, Bell RM. A phorbol ester binding domain of protein kinase Cγ: deletion analysis of the cys2 domain defines a minimal 43-amino acid peptide. J Biol Chem 1994; 269:2961–2970.
26. Zhang G, Kazanietz MG, Blumberg PM, Hurley JH. Crystal structure of the Cys2 activator-binding domain of protein kinase Cδ in complex with phorbol ester. Cell 1995; 81:917–924.
27. Ahmed S, Kozma R, Lee J, Monfries C, Harden N, Lim L. The cysteine-rich domain of human proteins, neuronal chimaerin, protein kinase C and diacylglycerol kinase binds zinc. Biochem J 1991; 280:233–241.
28. Slater SJ, Ho C, Kelly MB, Larkin JD, Taddeo FJ, Yeager MD, Stubbs CD. Protein kinase Cα contains two activator binding sites that bind phorbol esters and diacylglycerols with opposite affinities. J Biol Chem 1996; 271:4627–4631.
29. Szallasi Z, Bogi K, Gohari S, Biro T, Acs P, Blumberg PM. Nonequivalent roles for the first and second zinc fingers of protein kinase Cδ. J Biol Chem 1996; 271:18299–18301.
30. Kazanietz MG, Bustelo XR, Barbacid M, Kolch W, Mischak H, Wong G, Petit GR, Bruns JD, Blumberg PM. Zinc finger domains and phorbol ester pharmacophore: analysis of binding to mutated form of protein kinase Cζ and the vav and c-raf proto-oncogene products. J Biol Chem 1994; 269:11590–11594.
31. Sutton RB, Davletov BA, Berghuis AM, Sudhof TC, Sprang SR. Structure of the first C_2 domain of synaptotagmin I: a novel Ca^{2+}/phospholipid-binding fold. Cell 1995; 80:929–938.
32. Shao X, Davletov BA, Sutton RB, Sudhof TC, Rizo J. Bipartite Ca^{2+}-binding motif in C_2 domains of synaptotagmin and protein kinase C. Science 1996; 273:248–251.
33. Luo J-H, Kahn S, O'Driscoll K, Weinstein IB. The regulatory domain of protein kinase C β1 contains phosphatidylserine- and phorbol ester-dependent calcium binding activity. J Biol Chem 1993; 268:3715–3719.

34. Parker PJ, Coussens L, Totty N, Rhee L, Young S, Chen E, Stabel S, Waterfield MD, Ullrich A. The complete primary structure of protein kinase C—the major phorbol ester receptor. Science 1986; 233:853–859.
35. Ward NE, Gravitt KR, O'Brian CA. Irreversible inactivation of protein kinase C by a peptide-substrate analog. J Biol Chem 1995; 270:8056–8060.
36. Su L, Parissenti AM, Riedel H. Functional carboxyl terminal deletion map of protein kinase C alpha. Receptors Channels 1992; 1:1–9.
37. Kishimoto A, Mikawa K, Hashimoto K, Yasuda I, Tanaka S-I, Tommaga M, Kuroda T, Nishizuka Y. Limited proteolysis of protein kinase C subspecies by calcium-dependent neutral protease (calpain). J Biol Chem 1989; 264:4088-4092.
38. Freisewinkel I, Riethmacher D, Stabel S. Downregulation of protein kinase C-γ is independent of a functional kinase domain. FEBS Lett 1991; 280:262–266.
39. Pears C, Parker PJ. Down-regulation of a kinase defective PKC-α. FEBS Lett 1991; 284:120–122.
40. Hata A, Akita Y, Suzuki K, Ohno S. Functional divergence of protein kinase C (PKC) family members. PKCγ differs from PKCα and βII and nPKCε in its competence to mediate 12-O-tetradecanoyl phorbol 13-acetate (TPA)-responsive transcriptional activation through a TPA-responsive element. J Biol Chem 1993; 268:9122–9129.
41. Kazanietz MG, Lewin NE, Bruns JD, Blumberg PM. Characterization of the cysteine-rich region of the *Caenorhabditiselegans* protein Unc-13 as a high affinity phorbol ester receptor. J Biol Chem 1995; 270:10777–10783.
42. Wilkinson SE, Hallam TJ. Protein kinase C: is its pivotal role in cellular activation over-stated? Trends Pharmacol Sci 1994; 15:53–57.
43. Martiny-Baron G, Kazanietz MG, Mischak H, Blumberg PM, Kochs G, Hug H, Marme D, Schachtele C. Selective inhibition of protein kinase C isozymes by the indolecarbazole Go 6976. J Biol Chem 1993; 268:9194–9197.
44. Davis PD, Hill CH, Keech E, Lawton G, Nixon JS, Sedgwick AD, Wadsworth J, Westmacott D, Wilkinson SE. Potent selective inhibitors of protein kinase C. FEBS Lett 1989; 259:61–63.
45. Meyer T, Regenass U, Fabbro D, Alteri E, Rosel J, Muller M, Caravatti G, Matter A. A derivative of staurosporine (CGP 41 251) shows selectivity for protein kinase C inhibition and in vitro antiproliferative as well as in vivo anti-tumor activity. Int J Cancer 1989; 43:851–856.
46. Toullec D, Pianetti P, Coste H, Bellevergue P, Grand-Perret T, Ajakane M, Baudet V, Boissin P, Boursier E, Loriolle F, Duhamel L, Charon D, Kirilovksy J. The bisindolylmaleimide GF 109203X is a potent and selective inhibitor of protein kinase C. J Biol Chem 1991; 266:15771–15781.
47. Bit RA, Davis PD, Elliott LH. Inhibitors of protein kinase C. 3. Potent and highly selective bisindolymaleimides by conformational restriction. J Med Chem 1993; 36:21–29.
48. Ishii H, Jirousek MR, Koya D, Takagi C, Xia P, Clermont A, Bursell S-E, Kern TS, Ballas LM, Heath WF, Stramm LE, Feener EP, King GL. Amelioration of vascular dysfunctions in diabetic rats by an oral PKCβ inhibitor. Science 1996; 272:728–731.

49. Ricouart A, Gesquiere JC, Tartar A, Sergheraert C. Design of potent protein kinase inhibitors using the bisubstrate approach. J Med Chem 1991; 34:73–78.
50. Rotenberg SA, Sun X-g. Photo-induced inactivation of protein kinase C by dequalinium identifies the RACK-1 binding domain as a recognition site. J Biol Chem 1998; 273:2390–2395.
51. Zhuo S, Allison WS. Inhibition and photoinactivation of the bovine heart mitochondrial F1-ATPase by the cytotoxic agent, dequalinium. Biochem Biophys Res Commun 1988; 152:968–972.
52. Rotenberg SA, Smiley S, Ueffing M, Krauss RS, Chen LB, Weinstein IB. Inhibition of rodent protein kinase C by the anticarcinoma agent dequalinium. Cancer Res 1990; 50:677–685.
53. Hait WN. Targeting calmodulin for the development of novel cancer chemotherapeutic agents. Anti-Cancer Drug Des 1987; 2:139–149.
54. Castle NA, Haylett DG, Morgan JM, Jenkinson DH. Dequalinium: a potent inhibitor of apamin-sensitive potassium channels in hepatocytes and of nicotinic responses in skeletal muscle. Eur J Pharmacol 1993; 236:201–207.
55. Borner C, Filipuzzi I, Wartmann M, Eppenberger U, Fabbro D. Biosynthesis and post-translational modifications of protein kinase C in human breast cells. J Biol Chem 1989; 264:13902–13909.
56. Mitchell FE, Marais RM, Parker PJ. The phosphorylation of protein kinase C as a potential measure of activation. Biochem J 1989; 261:131–136.
57. Olivier AR, Parker PJ. Bombesin, platelet-derived growth factor, and diacylglycerol induce selective membrane association and downregulation of protein kinase C isotypes in Swiss 3T3 cells. J Biol Chem 1994; 269:2758–2763.
58. Li W, Mischak H, Yu J-C, Wang L-M, Mushinski JF, Heidaran MA, Pierce JH. Tyrosine phosphorylation of protein kinase C-δ in response to its activation. J Biol Chem 1994; 269:2349–2352.
59. Denning MF, Dlugosz AA, Threadgill DW, Magnuson T, Yuspa SH. Activation of the epidermal growth factor receptor signal transduction pathway stimulates tyrosine phosphorylation of protein kinase Cδ. J Biol Chem1996; 271:5325–5331.
60. Keranen LM, Dutil EM, Newton AC. Protein kinase C is regulated in vivo by three functionally distinct phosphorylations. Curr Biol 1995; 5:1394–1403.
61. Flint AJ, Paladini RD, Koshland DE. Autophosphorylation of protein kinase C at three separate regions of its primary sequence. Science 1990; 249:408–411.
62. Zhang J, Wang L, Petrin J, Bishop WR, Bond RW. Characterization of site-specific mutants altered at protein kinase C isozyme autophosphorylation sites. Proc Natl Acad Sci USA 1993; 90:6130–6134.
63. Cazaubon SM, Parker PJ. Identification of the phosphorylated region responsible for the permissive activation of protein kinase C. J Biol Chem 1993; 268:17559–17563.
64. Orr JW, Newton AC. Requirement for negative charge on "activation loop" of protein kinase C. J Biol Chem 1994; 269:27715–27718.
65. Bornancin F, Parker PJ. Phosphorylation of threonine 638 critically controls the dephosphorylation and inactivation of protein kinase Cα. Curr Biol 1996; 6:1114–1123.

66. Lehel C, Olah Z, Jakab G, Szal Z. Protein kinase C subcellular localization domains and proteolytic degradation sites. J Biol Chem 1995; 270:19651–19658.
67. Ueda Y, Hirai S-i, Osada S-i, Suzuki A, Mizuno K, Ohno S. Protein kinase C δ activates the MEK-ERK pathway in a manner independent of Ras and dependent on raf. J Biol Chem 1996; 271:23512–23519.
68. Faux MC, Scott JD. More on target with protein phosphorylation: conferring specificity by location. Trends Biochem Sci 1996; 21:312–315.
69. Pawson T, Gish GD. SH2 and SH3 domains: from structure to function. Cell 1992; 71:359–362.
70. Musacchio A, Gibson T, Rice P, Lehto V-P, Saraste M. SH3—an abundant protein domain in search of a function. FEBS Lett 1992; 307:55–61.
71. Musacchio A, Gibson T, Rice P, Thompson J, Saraste M. The PH domain: a common piece in the structural patchwork of signaling proteins. Trends Biochem Sci 1993; 18:343–348.
72. Haslam RJ, Koide HB, Hemmings BA. Pleckstrin domain homology. Nature 1993; 363:309–310.
73. Mayer BJ, Ren R, Clark KL, Baltimore D. A putative modular domain present in diverse signaling proteins. Cell 1993; 73:629–630.
74. Geer Pvd, Pawson T. The PTB domain: a new protein module implicated in signal transduction. Trends Biochem Sci 1995; 20:277–280.
75. Chapline C, Ramsay K, Klauck T, Jaken S. Interaction cloning of protein kinase substrates. J Biol Chem 1993; 268:6858–6861.
76. Hyatt SL, Liao L, Chapline C, Jaken S. Identification and characterization of α-protein kinase C binding proteins in normal and transformed REF52 cells. Biochemistry 1994; 33:1223–1228.
77. Liao L, Hyatt SL, Chapline C, Jaken S. Protein kinase C domains involved in interactions with other proteins. Biochemistry 1994; 33:1229–1233.
78. Mochly-Rosen D. Localization of protein kinases by anchoring proteins: a theme in signal transduction. Science 1995; 268:247–250.
79. Dellambra E, Patrone M, Sparatore B, Luca MD, Cancedda FD. Stratifin, a keratinocyte specific 14-3-3 protein, harbors a pleckstrin homology (PHs) domain and enhances protein kinase C activity. J Cell Sci 1995; 108:3569–3580.
80. Mochly-Rosen D, Khaner H, Lopez J. Identification of intracellular receptor proteins for activated protein kinase C. Proc Natl Acad Sci USA 1991; 88:3997–4000.
81. Ron D, Mochly-Rosen D. An autoregulatory region in protein kinase C: the pseudoanchoring site. Proc Natl Acad Sci USA 1995; 92:492–496.
82. Smith BL, Mochly-Rosen D. Inhibition of protein kinase C function by injection of intracellular receptors for the enzyme. Biochem Biophys Res Commun 1992; 188:1235–1240.
83. Fields S, Song O-K. A novel genetic system to detect protein-protein interactions. Nature 1989; 340:245–246.
84. Staudinger J, Zhou J, Olson EN. A perinuclear binding protein and substrate for protein kinase C isolated by the yeast two-hybrid system. J Cell Biol 1995; 128:263–271.
85. James G, Olson E. Deletion of the regulatory domain of protein kinase Cα exposes

regions in the hinge and catalytic domains that mediate nuclear targeting. J Cell Biol 1992; 116:863–874.
86. Stamler JS, Singe DJ, Loscalzo J. Biochemistry of nitric oxide and its redox-activated forms. Science 1992; 258:1898–1902.
87. Zou M-H, Ullrich V. Peroxynitrite formed by simultaneous generation of nitric oxide and superoxide selectively inhibits bovine aortic prostacyclin synthase. FEBS Lett 1996; 382:101–104.
88. Hausladen A, Fridovich I. Superoxide and peroxynitrite inactivate aconitases, but nitric oxide does not. J Biol Chem 1994; 269:29405–29408.
89. Castro L, Rodriguez M, Radi R. Aconitase is readily inactivated by peroxynitrite, but not by its precursor, nitric oxide. J Biol Chem 1994; 269:29409–29415.
90. Rouault TA, Klausner RD. Iron-sulfur clusters as biosensors of oxidants and iron. Trends Biochem Sci 1996; 21:174–177.
91. Lepoivre M, Fieschi F, Coves J, Thelander L, Fontecave M. Inactivation of ribonucleotide reductase by nitric oxide. Biochem Biophys Res Commun 1991; 179:442–448.
92. Radi R, Beckman JS, Bush KM, Freeman BA. Peroxynitrite oxidation of sulfhydryls. J Biol Chem 1991; 266:4244–4250.
93. Stamler JS. Redox signaling: nitrosylation and related target interactions of nitric oxide. Cell 1994; 78:931–936.
94. Powis G, Briehl M, Oblong J. Redox signalling and the control of cell growth and death. Pharmacol Ther 1995; 68:149–173.
95. Heales SJR, Bolanos JP, Clark JB. Glutathione deletion is accompanied by increased neuronal nitric oxide synthase activity. Neurochem Res 1996; 21:35–40.
96. Hentze MW, Weiss G, Werner-Felmayer G. Iron regulates nitric oxide synthase activity by controlling nuclear transcription. J Exp Med 1994; 180:969–976.
97. Pantopoulos K, Hentze MW. Nitric oxide signaling to iron-regulatory protein: direct control of ferritin mRNA translation and transferrin receptor mRNA stability in transfected fibroblasts. Proc Natl Acad Sci USA 1995; 92:1267–1271.
98. Goldstein BD, Witz G, Amoruso M, Stone DS, Troll W. Stimulation of human polymorphonuclear leukocyte superoxide anion radical production by tumor promoters. Cancer Lett 1981; 11:257–262.
99. Gopalakrishna R, Anderson WB. Susceptibility of protein kinase C to oxidative inactivation: loss of both phosphotransferase activity and phorbol ester binding. FEBS Lett 1987; 225:233–237.
100. Gopalakrishna R, Anderson WB. Ca^{2+}- and phospholipid-independent activation of protein kinase C by selective oxidation modification of the regulatory domain. Proc Natl Acad Sci USA 1989; 86:6758–6762.
101. Yan SR, Berton G. Regulation of Src family tyrosine kinase activities in adherent human neutrophils. Evidence that reactive oxygen intermediates produced by adherent neutrophils increase the activity of the $p58^{c\text{-}fgr}$ and $p53/56^{lyn}$ tyrosine kinases. J Biol Chem 1996; 271:23464–23471.
102. Ward NE, Gravitt KR, O'Brian CA. Covalent modification of protein kinase C isozymes by the inactivating peptide-substrate analog N-biotinyl-Arg-Arg-Arg-Cys-Leu-Arg-Arg-Leu: evidence that the biotinylated peptide is an active-site label. J Biol Chem 1996; 271:24193–24200.

103. Gopalakrishna R, Zhen ZH, Gundimeda U. Nitric oxide and nitric oxide-generating agents induce a reversible inactivation of protein kinase C activity and phorbol ester binding. J Biol Chem 1993; 268:27180–27185.
104. Ono Y, Fujii T, Ogita K, Kikkawa U, Igarashi K, Nishizuka Y. The structure, expression, and properties of additional members of the protein kinase C family. J Biol Chem 1988; 263:6927–6932.
105. Hardie G, Hanks S. The Protein Kinase Facts Book. Vol. I. San Diego: Academic Press, 1995:81.
106. Hortelano S, Genaro AM, Bosca L. Phorbol esters induce nitric oxide synthase in rat hepatocytes. J Biol Chem 1992; 267:24937–24940.
107. Busse R, Mulsch A. Induction of nitric oxide synthase by cytokines in vascular smooth muscle cells. FEBS Lett 1990; 275:87–90.
108. Sunyer T, Rothe L, Collin-Osdoby P. Proinflammatory agents, IL-8 and IL-10, upregulate inducible nitric oxide synthase expression and nitric oxide production in avian osteoclast-like cells. J Cell Biochem 1996; 60:469–483.
109. Weisz A, Cicatiello L, Esumi H. Regulation of the mouse inducible-type nitric oxide synthase gene promoter by interferon-γ, bacterial lipopolysaccharide and N^G-monomethyl-L-arginine. Biochem J 1996; 316:209–215.
110. Geller DA, Nussler AK, Di Silvio M, Lowenstein CJ, Shapiro RA, Wang SC, Simmons RL, Billiar TR. Cytokines, endotoxin, and glucocorticoids regulate the expression of inducible nitric oxide synthase in hepatocytes. Proc Natl Acad Sci USA 1993; 90:522–526.
111. Simmons ML, Murphy S. Roles for protein kinases in the induction of nitric oxide synthase in astrocytes. Glia 1994; 11:227–234.
112. Jorens PG, Overveld FJV, Vermeire PA. Synergism between interleukin-1β and interferon-γ, an inducer of nitric oxide synthase, in rat lung fibroblasts. Eur J Pharmacol 1992; 224:7–12.
113. Kengatharan M, Kimpe SJD, Thiemermann C. Analysis of the signal transduction in the induction of nitric oxide synthase by lipoteichoic acid in macrophages. Br J Pharmacol 1996; 117:1163–1170.
114. Pfeilschifter J, Rob P, Mulsch A, Fandrey J, Vosbeck K, Busse R. Interleukin 1β and tumor necrosis factor α induce a macrophage-type of nitric oxide synthase in rat mesangial cells. Eur J Biochem 1992; 203:251–255.
115. Kunz D, Muhl H, Walker G, Pfeilschifter J. Two distinct signaling pathways trigger nitric oxide synthase in rat renal mesangial cells. Proc Natl Acad Sci USA 1994; 91:5387–5391.
116. Davda RK, Chandler LJ, Guzman NJ. Protein kinase C modulates receptor-independent activation of endothelial nitric oxide synthase. Eur J Pharmacol 1994; 266:237–244.
117. Balligand J-L, Ungureanu-Longrois D, Simmons WW, Pimental D, Malinski TA, Kapturczak M, Taha Z, Lowenstein CJ, Davidoff AJ, Kelly RA, Smith TW, Michel T. Cytokine-inducible nitric oxide synthase (iNOS) expression in cardiac myocytes. Characterization and regulation of iNOS expression and detection of iNOS activity in single cardiac myocytes in vitro. J Biol Chem 1994; 269:27580–27588.
118. McKenna TM, Li S, Tao S. PKC mediates LPS- and phorbol ester-induced car-

diac cell nitric oxide synthase activity and hypocontractility. Am J Physiol 1995; 269:H1891–1898.
119. LaPointe MC, Sitkins JR. Mechanisms of IL1β regulation of nitric oxide synthase in cardiac myocytes. Hypertension 1996; 27:709–714.
120. Saura M, Martinez-Dalmau R, Minty A, Perez-Sala D, Lamas S. Interleukin-13 inhibits inducible nitric oxide synthase expression in human mesangial cells. Biochem J 1996; 313:641–646.
121. Perrella MA, Yoshizumi M, Fen Z, Tsai JC, Hsieh CM, Kourembanas S, Lee ME. Transforming growth factor β1, but not dexamethasone, down-regulates nitric oxide synthase mRNA after its induction by interleukin-1β in rat smooth muscle cells. J Biol Chem 1994; 269:14595-14600.
122. Pfeilschifter J. Platelet-derived growth factor inhibits cytokine induction of nitric oxide synthase in rat renal mesangial cells. Eur J Pharmacol 1991; 208:339–340.
123. Di Rosa M, Radomski M, Carnuccio R, Moncada S. Glucocorticoids inhibit the induction of nitric oxide synthase in macrophages. Biochem Biophys Res Commun 1990; 172:1246–1252.
124. Hartmut K, Euchenhofer C, Forstermann U. Glucocorticoids inhibit the induction of nitric oxide synthase II by down-regulating cytokine induced activity of transcription factor nuclear factor-κB. Mol Pharmacol 1996; 49:15–21.
125. Kunz D, Walker G, Eberhardt W, Pfeilschifter J. Molecular mechanisms of dexamethasone inhibition of nitric oxide synthase expression in interleukin 1β-stimulated mesangial cells: evidence for the involvement of transcriptional and posttranslational regulation. Proc Natl Acad Sci USA 1996; 93:255–259.
126. Miller L, Alley EW, Hunt JS. Progesterone inhibits inducible nitric oxide synthase gene expression and nitric oxide production in murine macrophages. J Leukoc Biol 1996; 59:442–450.
127. Xie Q-W, Kashiwabara Y, Nathan C. Role of transcription factor NF-κB/Rel in induction of nitric oxide synthase. J Biol Chem 1994; 269:4705–4708.
128. Menegazzi M, Guerriero C, Prati ACd, Cardinale C, Suzuki H, Armato U. TPA and cycloheximide modulate the activation of NF-κB and the induction and stability of nitric oxide synthase transcript in primary neonatal rat hepatocytes. FEBS Lett 1996; 379:279–285.
129. Hecker M, Preis C, Schini-Kerth VB, Busse R. NFκB-mediated nitric oxide synthase in muscle cells. FEBS Lett 1996; 380:224–228.
130. Igarashi K-i, Garotta G, Ozmen L, Ziemiecki A, Wilks AF, Harpur AG, Larner AC, Finbloom DS. Interferon-gamma induces tyrosine phosphorylation of interferon-gamma receptor and regulated association of protein tyrosine kinases, Jak1 and Jak2, with its receptor. J Biol Chem 1994; 269:14333–14336.
131. Kitamura Y, Takahashi H, Taniguchi T. Possible involvement of Janus Jak2 in interferon-γ induction of nitric oxide synthase in rat glial cells. Eur J Pharmacol 1996; 306:297–306.
132. Paul A, Pendreigh RH, Plevin R. Protein kinase C and tyrosine kinase pathways regulate lipopolysaccharide-induced nitric oxide synthase activity in RAW 264.7 murine macrophages. Br J Pharmacol 1995; 114:484–488.
133. Severn A, Wakelam MJO, Liew FY. The role of protein kinase C in the induction

of nitric oxide synthesis by murine macrophages. Biochem Biophys Res Commun 1992; 188:997–1002.

134. Fujihara M, Connolly N, Ito N, Suzuki T. Properties of protein kinase C isoforms (βII, ε, and ζ) in a macrophage cell line (J774) and their roles in LPS-induced nitric oxide production. J Immunol 1994; 152:1898–1906.
135. Jun C-D, Choi B-M, Ryu H, Um J-Y, Kwak H-J, Lee B-S, Paik S-G, Kim H-M, Chung H-T. Synergistic cooperation between phorbol ester and IFN-γ for induction of nitric oxide synthesis in murine peritoneal macrophages. J Immunol 1994; 153:3684–3690.
136. Singh K, Balligand J-L, Fischer TA, Smith TW, Kelly RA. Regulation of cytokine-inducible nitric oxide synthase in cardiac myocytes and microvascular endothelial cells. Role of extracellular signal-regulated kinases 1 and 2 (ERK1/ERK2) and STAT1α. J Biol Chem 1996; 271:1111–1117.
137. Ventura C, Pintus G, Vaona I, Bennardini F, Pinna G, Tadolini B. Phorbol ester regulation of opioid peptide gene expression in myocardial cells. J Biol Chem 1995; 270:30115–30120.
138. Maiese K, Boniece IR, Skurat K, Wagner JA. Protein kinases modulate the sensitivity of hippocampal neurons to nitric oxide toxicity and anoxia. J Neurosci Res 1993; 36:77–87.
139. Yavin E, Inserte-Igual J, Gil S. Ischemia-triggered translocation and inactivation of protein kinase C isoforms in the fetal brain. J Neurochem 1995; 65:2594–2602.
140. Okada D. Protein kinase C modulates calcium sensitivity of nitric oxide synthase in cerebellar slices. J Neurochem 1995; 64:1298–1304.
141. Okada D. Differential effects of protein kinase C on neuronal nitric oxide synthase activity in rat cerebellar slices and in vitro. J Chem Neuroanat 1996; 10:213–220.
142. Sheu F-S, Mahoney CW, Seki K, Huang K-P. Nitric oxide modification of rat brain neurogranin affects its phosphorylation by protein kinase C and affinity for calmodulin. J Biol Chem 1996; 271:22407–22413.
143. Nathan C, Xie Q-W. Regulation of biosynthesis of nitric oxide. J Biol Chem 1994; 269:13725–13728.
144. Pan J, Burgher KL, Szczepanik AM, Ringheim GE. Tyrosine phosphorylation of inducible nitric oxide synthase: implications for potential post-translational regulation. Biochem J 1996; 314:889–894.
145. Michel T, Li GK, Busconi L. Phosphorylation and subcellular translocation of endothelial nitric oxide synthase. Proc Natl Acad Sci USA 1993; 90:6252–6256.
146. Goodnight J, Mischak H, Kolch W, Mushinski JF. Immunochemical localization of eight protein kinase C isozymes overexpressed in NIH 3T3 fibroblasts. J Biol Chem 1995; 270:9991–10001.
147. Olivier AR, Parker PJ. Bombesin, a platelet-derived growth factor, and diacylglycerol induce selective membrane association and down-regulation of protein kinase C isotypes in Swiss 3T3 cells. J Biol Chem 1994; 269:2758–2763.
148. Leach KL, Ruff VA, Jarpe MB, Adams LD, Fabbro D, Raben DM. α-Thrombin stimulates nuclear diglyceride levels and differential nuclear localization of protein kinase C isozymes in IIC9 cells. J Biol Chem 1992; 267:21816–21822.
149. Ha KS, Exton JH. Differential translocation of protein kinase C isozymes by

thrombin and platelet-derived growth factor. A possible function for phosphatidylcholine-derived diacylglycerol. J Biol Chem 1993; 268:10534–10539.
150. Dang PM-C, Rais S, Hakim J, Perianin A. Redistribution of protein kinase C isoforms in human neutrophils stimulated by formyl peptides and phorbol myristate acetate. Biochem Biophys Res Commun 1995; 212:664–672.
151. Kiley SC, Parker PJ, Fabbro D, Jaken S. Differential regulation of protein kinase C isozymes by thyrotropin-releasing hormone in GH4C1 cells. J Biol Chem 1991; 266:23761–23768.
152. Dekker LV, Parker PJ. Protein kinase C—a question of specificity. Trends Biochem Sci 1994; 19:73–77.

8
Nitric Oxide and Inflammation in Autoimmune Disorders

J. Brice Weinberg
Veterans Affairs Medical Center and Duke University Medical Center, Durham, North Carolina

I. INTRODUCTION

Nitric oxide may exhibit proinflammatory features. NO synthase (NOS) type 2 (NOS2) is overexpressed, and NO is overproduced in rodent models of inflammation. Blockage of NO production by administration of NOS inhibitors prevents or reduces various types of induced inflammation in rodents. Humans with the autoimmune disease rheumatoid arthritis have increased levels of NOS2 and NO production in their diseased synovia and in blood leukocytes, and the levels of NOS2 correlate significantly with disease activity. We have shown that autoimmune MRL-*lpr*/*lpr* mice overexpress NOS2 and overproduce NO in an age-dependent fashion that parallels expression of arthritis, glomerulonephritis, and vasculitis. The mice have high levels of nitrosylhemoglobin in their blood and high levels of a nitrosyl nonheme, iron protein in their diseased kidneys. Blocking NO production by oral administration of the NOS inhibitor N^G-monomethyl-L-arginine reduced the arthritis, glomerulonephritis, and vasculitis, while it did not modify serum anti-DNA antibody levels or glomerular deposition of immune complexes. When mice with genetically disrupted NOS2 were backcrossed to MRL-*lpr*/*lpr* mice, the resultant –/– mice expressed no NOS2 and produced no NO, the wild-type +/+ mice overexpressed NOS2 and overproduced NO (in comparison to normal, control mice), and the heterozygous +/– mice expressed and produced intermediate levels. Nephritis and arthritis in the –/– mice were comparable to that in MRL-*lpr*/*lpr* mice, but vasculitis was markedly decreased. Levels of anti-DNA antibodies were comparable in all mice, but IgG rheumatoid factor production was markedly reduced in the –/– mice. These results of studies in MRL-*lpr/lpr* mice with genetically disrupted NOS2 highlight the heterogeneity and complexity of the role NOS2 and NO in inflammation. Further studies in animal models and in humans with inflammatory diseases such as rheumatoid arthritis may lead to the development of novel therapies targeting NOS and NO.

II. OVERVIEW

Inflammatory joint tissue in RA is characterized by infiltration and accumulation of mononuclear phagocytes, lymphocytes, and plasma cells; proliferation of synovial cells; and expression of pro-inflammatory cytokines (1–3). While some cytokines are undetectable or are present in only low levels in rheumatoid synovia, synovial macrophages and fibroblasts in this disease are a good source of cytokines such as IFN-α, IL-1, TNF-α, IL-6, IL-8, and GM-CSF (4–7). In addition to these protein mediators, arachidonic acid metabolites, reactive oxygen species (ROS, O_2^-, hydrogen peroxide, and hydroxyl radical), and reactive nitrogen species (NO and related molecules such as peroxynitrite) likely contribute to the pathology. The combined actions of these mediators (along with certain inherent and induced anti-inflammatory mediators) contribute to the eventual pathology—accumulation of inflammatory cells, modification of synovial vascular cells, proliferation of synovial fibroblasts, disruption of the general synovial architecture, and destruction of cartilage and bone. NO, O_2^-, and peroxynitrite appear to be central to the inflammatory process. Antioxidant enzymes such as catalase, superoxide dismutase (SOD), and glutathione peroxidase may be critical as endogenous defenses against inflammation. Various cells in the joint may participate in the inflammation. These include mononuclear phagocytes and chondrocytes.

III. NO AND NITRIC OXIDE SYNTHASES

The simple gas NO has multiple important physiologic and pathologic functions (8–10 for reviews). These include roles in (to mention only a few) host resistance to tumors and microbes, regulation of blood pressure and vascular tone, neurotransmission, learning, neurotoxicity, carcinogenesis, and control of cellular growth and differentiation (11,12). In the presence of oxygen, NO rapidly (seconds) is converted to nitrite and nitrate, substances that are generally not bioactive (13 for review). NO binds with high affinity to iron in heme groups of proteins such as hemoglobin (Hb), myoglobin (Mb), and guanylyl cyclase; Hb and Mb are very effective quenchers of NO action. NO also reacts with O_2^-, and

SOD prolongs NO life by eliminating O_2^-. On reacting with O_2^-, NO may form peroxynitrite ($OONO^-$), a very toxic/reactive molecule that may be the most important effector toxic molecules when one thinks of NO toxicity in oxygenated systems.

There are three forms of the enzyme nitric oxide synthase encoded by three different genes. Neural NOS (nNOS or NOS1) and endothelial cell NOS (eNOS or NOS3) are constitutive enzymes, demonstrating low-level, constant transcription of mRNA. The enzymatic actions of NOS1 and NOS3 are modulated by regulation of cytoplasmic calcium levels, with agents inducing increases in calcium (e.g., calcium ionophores or ligands such as acetylcholine, or mechanical stress) and subsequent binding to calmodulin and activation of the enzyme. Inducible NOS (iNOS or NOS2) can be regulated at multiple levels, but induction of mRNA transcription by agents such as cytokines or lipopolysaccharide (LPS) appears to be of major importance (14 for review). The activity of NOS2 is generally thought to be calcium-independent. Although NOS2 was described initially in mononuclear phagocytes, it also is found in synoviocytes, chondrocytes, smooth muscle cells, hepatocytes, and others (8,9,15–17).

The NOS2 promoter contains two functionally important areas upstream of the 5′ end of the gene with at least 26 DNA consensus sequences for known DNA-binding proteins (14). Several of these DNA regions (e.g., IFN-α stimulatable RE [ISRE], IFN-γ RE [γ-IRE]) are targets for DNA-binding factors induced by IFN-α and IFN-γ (18,19).

Regulation of NOS2 can occur at multiple steps (14), including mRNA transcription, stability, and translation. At the protein level, NOS may be regulated by calmodulin binding, dimer formation (the functional enzyme exists as a dimer), substrate (L-arginine) depletion, substrate recycling (L-citrulline to L-arginine), tetrahydrobiopterin availability, end-product inhibition (NO interaction with NOS heme), phosphorylation, and subcellular localization. Important NOS co-factors include FAD, FMN, NADPH, tetrahydrobiopterin, and calmodulin-calcium. For NOS2, calmodulin is tightly bound to protein, making it relatively resistant to inhibition by calcium chelators. Activities of NOS can be influenced by tetrahydrobiopterin levels, and cytokines/LPS can enhance tetrahydrobiopterin production (20,21). Heme is a critical component of NOS; NO can act as a feedback inhibitor of NOS activity by binding to the iron in heme (22,23).

IV. HUMAN MONONUCLEAR PHAGOCYTE NOS2 EXPRESSION AND NO PRODUCTION

Several researchers have shown high-level NO production by murine macrophages, while others have difficulty showing that human monocytes or macrophages produce NO in vitro (24 for review). However, human hepatocytes, chondrocytes, and human DLD-1 colon tumor cells can express NOS2 mRNA and produce high levels of NO after treatment with cytokines and LPS (15,16,25–27). Some have shown human monocyte NO production, but levels have been low when compared to murine mononuclear phagocytes (24,28–30). In a study of normal human monocytes and peritoneal macrophages, we showed that these cells can produce NOS2 mRNA, protein, and NO, but levels were much less than that of mouse macrophages (31). We examined a large array of culture conditions and cytokines. IFN-γ and LPS showed only slight activity. Human alveolar macrophages have NOS2 antigen (32,33), and patients with tuberculosis have increased numbers of alveolar macrophages that express NOS2 antigen (34). Humans can be induced to make high levels of NO in vivo via an NOS mechanism (35), but the cells responsible for NO production in vivo are not known. We have recently noted that children with mild malaria overproduce NO and have mononuclear cells that express NOS2 antigen (36). We have also noted that patients with RA overexpress NOS in synovial macrophages (unpublished), that RA patients have overexpression of NOS2 antigen in blood mononuclear cells, and that their monocytes have enhanced NO production in vitro and enhanced responsiveness to treatment with IFN-γ and LPS (37).

In recent work, we have shown that IFN-α can serve to activate normal human monocytes in vitro for NO production, NOS2 antigen expression, and NOS2 mRNA expression (38). To determine if IFN-α would activate blood cells for NOS expression when administered in vivo, we studied patients with hepatitis C. Peripheral blood mononuclear cells (PBMCs) from these patients at baseline were no different from those from normal individuals. However, after they received IFN-α, their PBMCs had enhanced NOS enzyme activity, enhanced NOS2 antigen content on immunoblot analysis, and enhanced expression of NOS2 mRNA. The degree of IFN-α-induced increase of NOS2 antigen correlated significantly with the degree of improvement in their hepatitis (38). Since NO can inhibit infection with DNA viruses (e.g., ectromelia, herpes simplex virus, vaccinia, and Epstein-Barr virus) and RNA viruses (e.g., vesicular stomatitis virus and Coxsackie virus) (39–46), we speculated that the IFN-α-induced NOS2 expression and NO production might be responsible (at least in part) for the antihepatitis C effects of IFN-α in vivo. It is interesting to note that some patients receiving IFN-α treatment may develop "autoimmune" illnesses with inflammation similar to rheumatoid arthritis (RA) and systemic lupus erythematosus (SLE) (47–50). We have speculated that IFN-α-induced NO production my contribute to this (38).

Previous clinical trials of RA patients using TNF-α-neutralizing antibodies or a soluble recombinant TNF receptor fusion protein (p75) demonstrate that TNF-α blockade leads to reductions in clinical and laboratory disease activity in the majority of patients (51–53). The molecular mechanism(s) responsible for the beneficial effects of anti-TNF therapy are unknown. We have recently studied RA patients receiving the chimeric, monoclonal anti-TNF-α antibody cA2 (54). NOS2 mRNA and protein, and NOS enzyme activity were measured in PBMC at baseline and 4 weeks following cA2 therapy. At baseline, elevated levels of NOS2 mRNA and protein, and NOS enzyme activity were detected in the PBMC of RA patients but not in healthy controls. Treatment of the RA patients with cA2 led to significant reductions in NOS2 antigen expression and NOS enzyme activity in PBMC. The changes in NOS activity and NOS2 antigen expression correlated significantly with changes in the number of tender joints (54). These results documented again that NOS2 is increased in human mononuclear cells and monocytes in human disease. Importantly, this study also demonstrated that a treatment that decreases inflammation results in a decrease in NOS2 overexpression.

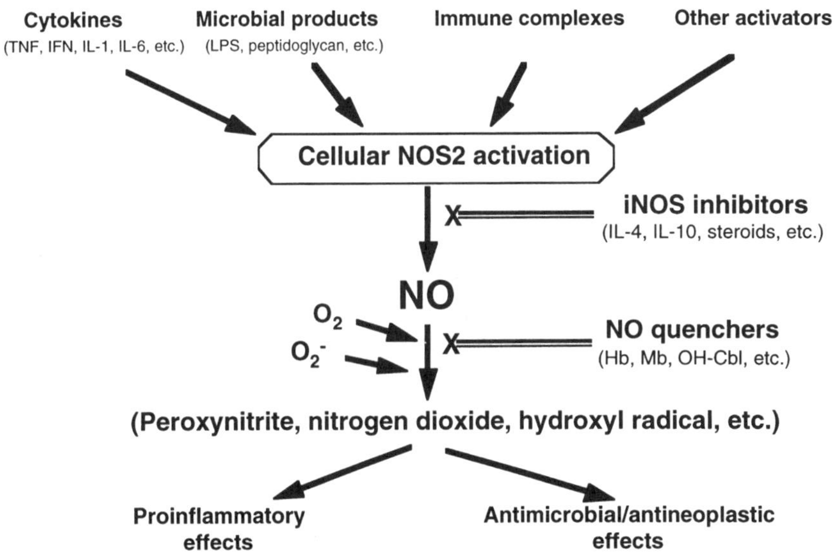

Figure 1 Nitric oxide and inflammation. Cellular NOS2 can be activated by a variety of cytokines (e.g., IFN-γ, IFN-α, and TNF), microbial products (e.g., endotoxin), immune complexes, and others. The formed NO can (either alone, or through interaction with superoxide and other, more reactive molecules) initiate and propagate inflammation. NOS2 activity can by reduced by various means (e.g., L-arginine analogs which act as competitive inhibitors of NOS, tetrahydrobiopterin depletion, and L-arginine depletion). NO effects can be quenched by a variety molecules, most of which contain metals such as iron or cobalt (e.g., hemoglobin and hydroxocobalamin). NO and certain of its catabolites may contribute to inflammation or may produce beneficial effects (e.g., antimicrobial and antineoplastic effects).

V. NO AND INFLAMMATION

NO has many actions appropriate for a pro-inflammatory agent; it is made by numerous cell types in sites of inflammation, and it increases blood flow and vascular permeability. NO has cell/tissue destructive abilities, and it can induce cyclo-oxygenase, cause pain, destroy certain protease inhibitors, and enhance production of IL-1 and TNF, and NADPH oxidase activity in myeloid cells (11,55). NO production may result from the actions of several substances including cytokines, immune complexes, and bacterial products. Since O_2^- may convert NO to peroxynitrite, coincident production of NO and O_2^- sets the stage for a severe inflammatory state. In the joint, several cell types may produce NO—macrophages, chondrocytes, endothelial cells, and possibly others.

Macrophages and chondrocytes are the most likely contributors. Human articular chondrocytes produce relatively high levels of NO, and cytokines can augment this production (15,56,57). Since several paths might lead to increased NO production, blocking only one pathway (e.g., blocking an interferon alone) might not fully blunt NO production since alternate paths could compensate.

Researchers have noted that in addition to pro-inflammatory effects, NO may also be anti-inflammatory. The "double-edged sword" phenomenon in NO biology thus applies to inflammation as well as other areas in NO biology (8,58). NO modifies adhesiveness and chemotaxis of PMN and monocytes (59,60), inhibits platelet aggregation and secretion (61), and inhibits cell proliferation (including lymphocytes [8]). We have shown that NO inhibits growth of normal and leukemic human blood cells, and induces monocytic differentiation of these normal or malignant cells (11). NO produced in murine bone marrow cells in response to LPS-IFN-γ can inhibit bone marrow cell growth; inhibitors of NO formation protect against the inhibition (12). NO inhibits human bone marrow colony formation in vitro (62,63) Importantly, we have found that NO preferentially reduces erythroid colony formation (63). NO regulates the RNA-binding protein "iron response element-binding protein" (IRE-BP). This cytoplasmic protein binds to mRNA of genes such as ferritin, erythroid aminolevulinic acid synthase, and transferrin receptor (TfR), with subsequent modulation of mRNA stability and/or translation. IRE-BP is cytoplasmic aconitase, an iron-sulfur cluster protein. NO reacts with the iron in aconitase and modulates its enzymatic and nucleic acid-binding functions. NO produced by mouse macrophages can modify IRE-BP activity and influence ferritin synthesis; also, NMMA prevents IFN-γ and LPS-induced "activation" of IRE-BP, and IRE-BP binding to the appropriate mRNAs (64,65). This is important as a clue to understanding anemia and abnormalities in iron metabolism noted in inflammatory states such as RA. We postulate that this anemia in RA is mediated in part by NO or peroxynitrite.

Arachidonic acid metabolites play important roles in inflammation, and cyclo-oxygenase inhibitors are drugs useful in the management of inflammatory disease (66). There is significant cross-talk between NO and cyclo-oxygenases. Eicosanoids can reduce NOS2 expression and NO production (67–69), and NO modulates prostaglandin E2 formation (70). Stimuli that enhance NOS2 and NO formation also may induce COX2 expression, but the time course for induction differs (71–73). Arginine analogs such as NMMA may be anti-inflammatory by inhibiting both COX2 and NOS (74). Furthermore, aspirin (in high doses) inhibits both cyclo-oxygenase and NOS2 (75).

NO is important in animal models of arthritis that mimic human RA. These include adjuvant arthritis, collagen-induced arthritis, and spontaneous arthritis in MRL mice (55, 57, 76 for reviews). Likewise, NO participates in the pathogenesis of spontaneous myositis in the SJL mouse (77). Mice with genetically disrupted transforming growth factor β1 have severe multifocal inflammation, and they die after only 2 to 3 weeks of life. These mice have overexpression of NOS2 and overproduction of NO (78). The overproduction of NO is inhibited by treatment with N^G-monomethyl-L-arginine (NMMA) in vivo. The absence of TGF-β, an inhibitor of NOS2 transcription and translation (79), appears to cause a systemic lethal inflammatory condition that results from NO overproduction.

NO may be increased in synovial fluid and serum of patients with RA (80–82). Kaur and Halliwell showed increased levels of nitrotyrosine (a product resulting from peroxynitrite action) in serum and synovial fluid from arthritis patients (83). Sakurai et al. found that synovia from patients with RA and OA produce NO in vitro and express NOS2 mRNA (RT-PCR) and protein (immunoblot and immunohistology). The NOS2 was associated primarily with CD14+ cells (mononuclear phagocytes) (84). Other investigators have also found that synovial fibroblasts (as well as macrophages) from patients with RA and OA produced NO and expressed NOS2 (85). Ueki and coworkers showed that RA patients have higher serum NO than do osteoarthritis (OA) patients, that RA synovial fluid NO was much higher than serum NO, and that serum NO levels correlated significantly with clinical parameters of disease activity (duration of morning stiffness, number of swollen and tender joints, and serum levels of C-reactive protein, TNF, and IL-6) (86). We had earlier noted comparable associations (37).

VI. CYTOKINES, NO, AND RHEUMATOID ARTHRITIS

As noted above, several cytokines can modulate NOS expression and NO production. IFN-γ is the prototypic activator of NOS2 transcription (at least with regard to murine cells), and TNF enhances NOS2 expression and appears to be important in RA (51,52,87). We think that IFN-α, IL-10, and IL-4 are very important. This relates in part to some of our recent findings regarding IFN-α as an activator of human mononuclear phagocyte NOS2.

A. Interferon-α

IFN-α is increased in sera and synovial fluids of people with autoimmune diseases such as RA and SLE (6,7,88). In the 1980s and 1990s, most investigators have focused on IFN-γ, and immunoreagents and immunoassays for IFN-α measurement have not been readily available or highly sensitive. Therefore, there is a paucity of quantitative data regarding IFN-α and RA. We believe that a more detailed analysis of this is warranted now. IFN-α (also called interferon type I) is produced primarily by mononuclear phagocytes. IFN-α and IFN-β share receptors, while IFN-γ has a distinct receptor (89). IFN-α and IFN-β can activate mouse macrophages for NO production, and macrophage-elaborated IFN-α serves in an autocrine fashion to modulate mononuclear phagocyte function (90,91). IFN-α can increase production of itself and IL-10, and certain complement components by macrophages, and it can decrease macrophage expression of CD14, IL-8 and GM-CSF (92–97). IFN-α also induces expression of other relatively specific IFN-related products such as 2′-5′ oligoA synthase, Mx, guanylate-binding protein, and 6–16 and 9–27 proteins. After IFN-α binds to surface receptors, signal transduction is accomplished by a complicated series of phosphorylations and interactions involving cytoplasmic factors (signal transducers and activators of transcription [STAT] factors) (18,19). We have found that IFN-α is a potent activator of NOS2 expression and NO production in human monocytes when used either in vitro or in vivo in humans (38).

B. Interleukin-10

IL-10 will likely be anti-inflammatory in man. Joint inflammation in RA displays many features of a cell-mediated immune response. IL-10, a product of Th2 cells and macrophages, exhibits potent anti-inflammatory effects and has been considered as a potential therapeutic agent for disorders provoked by cell-mediated immunity. It reciprocally suppresses Th1 cytokine production and decreases many functions of mononuclear phagocytes, including the expression of class II major histocompatibility complex (MHC) molecules, antigen-presenting cell function, and secretion of IL-1, IL-6, IL-8, TNF-α, and GM-CSF (98). IL-10-deficient mice develop a chronic enterocolitis resembling inflammatory bowel disease in humans (99). Treatment of diabetic-prone NOD mice with IL-10 prevents the development of hyperglycemia and attenuates islet cell inflammation (100). IL-10 can also prevent experimental allergic encephalomyelitis (101) and reduce the severity of established collagen-induced arthritis. IL-10 is known to be a potent inhibitor of NOS2 expression (102–105). IL-10 is an important immunoregulator in RA. IL-10 mRNA and protein are overexpressed in joint issue by synovial macrophages and T cells, which also spontaneously release IL-10 in synovial cell cultures (106). Neutralization of IL-10 in synovial culture supernatants increases TNF-α and IL-1 production, whereas addition of exogenous IL-10 has an opposite effect (106). Sera from patients with RA contain elevated levels of IL-10 (107). Taken together, these results suggest that insufficient levels of IL-10 may be present in the joint to control inflammation and that supplemental therapy with this cytokine may benefit patients with RA. Furthermore, we suggest that NO could be involved in the control of inflammation by IL-10 in RA.

C. Interleukin-4

IL-4 has numerous effects on mononuclear phagocytes and synoviocytes. Most of the effects are anti-inflammatory. These include decreases in cartilage degradation; immunoglobulin synthesis; COX2 expression and prostaglandin production; decreases in TNF, TNF-R, IL-6, and IL-1; and metalloproteinase production (108–112). Also, IL-4 causes increases IL-1 receptor antagonist (113). As noted before, IL-4 also decreases NOS2 expression and NO production. Thus, many anticipate that IL-4 treatment will benefit patients with RA. Some researchers have found that IL-4 can actually increase human mononuclear phagocyte NO production under certain circumstances (114). They feel that IL-4 increases cell expression of CD23 (the low-affinity Fc receptor for IgE—FcεR-II) and that then stimulation of cells with anti-CD23 or aggregated IgE increases NO production. However, most workers have found that IL-4 inhibits expression of mouse macrophage NO production. We have not studied CD23 and human monocytes.

D. Tumor Necrossi Factor-α

TNF-α is a likely an important factor in the development of inflammation in RA. A chronic inflammatory polyarthritis develops spontaneously by 4 weeks in transgenic mice that overexpress human TNF–α; this lesion is abrogated by administration of anti-TNF-α monoclonal antibody (115). Inflammation and tissue destruction are also attenuated in DBA/1 mice with type II collagen-induced arthritis treated with hamster antimurine TNF monoclonal antibody (116). In addition, clinical trials of RA patients using TNF-α neutralizing antibodies or a soluble recombinant TNF receptor fusion protein (p75) demonstrate that TNF-α blockade leads to reductions in clinical and laboratory disease activity (51–53). As noted above, we have recently documented that RA patients receiving the humanized monoclonal anti-TNF-α antibody cA2 have reductions in the levels of their overexpression of PBMC NOS2 antigen and activity to correlate significantly with the degree of cA2-induced clinical improvement (54). This suggests that TNF is involved in the pathogenesis of RA, and that TNF may function (at least in part, directly or indirectly) by enhancing NOS2 expression.

VII. NO, CARTILAGE, AND CHONDROCYTES

Chondrocytes can produce NO via NOS action (15,56,117). NO can influence cartilage and chondrocyte metabolism and function in a variety of ways including inhibition of chondrocyte proliferation, induction of apoptosis, and modification of proteoglycan synthesis and turnover (118–124). Also, NO activates latent metalloproteinases in cartilage (125). Furthermore, investigators have noted that cartilage from patients with OA has enhanced production of NO in vitro (126, 127). Amin and coworkers have found that cartilage from OA patients may contain a unique NOS isoform (127). Using immunoblot studies, they noted a NOS1-like protein (mass of 150 kD as opposed to the expected 133 kD NOS2) that was increased in content in OA patient cartilage. Activity of this enzyme was inhibited by inhibitors of protein synthesis or NF-κB expression, but not by TGF-β or dexamethasone. Thus, it did not have the characteristics of the known isoforms. These investigators have not published a characterization of its DNA or RNA composition.

VIII. THE MRL-*lpr/lpr* MODEL OF AUTOIMMUNITY

These mice develop a disease that resembles human SLE and RA (128). They have autoantibodies, rheumatoid factor, arthritis, nephritis, and vasculitis, and they die prematurely from disease. They have lymphadenopathy and splenomegaly, and a defect in apoptosis caused by a mutated Fas gene (an insert of the early transposable element [ETn] resulting in aberrant splicing and premature termination of transcription) (129). Fas protein is a membrane protein that, after ligation with an antibody or with its ligand FasL, causes cell death by apoptosis. C_3H-*gld* mice are similar to MRL-*lpr/lpr* mice in that they have generalized lymphadenopathy and a defect in apoptosis. The *gld* defect is a defect in the FasL gene, which encodes the ligand for Fas. Mice of both strains have increased numbers of CD4–, CD8–, CD3+ cells; these cells are polyclonal and nonmalignant. Fas is essential for the activation-induced death of mature T-cells in the periphery, but not in the thymus. In MRL-*lpr/lpr* mice, failure of apoptosis results in persistence of autoreactive T-cells which help autoreactive B-cells that are not eliminated. Genetic factors in addition to *lpr* are also important in determination of disease. MRL mice without the Fas abnormality still develop an autoimmune disorder (albeit less severe). Our work showed that while enhanced NOS2 expression and NO production is critical for disease development, disease likely does not result from an abnormality in the NOS2 gene, per se (130). Since these mice have high serum and tissue levels of immune complexes and various cytokines (including IFN-γ and TNF), activation may result from these factors. We and others have proposed potential final effectors of disease; these include arachidonic acid metabolites, ROS, and NO (130–132). Humans with Fas gene mutations, defective apoptosis, and autoimmune abnormalities have been recently described (133).

IX. NO, PEROXYNITRITE, INFLAMMATION, AND ARTHRITIS

We are currently studying NO biology in MRL-*lpr*/*lpr* mice. We noted before that their macrophages are "activated," including enhanced ROS generation (131). We hypothesized that they might also overproduce NO (130), and we demonstrated that they spontaneously excrete 5 to 10 times more urinary nitrite/nitrate (stable oxidation products of NO) than do normal mice; they have enhanced expression of NOS2 mRNA and protein in macrophages, liver, kidney, and spleen; NOS2 maps to mouse chromosome 11 (corresponding to human 17p21); this location is different from sites previously determined to be linked to disease susceptibility (chromosome 19 [*fas*], 7, and 12), making it unlikely that a defect in the NOS2 gene is the cause of this autoimmune disorder; treatment of the mice with NMMA orally blocks the overproduction and histology and prevents or blunts the development of arthritis and nephritis, indicating that NO is causally related to the disease. Despite marked improvement in renal function and histology, drug treatment does not reduce the serum levels of anti-DNA antibody or alter deposition of immune complexes. Other workers have corroborated our findings of overproduction of NO in these autoimmune mice (134).

NO binds to Hb tightly to form nitrosylhemoglobin (NOHb). This NOHb can be accurately detected by electron paramagnetic resonance techniques (epr). Nonhuman animals in septic shock and humans receiving IV nitroglycerin have detectable levels of NOHb (135). We measured NOHb in whole blood. MRL-*lpr*/*lpr* mice had statistically significant elevated levels of NOHb in blood (136). This increased with age and paralleled the course of NO overproduction. We also examined the kidneys using epr, and found that MRL kidneys have a nitrosyl nonheme iron protein signal as well as some NOHb (probably from blood trapped in the kidney). The nitrosyl nonheme, iron protein at $g = 2.04$ may be an iron-sulfur cluster protein such as one of the mitochondrial electron transport enzymes.

Our studies showed that kidneys from MRL mice have an increase in the amount of nitrotyrosine-containing proteins. Nitrotyrosine is formed as a consequence of action of peroxynitrite on tyrosine in proteins, and thus may serve as a marker of NO and peroxynitrite action in tissues (137–139). On immunoblots using specific antinitrotyrosine antibody, extracts from kidneys from normal mice were essentially negative, while those from MRL mice had two major bands of immunoreactivity (Mr ≈ 60,000 and 48,000) and three minor bands (140). Reactivity was eliminated by omitting the primary antibody, or by coincubating the primary antibody reaction mixture with 10 mM nitrotyrosine. The identity of the nitrated proteins in the tissues from the diseased kidneys is unknown. NO and peroxynitrite can react with numerous different proteins, and these reactions can alter functions of some (9,10). In an attempt to identify one of the target proteins for NO and peroxynitrite, we measured catalase activity in the mouse kidneys. Catalase levels were diminished in MRL mice, and this decrease was prevented by treating them with NMMA in vivo. In in vitro studies, we noted peroxynitrite would inactivate purified catalase. This suggested that catalase is one of the target proteins inactivated by NO and/or peroxynitrite. All this indicates that catalase can serve as a target for peroxynitrite-mediated modification and that the modified protein has decreased activity. Thus, in the MRL mice, there is overproduction of oxidants (NO, O_2^-, and peroxynitrite) with depletion of the antioxidant enzyme catalase. This produces the setting of extreme oxidant stress. Evidence of increased lipid peroxidation and oxidant stress in these mice has been reported (141).

X. TARGETED DISRUPTION OF THE GENES FOR NOS2

Studies using homologous recombination between incoming DNA and a chosen target gene (gene targeting) in embryonic stem cells to make planned changes in the mouse germ line have allowed production of desired "knockout" (KO) mice. One can study the effects of the absence of genes and their products. Knockout models for the three isoforms of NOS (NOS1 [142], NOS2 [143–145], and NOS3 [146]) have been reported. Mice of each of these KOs develop normally, reproduce, and appear grossly normal. The NOS1 KO mice have gastromegaly due to absence of NO-releasing neurons in the pylorus of the stomach, and they are overly aggressive. NOS3 KO mice have hypertension (146). NOS2 KO mice do not elevate levels of NO after immune stimulation or after LPS injection, and their macrophages cannot make NOS2 protein or NO (143,145). They have enhanced susceptibility to disseminated infection with *Leishmania* and *Listeria monocytogenes*, and anesthetized mice may have increased lethality after challenge with high doses of LPS (especially those mice that have previously been injected with *Corynebacterium parvum*). One group has shown that awake, untreated NOS2 KO mice are not resistant to LPS-induced death (144), suggesting that other factors are operative in this complex model of shock. NOS2 KO mice have diminished paw swelling after injection of carrageenin, and their lymphocytes show increased production of Th2 cytokines after stimulation in vitro (145). NOS2 KO mice have reduced resistance to infection with *Leishmania major* and *Mycobacteria tuberculosis* (145, 147).

XI. NOS2 KNOCKOUT MOUSE STUDIES IN AUTOIMMUNE MICE

We have crossed the NOS2 knockout mice with MRL mice, and these are to four back-crosses (N4). Since the embryonic stem cells are the 129 strain, the targeted mice are not genetically homogeneous, but N4 mice have diluted out most of 129 strain of the embryonic stem cells, and express essentially the MRL background. MRL-*lpr/lpr* littermates homozygous for disrupted NOS2 [–/–], heterozygous for disrupted NOS2 [+/–], or no disruption of NOS2 [+/+] were derived for this study (148).

The +/+ mice excreted large amounts of nitrite/nitrate, confirming our prior observations (130). The +/– mice excreted intermediate levels, while –/– mice excreted very low levels of nitrite/nitrate (comparable to those of normal BALB/c mice). Nitrite/nitrate levels in sera from 20-week-old animals paralleled the urinary measures, with very low levels in the –/– mice, high levels in the +/+ mice, and intermediate levels in the +/– mice. Levels of urinary and serum nitrite/nitrate in +/+ or +/– mice were significantly higher than those in the –/– mice (148).

To assess in vitro NO production by cells from mice of the three groups, peritoneal macrophages were cultured without additives and with IFN-γ (50 units/mL) and LPS (10 ng/mL). Nitrite/nitrate levels were significantly lower in the tissue culture supernatants of macrophages from the –/– mice than +/+ mice or the +/– mice, both at baseline and following stimulation. Similarly, NOS2 enzyme activity, as measured by the conversion of L-arginine to L-citrulline, was significantly less in the cells from –/– mice than those from +/– or +/+ mice (28 ± 8 vs. 493 ± 93 vs. 990 ± 32 pmol citrulline/mg for LPS/IFN-γ-treated cells, respectively). NOS activities were reduced by more than 90% by inclusion of 2 mM NMMA in the reaction mixtures. These studies confirm lack of detectable NOS activity in –/– mice. Differences in peritoneal macrophage NOS activity between –/– and the other two groups were significant at $p = 0.0002$ (148).

Immunoblots were performed on protein extracts from spleens, kidneys, liver, and peritoneal macrophages from the mice using an anti-NOS2 antibody. The macrophage cell line J774 (0 Rx or treated with LPS/IFN-γ [L/I]) served as control. As noted earlier, tissue from BALB/c (B) mice had no NOS2. There was minimal or no detectable NOS2 in the splenic, kidney, and macrophage protein extracts from –/– mice; extracts from +/+and +/– mice contained NOS2 protein, with those of the +/– mice having approximately half the amount of the +/+ mice (148).

Based on the effect of in vivo administration of NMMA on renal disease and arthritis in MRL-*lpr*/*lpr* mice (130), we predicted that –/– mice would develop less renal disease and arthritis than mice of the other two groups. Twenty-four-hour urinary protein excretion at 20 weeks of age was less in the –/– mice than in the other two groups, suggesting a possible difference in renal disease. Pathologic examination, however, indicated that glomerulonephritis in the –/– mice was similar in severity to that of the other two groups. Proliferative glomerulonephritis was present in all mice regardless of NOS2 genotype, with overall glomerular scores similar among the groups. Crescentic glomerulonephritis and interstitial disease was present in a small number of mice in each group. Synovitis was present in most of the mice, with overall synovial scores similar in the three groups. Synovial hypertrophy, synovial inflammation, and erosive disease were present to a similar degree in the MRL-*lpr*/*lpr* mice regardless of NOS2 genotype (148).

Infiltrates of lymphocytes and perivascular collection of lymphocytes occur in the kidneys of all *lpr* mice regardless of genetic background (128,149); true medium to large vessel vasculitis is found, however, only in MRL-*lpr*/*lpr* mice (149–151). Mononuclear cell infiltrates were present in the kidneys of all mice in this study. However, in contrast to glomerulonephritis and arthritis, there was a significant difference in the presence and severity of medium to large vessel vasculitis depending on NOS2 genotype. Indeed, four of six (4/6) +/+ mice had prominent vasculitis of medium to large vessels in the kidney, while only 1/9 of the +/– mice and 0/9 of –/– mice had medium to large vessel vasculitis. The incidence of vasculitis in the +/+ mice was similar to that in 20-week-old female MRL-*lpr*/*lpr* mice (80%). The difference in the occurrence of vasculitis, between the +/+ mice and the –/– mice was significant at $P = .011$. Pathologic examination of the brain, liver, lymph nodes, spleen, and lung revealed similar, mild lymphocytic infiltration in all three groups. Although the combined spleen and lymph node weights were less in the –/– mice than in the +/+ mice, the differences were not significant. Hematocrits were slightly higher in the –/– mice than in the other two groups, but these differences were not significant (148).

MRL-*lpr*/*lpr* mice are notable for autoantibody production (128,149). We therefore determined if MRL-*lpr*/*lpr* mice of various NOS2 genotypes displayed qualitative or quantitative differences in autoantibody production. Serum levels of antibodies to ss or dsDNA did not differ in the various mice. These results are comparable to the lack of effect of NMMA treatment on anti-DNA production we noted before in MRL-*lpr*/*lpr* mice (130). However, there was a shift in the isotype of anti-DNA antibodies produced. The IgG1/IgG3 ratio of anti-DNA antibodies was higher in the –/– mice (0.39 vs. 1.31). In contrast to anti-DNA production, rheumatoid factor production differed among the MRL-*lpr/lpr* mice of various NOS2 genotypes. The –/– mice produced significantly less IgG RF than did +/+ mice. IgM RF levels were also lower in the –/– mice, but levels were not significantly different from those in +/+ mice. IgG_3 RF activity (known to be associated with small-vessel, but not medium-vessel vasculitis in MRL-*lpr/lpr* mice [152]), was similar in the three groups of mice. Total serum IgG and IgM were similar in the three groups as were serum levels of anti-Sm and anti-La antibodies (results not shown). The different antibody levels in NOS2-disrupted mice may highlight the fact that small amounts of NO can modulate B cell function by modifying bcl-2 levels and apoptosis (153).

The mechanism(s) for the contrasting effects on renal and synovial diseases of a genetically disrupted NOS2, as opposed to pharmacological inhibition of NOS activity with N^G-monomethyl-L-arginine are not clear. It is unlikely that there were compensatory increases in NOS1 or NOS3 in the –/– mice, since we noted very low total body NO production in the –/– mice. N^G-monomethyl-L-arginine is an isoform-nonspecific NOS inhibitor, blocking all three isoforms of the NOS enzymes (154). Inhibiting all NOS isoforms (and hence potentially all NO production) with N^G-monomethyl-L-arginine may be more effective in disease prevention than genetically disrupting only NOS2. Alternative inflammatory pathways may not be active when NO production is acutely blocked by N^G-monomethyl-L-arginine; however, these pathways might become active over time, when NOS2 is genetically disrupted and absent the entire life of the animal.

XII. CONCLUSIONS

Studies using N^G-monomethyl-L-arginine in MRL-*lpr/lpr* mice indicate that NO is important in the pathogenesis of glomerulonephritis, arthritis, and vasculitis (130,155). However, studies of inflammation in MRL-*lpr/lpr* mice with genetically disrupted NOS2 highlight the heterogeneity and complexity of the role NOS2 and NO. Further studies in animal models of inflammation and in humans with inflammatory diseases such as RA and SLE should help us determine whether NOS2 and NO are suitable targets for new treatments in these diseases.

ACKNOWLEDGMENTS

Studies discussed above were funded in part by the Veterans Affairs Research Service, the James Swiger Hematology Research Fund, and NIH grant AR-39162.

REFERENCES

1. Cush JJ, Lipsky PE. Cellular basis for rheumatoid inflammation. Clin Ortho Rel Res 1991; 265:9–22.
2. Harris ED Jr. Rheumatoid arthritis. Pathophysiology and implications for therapy. New Eng J Med 1990; 322:1277–1289.
3. Weyand CM, Goronzy JJ. Pathogenesis of rheumatoid arthritis. Adv Rheumat 1997; 81:29–55.
4. Brennan FM, Feldmann M. Cytokines in autoimmunity. Curr Opin Immunol 1992; 4:754–759.
5. Alvaro-Gracia JM, Zvaifler NJ, Brown CB, Kaushansky K, Firestein GS. Cytokines in chronic inflammatory arthritis. VI. Analysis of the synovial cells involved in granulocyte-macrophage colony-stimulating factor production and gene expression in rheumatoid arthritis and its regulation by IL-1 and tumor necrosis factor-alpha. J Immunol 1991; 146:3365–3371.
6. Talal N. Interleukins, interferon and rheumatic disease. Clin Rheum Dis 1985; 11:633–644.
7. Hooks JJ, Jordan GW, Cupps T, Moutsopoulos HM, Fauci AS, Notkins AL. Multiple interferons in the circulation of patients with systemic lupus erythematosus and vasculitis. Arthritis Rheum 1982; 25:396–400.
8. Moncada S, Higgs A. The L-arginine-nitric oxide pathway. N Engl J Med 1993; 329:2002–2012.
9. Nathan C. Nitric oxide as a secretory product of mammalian cells. FASEB J 1992; 6:3051–3064.
10. Beckman JS, Crow JP. Pathological implications of nitric oxide, superoxide and peroxynitrite formation. (Review.) Biochem Soc Trans 1993; 21:330–334.
11. Magrinat G, Mason SN, Shami PJ, Weinberg JB. Nitric oxide modulation of human leukemia cell differentiation and gene expression. Blood 1992; 80:1880–1884.
12. Punjabi CJ, Laskin DL, Heck DE, Laskin JD. Production of nitric oxide by murine bone marrow cells. Inverse correlation with cellular proliferation. J Immunol 1992; 149:2179.
13. Stamler JS, Singel DJ, Loscalzo J. Biochemistry of nitric oxide and its redox-activated forms. (Review.) Science 1992; 258:1898–902.
14. Nathan C, Xie Q-W. Regulation of biosynthesis of nitric oxide. J Biol Chem 1994; 269:13725–13728.
15. Palmer RM, Hickery MS, Charles IG, Moncada S, Bayliss MT. Induction of nitric oxide synthase in human chondrocytes. Biochem Biophys Res Commun 1993; 193:398–405.
16. Charles IG, Palmer RMJ, Hickery MS, Bayliss MT, Chubb AP, Hall VS, Moss DW, Moncada S. Cloning, characterization, and expression of a cDNA encoding an inducible nitric oxide synthase from the human chondrocyte. Proc Natl Acad Sci USA 1993; 90:11419–11423.
17. Stefanovic-Racic M, Stadler J, Georgescu HI, Evans CH. Nitric Oxide Synthesis and Its Regulation By Rabbit Synoviocytes. J Rheumatol 1994; 21:1892–1898.
18. Fields NB, Knipe DM, Howley PM. Virology. Third ed. Philadelphia: Lippincott-Raven, 1996.

19. Darnell JE Jr, Kerr IM, Stark GR. Jak-STAT pathways and transcriptional activation in response to IFNs and other extracellular signaling proteins. Science 1994; 264:1415–1421.
20. Gross SS, Levi R. Tetrahydrobiopterin synthesis. An absolute requirement for cytokine-induced nitric oxide generation by vascular smooth muscle. J Biol Chem 1992; 267:25722–25729.
21. Rosenkranz-Weiss P, Sessa WC, Milstien S, Kaufman S, Watson CA, Pober JA. Regulation of nitric oxide synthesis by proinflammatory cytokines in human umbilical vein endothelial cells: elevations in tetrahydrobiopterin levels enhance endothelial nitric oxide synthase specific activity. J Clin Invest 1994; 93:2236–2243.
22. Assreuy J, Cunha FQ, Liew FY, Moncada S. Feedback inhibition of nitric oxide synthase activity by nitric oxide. Br J Pharmacol 1993; 108:833–837.
23. Rogers NE, Ignarro LJ. Constitutive nitric oxide synthase from cerebellum is reversibly inhibited by nitric oxide formed from L-arginine. Biochem Biophys Res Commun 1992; 189:242.
24. Denis M. Human monocytes/macrophages: NO or no NO? J Leukoc Biol 1994; 55:682–684.
25. Curran RD, Billiar TR, Stuehr DJ, Ochoa JB, Harbrecht BG, Flint SG, Simmons RL. Multiple cytokines are required to induce hepatocyte nitric oxide production and inhibit total protein synthesis. Ann Surg 1990; 212:462–469.
26. Sherman PA, Laubach VE, Reap BR, Wood ER. Purification and cloning of a cytokine-induced nitric oxide synthase from a human tumor cell line. Biochemistry 1993; 32:11600–11605.
27. Geller DA, Lowenstein CJ, Shapiro RA, Nussler AK, Di Silvio M, Wang SC, Nakayama DK, Simmons RL, Snyder SH, Billiar TR. Molecular cloning and expression of inducible nitric oxide synthase from human hepatocytes. Proc Natl Acad Sci USA 1993; 90:3491–3495.
28. Denis M. Tumor necrosis factor and granulocyte macrophage-colony stimulating factor stimulate human macrophages to restrict growth of virulent *Mycobacterium avium* and to kill avirulent *M. avium*: killing effector mechanism depends on the generation of reactive nitrogen intermediates. J Leukoc Biol 1991; 49:380–387.
29. Paul-Eugene N, Mossalayi D, Sarfati M, Yamaoka K, Aubry JP, Bonnefoy JY, Dugas B, Kolb JP. Evidence for a role of Fc-Epsilon-RII/CD23 in the IL-4-Induced nitric oxide production by normal human mononuclear phagocytes. Cell Immunol 1995; 163:314–318.
30. De Maria R, Cifone MG, Trotta R, Rippo MR, Festuccia C, Santoni A, Testi R. Triggering of human monocyte activation through CD69, a member of the natural killer cell gene complex family of signal transducing receptors. J Exp Med 1994; 180:1999–2004.
31. Weinberg JB, Misukonis MA, Shami PJ, Mason SN, Sauls DL, Dittman WA, Wood ER, Smith GK, McDonald B, Bachus KE, Haney AF, Granger DL. Human mononuclear phagocyte inducible nitric oxide synthase (iNOS). Analysis of iNOS mRNA, iNOS protein, biopterin, and nitric oxide production by blood monocytes and peritoneal macrophages. Blood 1995; 86:1184–1195.
32. Kobzik L, Bredt DS, Lowenstein CJ, Drazen J, Gaston B, Sugarbaker D, Stamler

JS. Nitric oxide synthase in human and rat lung: immunocytochemical and histochemical localization. Am J Respir Cell Mol Biol 1993; 9:371–377.
33. Tracey WR, Xue C, Klinghofer V, Barlow J, Pollock JS, Forstermann U, Johns RA. Immunochemical detection of inducible NO synthase in human lung. Am J Physiol 1994; 266:L722–L727.
34. Nicholson S, Bonecinialmeida MDG, Silva LE Jr, Nathan C, Xie QW, Mumford R, Weidner JR, Calaycay J, Geng J, Boechat N, Linhares C, Rom W, Ho JL. Inducible nitric oxide synthase in pulmonary alveolar macrophages from patients with tuberculosis. J Exp Med 1996; 183:2293–2302.
35. Hibbs JB, Jr., Westenfelder C, Taintor R, Vavrin Z, Kablitz C, Baranowski RL, Ward JH, Menlove RL, McMurry MP, Kushner JP, Samlowski WE. Evidence for cytokine-inducible nitric oxide synthesis from L-arginine in patients receiving interleukin-2 therapy [published erratum appears in J Clin Invest 1992; 90(1):295]. J Clin Invest 1992; 89:867–877.
36. Anstey NM, Weinberg JB, Hassanali M, Mwaikambo ED, Manyenga D, Misukonis MA, Arnelle DR, Hollis D, McDonald MI, Granger DL. Nitric oxide in Tanzanian children with malaria. Inverse relationship between malaria severity and nitric oxide production/nitric oxide synthase type 2 expression. J Exp Med 1996; 184:557–567.
37. St. Clair EW, Wilkinson WE, Lang T, Sanders L, Misukonis MA, Gilkeson GS, Pisetsky DS, Granger DL, Weinberg JB. Increased expression of blood mononuclear cell nitric oxide synthase type 2 in rheumatoid arthritis patients. J Exp Med 1996; 184:1173–1178.
38. Sharara AI, Perkins DJ, Misukonis MA, Chan S, Weinberg JB. Interferon-alpha activation of human mononuclear cells in vitro and in vivo for nitric oxide synthase type 2 mRNA and protein expression—possible relationship of induced NOS2 to the anti-hepatitis C effects of IFN-α in vivo. J Exp Med 1997; 186:1495–1502.
39. Croen KD. Evidence for antiviral effect of nitric oxide. Inhibition of herpes simplex virus type 1 replication. J Clin Invest 1993; 91:2446–2452.
40. Bi Z, Reiss CS. Inhibition of vesicular stomatitis virus infection by nitric oxide. J Virol 1995; 69:2208–2213.
41. Karupiah G, Xie QW, Buller RM, Nathan C, Duarte C, MacMicking JD. Inhibition of viral replication by interferon-gamma-induced nitric oxide synthase. Science 1993; 261:1445–148.
42. Karupiah G, Harris N. Inhibition of viral replication by nitric oxide and its reversal by ferrous sulfate and tricarboxylic acid cycle metabolites. J Exp Med 1995; 181:2171–2179.
43. Mannick JB, Asano K, Izumi K, Kieff E, Stamler JS. Nitric oxide produced by human B lymphocytes inhibits apoptosis and Epstein-Barr virus reactivation. Cell 1994; 79:1137–1146.
44. Mannick JB. The antiviral role of nitric oxide. Res Immunol 1995; 146:693–697.
45. Lowenstein CJ, Hill SL, Lafond-Walker A, Wu J, Allen G, Landavere M, Rose NR, Herskowitz A. Nitric oxide inhibits viral replication in murine myocarditis. J Clin Invest 1996; 97:1837–1843.

46. Lin Y-L, Huang Y-L, Ma S-H, Yeh C-T, Chiou S-Y, Chen L-K, Liao C-L. Inhibition of Japanese encephalitis virus infection by nitric oxide: antiviral effect of nitric oxide on RNA virus replication. J Virol 1997; 71:5227.
47. Vial T, Descotes J. Immune-mediated side-effects of cytokines in humans. Toxicology 1995; 105:31–57.
48. Wandl UB, Nagel-Hiemke M, May D, Kreuzfelder E, Kloke O, Kranzhoff M, Seeber S, Niederle N. Lupus-like autoimmune disease induced by interferon therapy for myeloproliferative disorders. Clin Immunol Immunopathol 1992; 65:70–74.
49. Ronnblom LE, Alm GV, Oberg K. Autoimmune phenomena in patients with malignant carcinoid tumors during interferon-alpha treatment. Acta Oncol 1991; 30:537–540.
50. Nadir F, Fagiuoli S, Wright HI, Nadir A, Hopp E, Gavaler J, Van Thiel DH. Rheumatoid arthritis: a complication of interferon therapy. J Okla State Med Assoc 1994; 87:228–230.
51. Elliott MJ, Maini RN, Feldmann M, Long-Fox A, Charles P, Katsikis P, Brennan FM, Walker J, Bijl H, Ghrayeb J, et al. Treatment of rheumatoid arthritis with chimeric monoclonal antibodies to tumor necrosis factor alpha. Arthritis Rheum 1993; 36:1681–1690.
52. Elliott MJ, Maini RN, Feldmann M, Kalden JR, Antoni C, Smolen JS, Leeb B, Breedveld FC, Macfarlane JD, Bijl H, et al. Randomised double-blind comparison of chimeric monoclonal antibody to tumour necrosis factor alpha (cA2) versus placebo in rheumatoid arthritis. Lancet 1994; 344:1105–1110.
53. Moreland LW, Baumgartner SW, Schiff MH, Tindall EA, Fleischmann RM, Weaver AL, Ettlinger RE, Cohen S, Koopman WJ, Mohler K, Widmer MB, Blosch CM. Treatment of rheumatoid arthritis with a recombinant human tumor necrosis factor receptor (P75)-Fc Fusion Protein. N Engl J Med 1997; 337:141–147.
54. Perkins DJ, St. Clair WE, Misukonis MA, Weinberg JB. Reduction of NOS2 overexpression in rheumatoid arthritis patients treated with anti-TNF-α monoclonal antibody (cA2). Arth Rheum, in press, 1998.
55. Clancy RM, Abramson SB. Nitric oxide—a novel mediator of inflammation [Review]. Proc Soc Exp Biol Med 1995; 210:93–101.
56. Stadler J, Stefanovic-Racic M, Billiar TR, Curran RD, McIntyre LA, Georgescu HI, Simmons RL, Evans CH. Articular chondrocytes synthesize nitric oxide in response to cytokines and lipopolysaccharide. J Immunol 1991; 147:3915–3920.
57. Evans CH, Stefanovic-Racic M, Lancaster JR Jr. Nitric oxide and its role in orthopaedic disease [Review]. Clin Ortho Rel Res 1995; 312:275–294.
58. Clancy RM, Leszczynska-Piziak J, Abramson SB. Nitric oxide, an endothelial cell relaxation factor, inhibits neutrophil superoxide anion production via a direct action on the NADPH oxidase. J Clin Invest 1992; 90:1116–1121.
59. Darius H, Grodzinska L, Meyer J. The effects of the nitric oxide donors molsidomine and SIN-1 on human polymorphonuclear leucocyte function in vitro and ex vivo. Eur J Clin Pharmacol 1992; 43:629–633.
60. Kubes P, Suzuki M, Granger DN. Nitric oxide: an endogenous modulator of leukocyte adhesion. Proc Natl Acad Sci USA 1991; 88:4651–4655.

61. Stamler JS, Loscalzo J. The antiplatelet effects of organic nitrates and related nitroso compounds in vitro and in vivo and their relevance to cardiovascular disorders [see comments]. J Am Coll Cardiol 1991; 18:1529–1536.
62. Maciejewski JP, Selleri C, Sato T, Cho HJ, Keefer LK, Nathan CF, Young NS. Nitric oxide suppression of human hematopoiesis in vitro—contribution to inhibitory action of interferon-gamma and tumor necrosis factor-alpha. J Clin Invest 1995; 96:1085–1092.
63. Shami PJ, Weinberg JB. Differential effects of nitric oxide on erythroid and myeloid colony growth from CD34(+) human bone marrow cells. Blood 1996; 87:977–982.
64. Weiss G, Goossen B, Doppler W, Fuchs D, Pantopoulos K, Werner-Felmayer G, Wachter H, Hentze MW. Translational regulation via iron-responsive elements by the nitric oxide/NO-synthase pathway. EMBO J 1993; 12:3651–3657.
65. Drapier JC, Hirling H, Wietzerbin J, Kaldy P, Kuhn LC. Biosynthesis of nitric oxide activates iron regulatory factor in macrophages. EMBO J 1993; 12:3643–3659.
66. Harris ED Jr. Textbook of rheumatology. 4th ed. Philadelphia: W.B. Saunders, 1993.
67. Gaillard T, Mulsch A, Klein H, Decker K. Regulation by prostaglandin E2 of cytokine-elicited nitric oxide synthesis in rat liver macrophages. Biol Chem Hoppe Seyler 1992; 373:897–902.
68. Marotta P, Sautebin L, Di Rosa M. Modulation of the induction of nitric oxide synthase by eicosanoids in the murine macrophage cell line J774. Br J Pharmacol 1992; 107:640–641.
69. Imai Y, Kolb H, Burkart V. Nitric oxide production from macrophages is regulated by arachidonic acid metabolites. Biochem Biophys Res Commun 1993; 197:105–109.
70. Stadler J, Harbrecht BG, Di Silvio M, Curran RD, Jordan ML, Simmons RL, Billiar TR. Endogenous nitric oxide inhibits the synthesis of cyclooxygenase products and interleukin-6 by rat Kupffer cells. J Leukoc Biol 1993; 53:165–172.
71. Salvemini D, Misko TP, Masferrer JL, Seibert K, Currie MG, Needleman P. Nitric oxide activates cyclooxygenase enzymes. Proc Natl Acad Sci USA 1993; 90:7240–7244.
72. Swierkosz TA, Mitchell JA, Warner TD, Botting RM, Vane JR. Co-induction of nitric oxide synthase and cyclo-oxygenase: interactions between nitric oxide and prostanoids. Br J Pharmacol 1995; 114:1335–1342.
73. Vane JR, Mitchell JA, Appleton I, Tomlinson A, Bishop-Bailey D, Croxtall J, Willoughby DA. Inducible isoforms of cyclooxygenase and nitric-oxide synthase in inflammation. Proc Natl Acad Sci USA 1994; 91:2046–2050.
74. Salvemini D, Manning PT, Zweifel BS, Seibert K, Connor J, Currie MG, Needleman P, Masferrer JL. Dual inhibition of nitric oxide and prostaglandin production contributes to the antiinflammatory properties of nitric oxide synthase inhibitors. J Clin Invest 1995; 96:301–308.
75. Amin AR, Vyas P, Attur M, Leszczynskapiziak J, Patel IR, Weissmann G, Abramson SB. The mode of action of aspirin-like drugs—effect on inducible nitric oxide synthase. Proc Natl Acad Sci USA 1995; 92:7926–7930.

76. Stefanovic-Racic M, Stadler J, Evans CH. Nitric oxide and arthritis. Arthritis Rheum 1993; 36:1036–1044.
77. Tamir S, Derojaswalker T, Gal A, Weller AH, Li XT, Fox JG, Wogan GN, Tannenbaum SR. Nitric oxide production in relation to spontaneous B-cell lymphoma and myositis in SJL mice. Cancer Res 1995; 55:4391–4397.
78. Vodovotz Y, Geiser AG, Chesler L, Letterio J, Campbell A, Lucia MS, Sporn MB, Roberts AB. Spontaneously increased production of nitric oxide and aberrant expression of the inducible nitric oxide synthase in vivo in the transforming growth factor beta-1 null mouse. J Exp Med 1996; 183:2337–2342.
79. Vodovotz Y, Bogdan C, Paik J, Xie QW, Nathan C. Mechanisms of suppression of macrophage nitric oxide release by transforming growth factor beta. J Exp Med 1993; 178:605–613.
80. Farrell AJ, Blake DR, Palmer RM, Moncada S. Increased concentrations of nitrite in synovial fluid and serum samples suggest increased nitric oxide synthesis in rheumatic diseases. Ann Rheum Dis 1992; 51:1219–1222.
81. Stichtenoth DO, Fauler J, Zeidler H, Frolich JC. Urinary nitrate excretion is increased in patients with rheumatoid arthritis and reduced by prednisolone. Ann Rheum Dis 1995; 54:820–824.
82. Grabowski PS, England AJ, Dykhuizen R, Copland M, Benjamin N, Reid DM, Ralston SH. Elevated nitric oxide production in rheumatoid arthritis—detection using the fasting urinary nitrate/creatinine ratio. Arthritis Rheum 1996; 39:643–647.
83. Kaur H, Halliwell B. Evidence for nitric oxide-mediated oxidative damage in chronic inflammation—nitrotyrosine in serum and synovial fluid from rheumatoid patients. FEBS Lett 1994; 350:9–12.
84. Sakurai H, Kohsaka H, Liu MF, Higashiyama H, Hirata Y, Kanno K, Saito I, Miyasaka N. Nitric oxide production and inducible nitric oxide synthase expression in inflammatory arthritides. J Clin Invest 1995; 96:2357–2363.
85. McInnes LB, Leung BP, Field M, Wei XQ, Huang FP, Sturrock RD, Kinninmonth A, Weidner J, Mumford R, Liew FY. Production of nitric oxide in the synovial membrane of rheumatoid and osteoarthritis patients. J Exp Med 1996; 184:1519–1524.
86. Ueki Y, Miyake S, Tominaga Y, Eguchi K. Increased nitric oxide levels in patients with rheumatoid arthritis. J Rheumatol 1996; 23:230–236.
87. Brennan FM, Maini RN, Feldmann M. TNF alpha—a pivotal role in rheumatoid arthritis? Br J Rheumatol 1992; 31:293–298.
88. Hopkins SJ, Meager A. Cytokines in synovial fluid. II. The presence of tumour necrosis factor and interferon. Clin Exp Immunol 1988; 73:88–92.
89. Vilcek J, Sen GC. Interferons and other cytokines. In: Fields BN, Knipe DM, Howley PM, eds. Virology. Vol. 1 3rd ed. Philadelphia; Lippincott-Raven, 1996: 375–399.
90. Zhou AQ, Chen ZF, Rummage JA, Jiang H, Kolosov M, Kolosova I, Stewart CA, Leu RW. Exogenous interferon-gamma induces endogenous synthesis of interferon-alpha and -beta by murine macrophages for induction of nitric oxide synthase. J Interferon Cytokine Res 1995; 15:897–904.
91. Zhang X, Alley EW, Russell SW, Morrison DC. Necessity and sufficiency of beta

interferon for nitric oxide production in mouse peritoneal macrophages. Infect Immun 1994; 62:33–40.

92. Aman MJ, Tretter T, Eisenbeis I, Bug G, Decker T, Aulitzky WE, Tilg H, Huber C, Peschel C. Interferon-alpha stimulates production of interleukin-10 in activated CD4+ T cells and monocytes. Blood 1996; 87:4731–4736.
93. Kolosov M, Kolosova I, Zhou A, Leu RW. Autocrine induction of macrophage synthesis of complement subcomponent C1q by endogenous interferon-alpha/beta. J Interferon Cytokine Res 1996; 16:209–215.
94. Landmann R, Wesp M, Obrecht JP. Cytokine regulation of the myeloid glycoprotein CD14. Pathobiology 1991; 59:131–135.
95. Meltzer MS, Baca L, Turpin JA, Kalter DC, Dieffenbach C, Friedman RM, Gendelman HE. Regulation of cytokine and viral gene expression in monocytes infected with the human immunodeficiency virus. Pathobiology 1991; 59:209–213.
96. Schnyder-Candrian S, Strieter RM, Kunkel SL, Walz A. Interferon-alpha and interferon-gamma down-regulate the production of interleukin-8 and ENA-78 in human monocytes. J Leukoc Biol 1995; 57:929–935.
97. Baca LM, Genis P, Kalvakolanu D, Sen G, Meltzer MS, Zhou A, Silverman R, Gendelman HE. Regulation of interferon-alpha-inducible cellular genes in human immunodeficiency virus-infected monocytes. J Leukoc Biol 1994; 55:299–309.
98. Moore KW, O'Garra A, de Waal Malefyt R, Vieira P, Mosmann TR. Interleukin-10. Annu Rev Immunol 1993; 11:165–190.
99. Kuhn R, Lohler J, Rennick D, Rajewsky K, Muller W. Interleukin-10-deficient mice develop chronic enterocolitis [see comments]. Cell 1993; 75:263–274.
100. Pennline KJ, Roque-Gaffney E, Monahan M. Recombinant human IL-10 prevents the onset of diabetes in the nonobese diabetic mouse. Clin Immunolol Immunopath 1994; 71:169–175.
101. Rott O, Fleischer B, Cash E. Interleukin-10 prevents experimental allergic encephalomyelitis in rats. Eur J Immunol 1994; 24:1434–1440.
102. Wu J, Cunha FQ, Liew FY, Weiser WY. IL-10 inhibits the synthesis of migration inhibitory factor and migration inhibitory factor-mediated macrophage activation. J Immunol 1993; 151:4325–4332.
103. Oswald IP, Wynn TA, Sher A, James SL. Interleukin 10 inhibits macrophage microbicidal activity by blocking the endogenous production of tumor necrosis factor alpha required as a costimulatory factor for interferon gamma-induced activation. Proc Natl Acad Sci USA 1992; 89:8676–8680.
104. Cunha FQ, Moncada S, Liew FY. Interleukin-10 (IL-10) inhibits the induction of nitric oxide synthase by interferon-gamma in murine macrophages. Biochem Biophys Res Commun 1992; 182:1155–1159.
105. Cenci E, Romani L, Mencacci A, Spaccapelo R, Schiaffella E, Puccetti P, Bistoni F. Interleukin-4 and interleukin-10 inhibit nitric oxide-dependent macrophage killing of *Candida albicans*. Eur J Immunol 1993; 23:1034–1038.
106. Katsikis PD, Chu CQ, Brennan FM, Maini RN, Feldmann M. Immunoregulatory role of interleukin 10 in rheumatoid arthritis. J Exp Med 1994; 179:1517–1527.
107. Cush JJ, Splawski JB, Thomas R, McFarlin JE, Schulze-Koops H, Davis LS, Fujita K, Lipsky PE. Elevated interleukin-10 levels in patients with rheumatoid arthritis. Arthritis Rheum 1995; 38:96–104.

108. van Roon JA, van Roy JL, Gmelig-Meyling FH, Lafeber FP, Bijlsma JW. Prevention and reversal of cartilage degradation in rheumatoid arthritis by interleukin-10 and interleukin-4. Arthritis Rheum 1996; 39:829–835.
109. Suzuki H, Sugiyama E, Tunru IS, Yamashita N, Matsuno H, Hamazaki T, Kobayashi M. Suppressive effect of interleukin-4 (IL-4) on IL-6 production by adherent rheumatoid synovial cells. Clin Immunol Immunopathol 1993; 66:67–72.
110. Briolay J, Dechanet J, Blanchard D, Banchereau J, Miossec P. Interleukin 4 inhibits polyclonal immunoglobulin secretion and cytokine production by peripheral blood mononuclear cells from rheumatoid arthritis patients. J Clin Immunol 1992; 12:36–44.
111. Seitz M, Loetscher P, Dewald B, Towbin H, Ceska M, Baggiolini M. Production of interleukin-1 receptor antagonist, inflammatory chemotactic proteins, and prostaglandin E by rheumatoid and osteoarthritic synoviocytes—regulation by IFN-gamma and IL-4. J Immunol 1994; 152:2060–2065.
112. Sugiyama E, Kuroda A, Taki H, Ikemoto M, Hori T, Yamashita N, Maruyama M, Kobayashi M. Interleukin 10 cooperates with interleukin 4 to suppress inflammatory cytokine production by freshly prepared adherent rheumatoid synovial cells. J Rheumatol 1995; 22:2020–2026.
113. Chomarat P, Vannier E, Dechanet J, Rissoan MC, Banchereau J, Dinarello CA, Miossec P. Balance of IL-1 receptor antagonist/IL-1 beta in rheumatoid synovium and its regulation by IL-4 and IL-10. J Immunol 1995; 154:1432–1439.
114. Paul-Eugene N, Mossalayi D, Sarfati M, Yamaoka K, Aubry JP, Bonnefoy JY, Dugas B, Kolb JP. Evidence for a role of Fc epsilon RII/CD23 in the IL-4-induced nitric oxide production by normal human mononuclear phagocytes. Cell Immunol 1995; 163:314–318.
115. Keffer J, Probert L, Cazlaris H, Georgopoulos S, Kaslaris E, Kioussis D, Kollias G. Transgenic mice expressing human tumor necrosis factor: a predictive genetic model of arthritis. EMBO J 1991; 10:4025–4031.
116. Williams RO, Feldmann M, Maini RN. Anti-tumor necrosis factor ameliorates joint disease in murine collagen-induced arthritis. Proc Natl Acad Sci USA 1992; 89:9784–9788.
117. Maier R, Bilbe G, Rediske J, Lotz M. Inducible nitric oxide synthase from human articular chondrocytes: cDNA cloning and analysis of mRNA expression. Biochim Biophys Acta 1994; 1208:145–150.
118. Blanco FJ, Lotz M. IL-1-induced nitric oxide inhibits chondrocyte proliferation via PGE2. Exp Cell Res 1995; 218:319–325.
119. Kondo S, Ishiguro N, Iwata H, Nakashima I, Isobe K. The effects of nitric oxide on chondrocytes and lymphocytes. Biochem Biophys Res Commun 1993; 197:1431–1437.
120. Stefanovic-Racic M, Taskiran D, Georgescu HI, Evans CH. Modulation of chondrocyte proteoglycan synthesis by endogeneously produced nitric oxide. Inflamm Res 1995; 44:S216–S217.
121. Tamura T, Nakanishi T, Kimura Y, Hattori T, Sasaki K, Norimatsu H, Takahashi K, Takigawa M. Nitric oxide mediates interleukin-1-induced matrix degradation and basic fibroblast growth factor release in cultured rabbit articular

chondrocytes—a possible mechanism of pathological neovascularization in arthritis. Endocrinology 1996; 137:3729–3737.

122. Stefanovic-Racic M, Morales TI, Taskiran D, McIntyre LA, Evans CH. The role of nitric oxide in proteoglycan turnover by bovine articular cartilage organ cultures. J Immunol 1996; 156:1213–1220.
123. Blanco FJ, Ochs RL, Schwarz H, Lotz M. Chondrocyte apoptosis induced by nitric oxide. Am J Pathol 1995; 146:75–85.
124. Hauselmann HJ, Oppliger L, Michel BA, Stefanovic-Racic M, Evans CH. Nitric oxide and proteoglycan biosynthesis by human articular chondrocytes in alginate culture. FEBS Lett 1994; 352:361–364.
125. Murrell GA, Jang D, Williams RJ. Nitric oxide activates metalloprotease enzymes in articular cartilage. Biochem Biophys Res Commun 1995; 206:15–21.
126. Pelletier JP, Mineau F, Ranger P, Tardif G, Martelpelletier J. The increased synthesis of inducible nitric oxide inhibits Il-1ra synthesis by human articular chondrocytes—possible role in osteoarthritic cartilage degradation. Osteoarthritis Cartilage 1996; 4:77–84.
127. Amin AR, Di Cesare PE, Vyas P, Attur M, Tzeng E, Billiar TR, Stuchin SA, Abramson SB. The expression and regulation of nitric oxide synthase in human osteoarthritis-affected chondrocytes: evidence for up-regulated neuronal nitric oxide synthase. J Exp Med 1995; 182:2097–2102.
128. Cohen PL, Eisenberg RA. *Lpr* and *gld*: single gene models of systemic autoimmunity and lymphoproliferative disease. Ann Rev Immunol 1991; 9:243–269.
129. Nagata S, Suda T. Fas and Fas ligand: lpr and gld mutations. Immunol Today 1995; 16:39–43.
130. Weinberg JB, Granger DL, Pisetsky DS, Seldin MF, Misukonis MA, Mason SN, Pippen AM, Ruiz P, Wood ER, Gilkeson GS. The role of nitric oxide in the pathogenesis of spontaneous murine autoimmune disease: increased nitric oxide production and nitric oxide synthase expression in MRL-lpr/lpr mice, and reduction of spontaneous glomerulonephritis and arthritis by orally administered N^G-monomethyl-L-arginine. J Exp Med 1994; 179:651–660.
131. Dang-Vu AP, Pisetsky DS, Weinberg JB. Functional alterations of macrophages in autoimmune MRL-lpr/lpr mice. J Immunol 1987; 138:1757–1761.
132. Spurney RF, Fan PY, Ruiz P, Sanfilippo F, Pisetsky DS, Coffman TM. Thromboxane receptor blockade reduces renal injury in murine lupus nephritis. Kidney Int 1992; 41:973–982.
133. Fisher GH, Rosenberg FJ, Straus SE, Dale JK, Middleton LA, Lin AY, Strober W, Lenardo MJ, Puck JM. Dominant interfering Fas gene mutations impair apoptosis in a human autoimmune lymphoproliferative syndrome. Cell 1995; 81:935–946.
134. Huang FP, Feng GJ, Lindop G, Stott DI, Liew FY. The role of interleukin 12 and nitric oxide in the development of spontaneous autoimmune disease in MRL:MP-lpr:lpr mice. J Exp Med 1996; 183:1447–1459.
135. Henry Y, Lepoivre M, Drapier JC, Ducrocq C, Boucher JL, Guissani A. EPR characterization of molecular targets for NO in mammalian cells and organelles. FASEB J 1993; 7:1124–1134.

136. Weinberg JB, Gilkeson GS, Mason RP, Chamulitrat W. Nitrosylation of blood hemoglobin and renal non-heme proteins in autoimmune MRL-*lpr/lpr* mice. Free Radic Biol Med 1998; 24:191–196.
137. Beckman JS, Ye YZ, Anderson PG, Chen J, Accavitti MA, Tarpey MM, White CR. Extensive nitration of protein tyrosines in human atherosclerosis detected by immunohistochemistry. Biol Chem Hoppe Seyler 1994; 375:81–88.
138. Kooy NW, Royall JA, Ye YZ, Kelly DR, Beckman JS. Evidence for in vivo peroxynitrite production in human acute lung injury. Am J Respir Crit Care Med 1995; 151:1250–1254.
139. Haddad IY, Pataki G, Hu P, Galliani C, Beckman JS, Matalon S. Quantitation of nitrotyrosine levels in lung sections of patients and animals with acute lung injury. J Clin Invest 1994; 94:2407–2413.
140. Privalle CT, Keng T, Gilkeson GS, Weinberg JB. The role of nitric oxide and peroxynitrite in the pathogenesis of spontaneous murine autoimmune disease. In: Stamler J, Gross SS, Moncada S, eds. The Biology of Nitric Oxide. Vol. 5 London: Portland Press, 1996; 20.
141. Venkatraman JT, Chandrasekar B, Kim JD, Fernandes G. Genotype effects on the antioxidant enzymes activity and mRNA expression in liver and kidney tissues of autoimmune-prone MRL/MpJ-lpr/lpr mice. Biochim Biophys Acta 1994; 1213:167–175.
142. Huang PL, Dawson TM, Bredt DS, Snyder SH, Fishman MC. Targeted disruption of the neuronal nitric oxide synthase gene. Cell 1993; 75:1273–1286.
143. MacMicking JD, Nathan C, Hom G, Chartrain N, Fletcher DS, Trumbauer M, Stevens K, Xie QW, Sokol K, Hutchinson N, Chen H, Mudgett JS. Altered responses to bacterial infection and endotoxic shock in mice lacking inducible nitric oxide synthase. Cell 1995; 81:641–650.
144. Laubach VE, Shesely EG, Smithies O, Sherman PA. Mice lacking inducible nitric oxide synthase are not resistant to lipopolysaccharide-induced death. Proc Natl Acad Sci USA 1995; 92:10688–10692.
145. Wei XQ, Charles IG, Smith A, Ure J, Feng CJ, Huang FP, Xu DM, Muller W, Moncada S, Liew FY. Altered immune responses in mice lacking inducible nitric oxide synthase. Nature 1995; 375:408–411.
146. Huang PL, Huang ZH, Mashimo H, Bloch KD, Moskowitz MA, Bevan JA, Fishman MC. Hypertension in mice lacking the gene for endothelial nitric oxide synthase. Nature 1995; 377:239–242.
147. Macmicking JD, North RJ, Lacourse R, Mudgett JS, Shah SK, Nathan CF. Identification of nitric oxide synthase as a protective locus against tuberculosis. Proc Natl Acad Sci USA 1997; 94:5243–5248.
148. Gilkeson GS, Mudgett J, Seldim MF, Ruiz P, Alexander AA, Pisetsky DS, Misukonis MA, Weinberg JB. Clinical and serological manifestations of autoimmune disease in MRL-lpr/lpr mice lacking nitric oxide synthase type 2. J Exp Med 1997; 365–374.
149. Hang L, Theofilopoulos AN, Dixon FJ. A spontaneous rheumatoid arthritis-like disease in MRL-l mice. J Exp Med 1982; 155:1690–1701.
150. Nose M, Nishimura M, Kyogoku M. Analysis of granulomatous arteritis in MRL/

Mp autoimmune disease mice bearing lymphoproliferative genes. The use of mouse genetics to dissociate the development of arteritis and glomerulonephritis. Am J Pathol 1989; 135:271–280.

151. Tarkowski A, Jonsson R, Sanchez R, Klareskog L, Koopman WJ. Features of renal vasculitis in autoimmune MRL lpr/lpr mice: phenotypes and functional properties of infiltrating cells. Clin Exp Immunol 1988; 72:91–7.
152. Takahashi S, Nose M, Sasaki J, Yamamoto T, Kyogoku M. IgG3 production in MRL/lpr mice is responsible for development of lupus nephritis. J Immunol 1991; 147:515–519.
153. Genaro AM, Hortelano S, Alvarez A, Martineza C, Bosca L. Splenic B lymphocyte programmed cell death is prevented by nitric oxide release through mechanisms involving sustained bcl-2 levels. J Clin Invest 1995; 95:1884–1890.
154. Southan GJ, Szabo C. Selective pharmacological inhibition of distinct nitric oxide synthase isoforms. Biochem Pharmacol 1996; 51:383–394.
155. Oates JC, Ruiz P, Alexander A, Pippen AMM, Gilkeson GS. Effect of late modulation of nitric oxide production on murine lupus. Clin Immunol Immunopathol 1997; 83:86–92.

9

Protective and Pathologic Roles of Nitric Oxide in Tissue Injury

Carol R. Gardner and Debra L. Laskin
Environmental and Occupational Health Sciences Institute, Rutgers University, and University of Medicine and Dentistry of New Jersey–Robert Wood Johnson Medical School, Piscataway, New Jersey

I. INTRODUCTION

Tissue injury induces an inflammatory response characterized by an accumulation of macrophages and neutrophils at sites of damage. These cells release pro-inflammatory cytokines and cytotoxic mediators to destroy invading pathogens and initiate tissue repair. However, recent studies have demonstrated that these mediators, which include pro-inflammatory cytokines, bioactive lipids, hydrolytic enzymes, and reactive oxygen and nitrogen intermediates, may also contribute to tissue damage (1–4). This process involves both direct damage to target tissue and amplification of the inflammatory response. Among the macrophage-derived mediators implicated in toxicity, nitric oxide has recently received considerable attention. It is reportedly produced in excess quantities during tissue injury associated with inflammation (5–7). Moreover, in various experimental models, modulating nitric oxide production modifies tissue injury. Whether nitric oxide plays a protective or pathological role in tissue injury appears to vary with the toxicant, the target tissue, and the extent to which toxicity is mediated by reactive oxygen intermediates. In this chapter, we review evidence implicating nitric oxide in liver and lung injury.

II. NITRIC OXIDE AND LIVER INJURY

The liver is a major target organ for a number of different toxicants. All blood-derived pathogens and orally ingested xenobiotics pass through the liver via the portal circulation. Moreover, metabolic activation of xenobiotics to toxic intermediates occurs in the liver. The response of the liver to toxicants involves both necrotic and apoptotic reactions. Excessive production of nitric oxide has been implicated in both of these processes (5–7).

Hepatocytes constitute the majority of the cells of the liver (>70%). They play a major role in the detoxification of xenobiotics and are involved in bilirubin and ammonia metabolism and urea, glutamine, and cholesterol biosynthesis (8). Hepatocytes are also the primary source of acute-phase proteins (8,9). Nonparenchymal cells consist predominantly of cells lining the hepatic sinusoids. These include Kupffer cells, endothelial cells and stellate cells. Kupffer cells comprise 80% to 90% of all the macrophages in the body and represent 29% of the sinusoidal cells in the liver (10). They are localized in the lumen of the hepatic sinusoids in periportal and central regions of the liver lobule. The major function of Kupffer cells is to clear particulate and foreign materials from the portal circulation, primarily through phagocytosis (11). Like other mononuclear phagocytes, Kupffer cells respond to chemotactic and inflammatory signals and produce a myriad of mediators that aid in antigen destruction and participate in various immunoregulatory and inflammatory activities (4,12–15).

Approximately 48% of the hepatic sinusoidal cells are endothelial cells. These cells are unique in that they possess pores or fenestrae that function as a selective barrier between the blood and hepatocytes (16). Hepatic endothelial cells have a well-developed endocytic capacity and function in the turnover and catabolism of glycoproteins, as well as lactoferrin, lipoproteins, albumin, and hyaluronic acid (17–19). Hepatic endothelial cells also release a variety of mediators that regulate the function of sinusoidal cells and hepatocytes as well as infiltrating leukocytes (4,12–14). Thus, these cells, like Kupffer cells, appear to be important in inflammatory activities in the liver.

Hepatic stellate cells, also called fat-storing cells, perisinusoidal cells, Ito cells, and lipocytes, are located between the endothelial cells and hepatocytes and represent approximately 20% of the sinusoidal cells in the liver. Stellate cells are the major storage site in the body for vitamin A (20,21). They also synthesize extracellular matrix proteins, and evidence indicates that they play a role in collagen biosynthesis in both normal and fibrotic liver (22,23). Recent studies have demonstrated that stellate cells can release proteins and inflammatory mediators that may contribute to liver injury (4,12,13,24,25).

Nitric oxide and its oxidation products, nitrate and nitrite, have been reported to be produced by both parenchymal and nonparenchymal liver cells (24–

42) and are thought to be involved in altered hepatic functioning following xenobiotic exposure and in liver injury (27,40,41,43). Liver cells produce nitric oxide in response to cytokines and bacterial derived products via the NADPH and L-arginine-dependent inducible form of nitric oxide synthase (NOS2) (2–6,27,44–47). Although mRNA for this enzyme is not detectable in hepatocytes or sinusoidal cells from the livers of untreated rats, small quantities of NOS2 protein have been identified both in situ in histological sections and in vitro in isolated cells (26–28,44). This is most likely due to continuous exposure of liver cells to endotoxin in the portal circulation. In the absence of exogenous stimulation hepatocytes produce only low levels of nitric oxide (27–29,45). Following treatment with lipopolysaccharide (LPS), interferon-γ (IFN-γ), or the combination of LPS and IFN-γ, hepatocytes release large quantities of this reactive nitrogen intermediate, a response that is blocked by nitric oxide synthase inhibitors such as N^G-monomethyl-L-arginine (L-NMMA), N^ω-nitro-L-arginine methyl ester (L-NAME) and aminoguanidine (26–29,45,46). Endotoxin-primed hepatocytes also respond to IFN-γ in combination with cytokines such as tumor necrosis factor-α (TNF-α), interleukin-1 (IL-1), interleukin-6, hepatocyte growth factor, granulocyte-macrophage colony-stimulating factor (GM-CSF), and macrophage colony-stimulating factor (M-CSF) (27,45). The NOS2 gene has been cloned from both human and rodent hepatocytes (47–49). NOS2 isolated from rat hepatocytes has been reported to have an amino acid sequence that is 94% identical to NOS2 from RAW264.7 mouse macrophages (49). Human hepatocytes express a 4.5-kb NOS2 mRNA that is induced by TNF-α, IL-1, IFN-γ, and LPS and has an amino acid sequence that is 80% identical to mouse macrophage NOS2 (47). In contrast to the NOS2 found in macrophages, human hepatocyte NOS2 requires calcium for activity (47).

Kupffer cells, endothelial cells, and stellate cells also express NOS2 protein and mRNA and produce nitric oxide in response to inflammatory mediators such as LPS and IFN-γ (24–26,30,31,42,50–52). This response is significantly increased in the presence of TNF-α, GM-CSF, M-CSF, and IL-1β (30). As observed in hepatocytes, nitric oxide production by these sinusoidal cells is dependent on L-arginine and inhibitable by L-NMMA, N^ω-nitro-L-arginine (L-NNA), aminoguanidine, and L-canavanine (26,53; C. Gardner, unpublished observations). In the presence of phorbol myristate acetate, which induces a respiratory burst, Kupffer cells can also produce peroxynitrite, an even more potent oxidant, that has been also implicated in tissue injury (4,31).

A. Endotoxin-Induced Hepatotoxicity

Endotoxin is derived from the cell wall of gram-negative bacteria. It is present in large quantities in the intestine as a result of bacterial death and is released during bacterial growth (54). Whereas small amounts of endotoxin are rapidly cleared from the portal circulation by Kupffer cells (11,19), exposure to large amounts or impairment of Kupffer cell functioning results in hepatic injury (54,55). A characteristic histologic feature observed after treatment of experimental animals with endotoxin is an increase in macrophages in the liver (55,56). Our laboratory has demonstrated that hepatic endothelial cells and stellate cells also increase in number during endotoxemia (25,56). Liver macrophages, as well as endothelial cells and stellate cells isolated from animals treated with endotoxin, display varying degrees of functional activation including altered morphology, increased proliferation, and increased generation of pro-inflammatory cytokines, as well as reactive oxygen and nitrogen intermediates (24–26,30,55-58). Hepatic macrophages from endotoxin-treated rats also exhibit enhanced phagocytosis, chemotaxis, and cytotoxicity (55,59). Thus, these cells appear to be highly responsive to endotoxin.

The role of nitric oxide in endotoxin-induced hepatotoxicity is controversial. Whereas some studies have suggested a pathologic role, others propose that nitric oxide is hepatoprotective. Following endotoxin administration to rodents, both parenchymal and nonparenchymal liver cells are sensitized to express increased amounts of NOS2 mRNA and protein and to produce increased amounts of nitric oxide in response to inflammatory mediators (25–27). Similar increases in NOS2 expression and nitric oxide production in isolated hepatocytes and Kupffer cells have been described in rats treated with the combination of endotoxin and *Corynebacterium parvum* (60). NOS2 mRNA and protein are observed in situ in histologic sections within 3 h after treatment of the animals with endotoxin in hepatocytes, as well as sinusoidal cells located throughout the liver lobule (27,52). This is correlated with nitric oxide production in the liver as measured by electron paramagnetic spin trapping (27,61) and with increases in serum nitrite and nitrate levels (35,62–64). To analyze the role of nitric oxide in hepatotoxicity, various nitric oxide synthase inhibitors have been used. In general these studies suggest that nitric oxide generated via NOS2 is cytotoxic, while activation of the constitutive forms of nitric oxide synthase, in particular NOS3, and subsequent nitric oxide production is hepatoprotective. Thus, whereas treatment of rats with L-NMMA or L-NAME, which block all isoforms of nitric oxide synthase, reportedly exacerbates tissue injury caused by endotoxin, inhibition of NOS2 with aminoguanidine, L-canavanine, or aminoethyl-isothiourea protects or reduces hepatic damage (35,52,62–65).

Several studies have demonstrated that pretreatment of animals with the macrophage activators *Bacillus* Calemette-Guérin or *C. parvum* exacerbates liver damage induced by endotoxin (33,35,65), a finding consistent with the suggestion that macrophages can contribute to tissue injury (1,4,55). Interestingly, in these models, treatment of animals with nonspecific nitric oxide inhibitors such as L-NAME or L-NMMA exacerbates liver injury (33,35,65). These data indicate that nitric oxide may also be hepatoprotective in these models. However, the source of the nitric oxide and the enzyme mediating nitric oxide production by liver cells following these treatments is unknown.

The timing of administration of the nitric oxide synthase inhibitor also seems to contribute to the hepatic response to endotoxin. For example, while rats pretreated with L-NAME are protected from tissue injury induced by endotoxin alone (27), simultaneous administration of the L-NMMA, L-NNA, or L-NAME and endotoxin augments liver injury (35,62,66,67). Similar increases in hepatic injury are observed when nonspecific nitric oxide synthase inhibitors are administered after endotoxin (68). In contrast, treatment of rats with L-canavanine or aminoguanidine following endotoxin administration results in a decrease in liver damage, as well as an improvement in survival (64,68). The results of these studies indicate that nitric oxide may be protective or pathologic depending on the kinetics of its generation and on the enzyme mediating its production in the liver.

The role of nitric oxide in endotoxin-induced tissue injury and in sepsis has also been investigated using knockout mice lacking the gene for NOS2. Although there are no detectable increases in serum nitrite and nitrate in these animals after endotoxin treatment, lactate dehydrogenase (LDH) and serum transaminase levels remain elevated (69). Nevertheless, NOS2 deficiency does confer partial resistance to lethality induced by higher doses of endotoxin (69). It may be that nitric oxide generated via constitutive nitric oxide synthases acts in concert with other inflammatory mediators to induce toxicity. These mediators may be produced in increased quantities in NOS2 knockout mice following endotoxin administration, leading to tissue injury.

B. Acetaminophen-Induced Hepatotoxicity

A cytotoxic role for nitric oxide in hepatic necrosis induced by acetaminophen has also recently been described (28,29). Acetaminophen is a mild analgesic and antipyretic agent that causes centrilobular necrosis in cases of overdose (70,71). This is thought to be due to the formation of a highly reactive intermediate, N-acetyl-*p*-benzoquinone-imine (NAPQI), via cytochrome P4502E1, which is predominantly localized in these regions of the liver (72). NAPQI binds to critical cellular macromolecules, leading to cytotoxicity. Studies from our laboratory have suggested that soluble products produced by hepatocytes and/or infiltrating macrophages and resident Kupffer cells, such as nitric oxide, may also contribute to the pathogenesis of acetaminophen-induced injury (28,29,71,73–75). We have found that treatment of rats with hepatotoxic doses of acetaminophen is associated with an accumulation of macrophages in centrilobular regions of the liver (74). We hypothesize that these cells become activated and release mediators that

immunologic line of defense in the lung (1–3,111). Also found within the lung are interstitial macrophages which reside in the lung parenchyma and are in direct contact with the connective tissue. While alveolar macrophages are primarily involved in removal and destruction of inhaled antigens, interstitial macrophages appear to protect the lung from bloodborne pathogens (112). Chemotaxis and phagocytosis are inflammatory functions important in the clearance of micro-organisms and endotoxin, as well as damaged cells and debris from the lungs. Both alveolar and interstitial macrophages respond chemotactically toward a variety of inflammatory mediators, including complement cleavage fragments, connective tissue breakdown products, and bacterially-derived products (112). Interstitial macrophages are more responsive to chemoattractants than alveolar macrophages (112). In contrast, alveolar macrophages have a greater capacity to phagocytize particles and release reactive oxygen intermediates.

Both alveolar and interstitial macrophages also produce nitric oxide in response to cytokines and LPS via NOS2 (1,2,113–118). This response is blocked by L-NMMA, L-NAME, and aminoguanidine (113,115,118–120). Alveolar macrophages produce significantly more nitric oxide than interstitial macrophages (114,115). Alveolar macrophages also respond to combinations of IFN-γ and IL-1β, TNF-α, or GM-CSF with nitric oxide generation (115,120). In contrast, interstitial macrophages do not produce substantial amounts of nitrite/nitrate in the presence of these cytokines (115). Interestingly, in the absence of exogenous stimulation interstitial, but not alveolar macrophages express low levels of NOS2 and produce nitric oxide (115). This may be due to activation of these cells in vivo by bacterially-derived pathogens present in the pulmonary circulation. Both alveolar and interstitial macrophages are highly sensitive to exogenously administered endotoxin. However, their response varies dramatically with the route of exposure. Thus, whereas only interstitial macrophages exhibit increased chemotaxis, phagocytosis, and oxidative metabolism after intravenous administration of endotoxin, both interstitial and alveolar macrophages are activated by endotoxin after intratracheal instillation or inhalation (112,121–124). These responses are consistent with the specialized functioning of these cells.

The epithelial lining of the alveolus contains two specialized cell types (type I and type II pneumocytes) that are crucial to proper lung functioning. Type I cells are large, flat cells that cover most of the gas exchange surface of the alveoli (125). They are particularly susceptible to destruction induced by oxidants such as ozone (126). In contrast, the smaller, cuboidal type II cells are relatively resistant to injury, presumably due to high intracellular levels of antioxidants (127,128). The major function of type II cells is to synthesize and release pulmonary surfactant (129). Recent studies have demonstrated that type II cells also produce interleukin-8, macrophage chemotactic protein-1, eicosanoids, interferon, complement proteins, hydrogen peroxide, and nitric oxide (130–137), suggesting that these cells may participate in local regulation of inflammatory and immunologic responses in the lung. Production of nitric oxide by type II cells appears to be mediated by an inducible form of nitric oxide synthase. Thus, these cells are highly sensitive to cytokines like IFN-γ and TNF-α, expressing NOS2 mRNA and protein and generating nitric oxide (131). Interestingly, in contrast to lung macrophages, type II cells do not produce large amounts of nitric oxide in response to LPS (131). Our laboratory has reported that nitric oxide production by type II cells, as well as alveolar macrophages, is blocked by pyrrolidinedithiocarbamate (PDTC), an inhibitor of the nuclear transcription factor, NF-κB (138). This is correlated with decreased NOS2 protein expression in the cells (139). The promoter region of the NOS2 gene is known to contain binding sites for several different transcription factors, including NF-κB, that regulate the activity of the gene (140). Our findings that PDTC blocks nitric oxide production and NOS2 expression in lung macrophages and epithelial cells implies that this is an important regulatory pathway for nitric oxide generation in lung cells.

In the lung, nitric oxide plays a critical role in regulating airway responsiveness and bronchiole relaxation (141,142). Nitric oxide is also involved in macrophage-mediated cytotoxicity and in the regulation of cellular proliferation (30,117,143–149). Thus, production of nitric oxide in the lung may be significant not only in tissue damage, but also in the regulation of alveolar epithelial regeneration following injury. Recent studies have suggested that nitric oxide released by lung macrophages, as well as epithelial cells, may contribute to tissue injury (2,113,115,150–152). Nitric oxide has been implicated in carrageenin-induced increases in vascular permeability and edema, as well as lung damage induced by leukotoxin, immune complexes, paraquat, oleic acid, and endotoxin (153–160). Our laboratory has demonstrated that acute exposure of rats to ozone results in increased production of nitric oxide by alveolar macrophages, as well as interstitial macrophages and type II cells (113,131,148–151,161). This is associated with increased expression of NOS2 protein and mRNA, which is observed in vitro in isolated cells, in situ in histologic sections, and in vivo as tissue nitric oxide measured by electroparamagnectic resonance spin trapping (113,151). We speculate that nitric oxide contributes to ozone-induced inflammation and tissue injury. In support of this possibility, we found that pretreatment of rats with the macrophage inhibitor gadolinium chloride suppresses ozone-induced nitric oxide production and alveolar epithelial damage (148). Moreover, treatment of rats with endotoxin, which is known to render rats resistant to oxidants (162,163), prior to ozone, also abrogates NOS2 expression, nitric oxide production, and tissue injury (151). Exposure of animals to a number of other toxicants, including endotoxin, silica, smoke, immune complexes, and inhaled nitric oxide is also associated with NOS2 expression in the lung (115,164–171). Lung macrophages, as well as type II cells, isolated from rats treated with endotoxin express NOS2 and produce increased amounts of nitric oxide in response to inflammatory mediators (115,151). Alveolar and interstitial macrophages from endotoxin-treated rats also produce peroxynitrite (115). These findings, together with the identification of nitrotyrosine residues in the lungs of endotoxemic rats (115), suggest that peroxynitrite is involved in the pathogenic response to this toxin. In a number of different models of lung injury, blocking nitric oxide production protects against tissue damage. For example, treatment of animals with nitric oxide synthase inhibitors such as L-NAME, L-NMMA, L-NNA, aminoguanidine, or S-methylisothiourea sulfate has been reported to reduce or totally prevent pulmonary injury induced by paraquat, endotoxin, leukotoxin, immune complexes, and oleic acid (154–156,164,171–175). These results suggest a destructive role for nitric oxide in the pathogenesis of tissue injury caused by these agents.

The mechanism by which nitric oxide induces pathological changes in the lung are likely to be similar to those observed in the liver. Nitric oxide produced by LPS- or IFN-γ-stimulated alveolar macrophages has been reported to induce DNA fragmentation in type II cells, and this may contribute to cytotoxicity in vivo (116). Nitric oxide may also act directly on lung parenchymal cells inducing enzyme inactivation, lipid peroxidation and/or mitochondrial damage (2,5,6,96,108,109). Recent evidence has suggested that the toxicity of nitric oxide in the lung may be due, in large part, to the generation of peroxynitrite. Alveolar macrophages are among the most active producers of superoxide anion in the body (3,112). Therefore, there is a significant potential for peroxynitrite production in the lung. In this regard, recent studies have demonstrated that peroxynitrite is produced during acute lung injury in both humans and in animals (115,154,171,174–179). Peroxynitrite has been implicated in immune complex-induced pulmonary edema, acute lung injury, obliterative bronchiolitis, emphysema, and damage to pulmonary surfactants (115,156,180–186).

IV. CONCLUSION

Tissue injury induced by xenobiotics is a complex process that involves a variety of cell types and soluble mediators. Although xenobiotics or their metabolites can directly injure the tissue, they may also activate parenchymal and nonparenchymal cells and infiltrating leukocytes, and indirectly augment tissue injury. Nitric oxide and other reactive mediators produced by these cells may act as primary mediators of tissue injury, and/or they may participate in the inflammatory response by initiating a cascade of additional immunologic reactions that result in tissue damage. Further studies on nitric oxide-induced effects on target tissues will be important for understanding the mechanisms of xenobiotic-induced injury to the liver and lung.

REFERENCES

1. Laskin DL, Pendino KJ. Macrophages and inflammatory mediators in tissue injury. Annu Rev Pharmacol Toxicol 1995; 35:655–677.
2. Laskin DL, Laskin JD. Macrophages, inflammatory mediators and lung injury. Methods (Companion to Methods Enzymol) 1996; 10:61–70.
3. Laskin DL, Laskin JD. Phagocytes. In: Lawrence DA, ed. Toxicology of the Immune System, Comprehensive Toxicology. Vol. 5. New York: Pergamon, 1997; pp. 97–111.
4. Laskin DL, Gardner CR. The role of nonparenchymal cells and inflammatory macrophages in hepatotoxicity. In: Plaa GL, Hewitt WR, eds. Toxicology of the Liver. 2nd ed. Washington, DC: Taylor & Francis, 1998; pp. 297–320.
5. Lyons CR. The role of nitric oxide in inflammation. Adv Immunol 1995; 60:323–371.
6. Gross SS, Wolin MS. Nitric oxide: pathological mechanisms. Annu Rev Physiol 1995; 57:737–769.
7. Laskin JD, Heck DE, Laskin DL. Multifunctional role of nitric oxide in inflammation. Trends Endocrinol Metab 1998; 5:377–382.
8. Arias IM, Boyer JL, Fausto N, Jakoby WB, Schacter DA, Shafritz DA. eds. The Liver: Biology and Pathobiology. 3rd ed. New York: Raven Press, 1994.
9. Kushner I, Baumann H, Mackiewicz A. eds. Acute Phase Proteins: Molecular Biology, Biochemistry, Clinical Applications. Boca Raton, FL: CRC Press, 1993.
10. Bouwens L, Baekeland M, DeZanger R, Wisse E. Quantitation, tissue distribution and proliferation kinetics of Kupffer cells in normal rat liver. Hepatology 1986; 6:718–722.
11. Mathison JC, Ulevitch RJ. The clearance, tissue distribution, and cellular localization of intravenously injected lipopolysaccharide in rabbits. J Immunol 1979; 123:2133–2143.
12. Laskin DL. Xenobiotic-induced inflammation and injury in the liver. In: McCuskey RS, Earnest DL, eds. Hepatic and Gastrointestinal Toxicology, Comprehensive Toxicology. Vol. 9. New York: Pergamon Press, 1997; pp. 151–164.
13. Laskin DL. Sinusoidal lining cells and hepatotoxicity. Toxicol Pathol 1996; 24:112–118.
14. Laskin DL. Nonparenchymal cells and hepatotoxicity. Semin Liver Dis 1990; 10:293–304.
15. Laskin DL. Potential role of activated macrophages in chemical and drug induced liver injury. In: Wisse E, Knook DL, Decker K, eds. Cells of the Hepatic Sinusoid. Vol. 2. Rijswijk, Netherlands: Kupffer Cell Foundation, 1989; pp. 284–287.
16. Wisse E, Knook DL. The investigation of sinusoidal cells: a new approach to the study of liver function. Prog Liver Dis 1979; 6:153–171.
17. Hahn E, Wick G, Pencev D, Timpl R. Distribution of basement membrane proteins in normal and fibrotic liver: collagen type IV, laminin and fibronectin. Gut 1980; 21:63–71.
18. De Leeuw AM, Praaning-Van Dalen DP, Brouwer A, Knook DL. Endocytosis in liver sinusoidal endothelial cells. In: Wisse E, Knook DL, Decker K, eds. Cells of

the Hepatic Sinusoid. Vol. 2. Rijswijk, Netherlands: Kupffer Cell Foundation, 1989; pp. 94–98.
19. Praaning-Van Dalen DP, Brouwer A, Knook DL. Clearance capacity of rat liver Kupffer, endothelial, and parenchymal cells. Gastroenterology 1981; 81:1036–1044.
20. Blomhoff R, Wake K. Perisinusoidal cells of the liver: important roles in retinol metabolism and fibrosis. FASEB J 1991; 5:271–277.
21. Hendriks HF, Verhoofstad WA, Brouwer A, De Leeuw AM, Knook DL. Perisinusoidal fat-storing cells are the main vitamin A storage cells in rat liver. Exp Cell Res 1985; 160:138–149.
22. Friedman SL. Cellular sources of collagen and regulation of collagen production in liver. Semin Liver Dis 1990; 10:20–29.
23. Friedman SL, Roll FJ, Boyles JK, Bissell DM. Hepatic lipocytes: the principle collagen-producing cells of normal liver. Proc Natl Acad Sci USA 1985; 82:8681–8685.
24. Helyar L, Bundschuh DS, Laskin JD, Laskin DL. Hepatic fat storing cells produce nitric oxide and hydrogen peroxide in response to bacterially-derived lipopolysaccharide. In: Knook DL, Decker K, eds. Cells of the Hepatic Sinusoid. Vol. 4. Amsterdam: Kupffer Cell Foundation, 1993; pp. 394–396.
25. Helyar L, Bundschuh DS, Laskin JD, Laskin DL. Induction of hepatic Ito cell nitric oxide production by acute endotoxemia. Hepatology 1994; 12:1509–1515.
26. Laskin DL, Heck DE, Gardner CR, Feder LS, Laskin JD. Distinct patterns of nitric oxide production in hepatic macrophages and endothelial cells following acute exposure of rats to endotoxin. J Leukoc Biol 1994; 56:751–758.
27. Laskin DL, Rodriguez del Valle M, Heck DE, Hwang SM, Ohnishi ST, Durham SK, Laskin JD. Hepatic nitric oxide production following acute endotoxemia in rats is mediated by increased inducible nitric oxide synthase gene expression. Hepatology 1995; 22:223–234.
28. Gardner CR, Heck DE, Yang CS, Thomas PE, Zhang X-J, DeGeorge GL, Laskin JD, Laskin DL. Role of nitric oxide in acetaminophen-induced hepatotoxicity in the rat. Hepatology 1998; 26:748–754.
29. Gardner CR, Durham SK, Barton D, Heck DE, Laskin JD, Laskin DL. Role of nitric oxide in hepatotoxicity. In: Wisse E, Knook DL, Balabaud C, eds. Cells of the Hepatic Sinusoid. Vol. 6. Leiden, Netherlands: Kupffer Cell Foundation, 1997; pp. 268–271.
30. Feder LS, Laskin DL. Regulation of hepatic endothelial cell and macrophage proliferation and nitric oxide production by GM-CSF, M-CSF, and IL-1β following acute endotoxemia. J Leukoc Biol 1994; 55:507–513.
31. Decker K. Biologically active products of stimulated liver macrophages (Kupffer cells). Eur J Biochem 1990; 192:245–261.
32. Spolarics Z, Spitzer JJ, Wang JF, Xie J, Kolls J, Greenberg S. Alcohol administration attenuates LPS-induced expression of inducible nitric oxide synthase in Kupffer and hepatic endothelial cells. Biochem Biophys Res Commun 1993; 197:606–611.
33. Wang GS, Liu GT. Role of nitric oxide in immunological liver damage in mice. Biochem Pharmacol 1995; 49:1277–1281.

34. Diez-Fernandez C, Sanz N, Bosca L, Hortelano S, Cascales M. Involvement of nitric oxide synthesis in hepatic perturbations induced in rats by a necrogenic dose of thioacetamide. Br J Pharmacol 1997; 121:820–826.
35. Billiar TR, Curran RD, Harbrecht BG, Stuehr DJ, Demetris AJ, Simmons RL. Modulation of nitrogen oxide synthesis in vivo: N^G-monomethyl-L-arginine inhibits endotoxin-induced nitrite/nitrate biosynthesis while promoting hepatic damage. J Leukoc Biol 1990; 48:565–569.
36. Billiar TR, Curran RD, Stuehr DJ, West MA, Bentz BG. An L-arginine-dependent mechanism mediates Kupffer cell inhibition of hepatocyte protein synthesis in vitro. J Exp Med 1989; 169:1467–1472.
37. Curran RD, Billiar TR, Stuehr DJ, Hofmann K, Simmons RL. Hepatocytes produce nitrogen oxides from L-arginine in response to inflammatory products from Kupffer cells. J Exp Med 1989; 170:1769–1774.
38. Miller DR, Collier JM, Billings RE. Protein tyrosine kinase activity regulates nitric oxide synthase induction in rat hepatocytes. Am J Physiol 1997; 272:G207–G214.
39. Duval DL, Miller DR, Collier JM, Billings RE. Characterization of hepatic nitric oxide synthase: identification as the cytokine-inducible form primarily regulated by oxidants. Mol Pharmacol 1996; 50:277–284.
40. Milbourne EA, Bygrave FL. Does nitric oxide play a role in liver function? Cell Signal 1995; 7:313–318.
41. Losser M-R, Payen D. Mechanisms of liver damage. Semin Liver Dis 1996; 16:357–367.
42. Rockey DC, Chung JJ. Regulation of inducible nitric oxide synthase and nitric oxide during hepatic injury and fibrogenesis. Am J Physiol 1997; 273:G124–G130.
43. Nussler AK, Beger HG, Liu ZZ, Billiar TR. Nitric oxide, hepatocytes and inflammation. Res Immunol 1995; 146:671–677.
44. Duval DL, Miller DR, Collier J, Billings RE. Characterization of hepatic nitric oxide synthase: identification as the cytokine-inducible form primarily regulated by oxidants. Mol Pharmacol 1996; 50:277–284.
45. Nussler AK, Di Silvio M, Liu ZZ, Geller DA, Freeswick P, Dorko K, Bartoli F, Billiar TR. Further characterization and comparison of inducible nitric oxide synthase in mouse, rat, and human hepatocytes. Hepatology 1995; 21:1552–1560.
46. Monshouwer M, Witkamp RF, Nijmeijer SM, Van Amsterdam JG, Van Miert ASJPAM. Suppression of cytochrome P450- and UDP glucuronosyl transferase-dependent enzyme activities by proinflammatory cytokines and possible role of nitric oxide in primary cultures of pig hepatocytes. Toxicol Appl Pharmacol 1996; 137:237–244.
47. Geller DA, Lowenstein CJ, Shapiro RA, Nussler AK, Di Silvio M, Wang SC, Nakayama DK, Simmons RL, Snyder SH, Billiar TR. Molecular cloning and expression of inducible nitric oxide synthase from human hepatocytes. Proc Natl Acad Sci USA 1993; 90:3491–3495.
48. Chartrain NA, Geller DA, Koty PP, Sitrin NF, Nussler AK, Hoffman EP, Billiar TR, Hutchinson NI, Mudgett JS. Molecular cloning, structure, and chromosomal localization of the human inducible nitric oxide synthase gene. J Biol Chem 1994; 269:6765–6772.

49. Wood ER, Berger H Jr, Sherman PA, Lapetina EG. Hepatocytes and macrophages express identical cytokine inducible nitric oxide synthase gene. Biochem Biophys Res Commun 1993: 191:767–774.
50. Spitzer JA. Cytokine stimulation of nitric oxide formation and differential regulation in hepatocytes and nonparenchymal cells of endotoxemic rats. Hepatology 1994; 19:217–228.
51. Gardner CR, Heck DE, Feder LS, McCloskey TW, Laskin JD, Laskin DL. Differential regulation of reactive nitrogen and reactive oxygen intermediate production by hepatic macrophages and endothelial cells. In: Jesaitis A, Dratz EA, eds. The Molecular Basis of Oxidative Damage by Leukocytes. Boca Raton, FL: CRC Press, 1992; pp. 267–272.
52. Aono K, Isobe KI, Kiuch K, Fan ZH, Ito M, Takeuchi A, Miyachi M, Nakashima I, Nimura Y. In vitro and in vivo expression of inducible nitric oxide synthase during experimental endotoxemia: involvement of other cytokines. J Cell Biochem 1997; 65:349–358.
53. Gaillard T, Mülsch A, Busse R, Klein H, Decker K. Regulation of nitric oxide production by stimulated rat Kupffer cells. Pathobiology 1991; 59:280–283.
54. Decker KF. Bacterial endotoxin. In: McCuskey RS, Earnest DL, eds. Hepatic and Gastrointestinal Toxicology, Comprehensive Toxicology. Vol. 9. New York: Pergamon Press, 1997; pp. 433–446.
55. Pilaro AM, Laskin DL. Accumulation of activated mononuclear phagocytes in the liver following lipopolysaccharide treatment of rats. J Leukoc Biol 1986; 40:29–41.
56. McCloskey TW, Todaro JA, Laskin DL. Lipopolysaccharide treatment of rats alters antigen expression and oxidative metabolism in hepatic macrophages and endothelial cells. Hepatology 1992; 16:191–203.
57. Feder LS, Todaro JA, Laskin DL. Characterization of interleukin-1 and interleukin-6 production by hepatic endothelial cells and macrophages. J Leukoc Biol 1993; 53:126–132.
58. Spolarics Z. Endotoxin stimulates gene expression of ROS-eliminating pathways in rat hepatic endothelial and Kupffer cells. Am J Physiol 1996; 270:G660–G666.
59. Gardner CR, Wasserman AJ, Laskin DL. Differential sensitivity of tumor targets to liver macrophage mediated cytotoxicity. Cancer Res 1987; 47:6686–6691.
60. Billiar TR, Curran RD, Stuehr DJ, Stadler J, Simmons RL, Murray SA. Inducible cytosolic enzyme activity for the production of nitrogen oxides from L-arginine in hepatocytes. Biochem Biophys Res Commun 1990; 168:1034–1040.
61. Kubrina LN, Caldwell WS, Mordvintcev PI, Malenkova IV, Vanin AF. EPR evidence for nitric oxide production from guanidino nitrogens of L-arginine in animal tissues in vivo. Biochim Biophys Acta 1992; 1099:233–237.
62. Ou J, Carlos TM, Watkins SC, Saavedra JE, Keefer LK, Kim Y-M, Harbrecht BG, Billiar TR. Differential effects of nonselective nitric oxide synthase (NOS) and selective NOS inhibition on hepatic necrosis, apoptosis, ICAM-1 expression, and neutrophil accumulation during endotoxemia. Nitric Oxide 1997; 1:404–416.
63. Thiemermann C, Ruetten H, Wu C-C, Vane JR. The multiple organ dysfunction syndrome caused by endotoxin in the rat: attenuation of liver dysfunction by inhibitors of nitric oxide synthase. Br J Pharmacol 1995; 116:2845–2851.

64. Wu CC, Ruetten H, Thiemermann C. Comparison of the effects of aminoguanidine and N^{ω}-nitro-L-arginine methyl ester on the multiple organ dysfunction caused by endotoxemia in the rat. Eur J Pharmacol 1996; 300:99–104.
65. Harbrecht BG, Stadler J, Demetris AJ, Simmons RL, Billiar TR. Nitric oxide and prostaglandins interact to prevent hepatic damage during murine endotoxemia. Am J Physiol 1994; 266:G1004–G1010.
66. Vos TA, Gouw ASH, Klok PA, Havinga R, Van Goor H, Huitema S, Roelofsen H, Kuipers F, Jansen PLM, Moshage H. Differential effects of nitric oxide synthase inhibitors on endotoxin-induced liver damage in rats. Gastroenterology 1997; 113:1323–1333.
67. Huang T-P, Nishida T, Kamiike W, Kosaka H, Seiyama A, Morimoto Y, Tanaka S, Obunai S, Takei Y, Shiga T, Matsuda H. Role of nitric oxide in oxygen transport in rat liver sinusoids during endotoxemia. Hepatology 1997; 26:336–342.
68. Liaudet L, Rosselet A, Schaller M-D, Markert M, Perret C, Feihl F. Nonselective versus selective inhibition of inducible nitric oxide synthase in experimental endotoxic shock. J Infect Dis 1998; 177:127–132.
69. MacMicking JD, Nathan C, Hom G, Chartrain N, Fletcher DS, Trumbauer M, Stevens K, Xie Q-W, Sokol K, Hutchinson N, Chen H, Mudgett JS. Altered responses to bacterial infection and endotoxic shock in mice lacking inducible nitric oxide synthase. Cell 1995; 81:641–650.
70. Price VF, Jollow DJ. Strain differences in susceptibility of normal and diabetic rats to acetaminophen hepatotoxicity. Toxicol Appl Pharmacol 1986; 101:687–695.
71. Laskin DL, Gardner CR, Price VF, Jollow DJ. Modulation of macrophage functioning abrogates the acute hepatotoxicity of acetaminophen. Hepatology 1995; 21:1045–1050.
72. Lee SST, Buters JTM, Pineau T, Fernandez-Salguero P, Gonzalez FJ. Role of CYP2E1 in the hepatotoxicity of acetaminophen. J Biol Chem 1996; 271:12063–12067.
73. Laskin DL, Pilaro AM. Potential role of macrophages in acetaminophen hepatotoxicity. I. Isolation and characterization of activated macrophages from rat liver. Toxicol Appl Pharmacol 1986; 86:204–215.
74. Laskin DL, Pilaro AM, Ji S. Potential role of macrophages in acetaminophen hepatotoxicity. II. Mechanism of macrophage accumulation and activation. Toxicol Appl Pharmacol 1986; 86:216–226.
75. Blazka ME, Wilmer JL, Holladay SD, Wilson RE, Luster MI. Role of proinflammatory cytokines in acetaminophen hepatotoxicity. Toxicol Appl Pharmacol 1995; 133:43–52.
76. Ohira H, Ueno T, Shakado S, Sakamoto M, Torimura T, Inuzuka S, Sata M, Tanikawa K. Cultured rat hepatic sinusoidal endothelial cells express intercellular adhesion molecule-1 (ICAM-1) by tumor necrosis factor-a or interleukin-1a stimulation. J Hepatol 1994; 20:729–734.
77. Schlayer HJ, Karck U, Ganter U, Hermann R, Decker K. Enhancement of neutrophil adherence to isolated rat sinusoidal endothelial cells by supernatants of lipopolysaccharide-activated monocytes. Role of tumor necrosis factor. J Hepatol 1987; 5:311–321.
78. Morita M, Watanabe Y, Akaike T. Inflammatory cytokines up-regulate intercel-

lular adhesion molecule-1 expression on primary cultured mouse hepatocytes and T-lymphocyte adhesion. Hepatology 1994; 19:426–431.
79. Fey GH, Hocke GM, Wilson DR, Ripperger JA, Juan TS, Cui M, Darlington GJ. Cytokines and the acute phase response of the liver. In: Arias IM, Boyer JL, Fausto N, Jakoby WB, Schachter DA, Shafritz DA, eds. The Liver: Biology and Pathobiology. 3rd ed. New York: Raven Press, 1994; pp. 113–143.
80. Andus T, Bauer J, Gerok W. Effects of cytokines on the liver. Hepatology 1991; 13:364–375.
81. Ghezzi P, Saccardo B, Bianchi M. Recombinant tumor necrosis factor depresses cytochrome P450-dependent microsomal drug metabolism in mice. Biochem Biophys Res Commun 1986; 136:316–321.
82. Peterson TC, Renton KW. Kupffer cell factor mediated depression of hepatic parenchymal cell cytochrome P-450. Biochem Pharmacol 1986; 35:1491–1497.
83. Ding AH, Nathan CF, Stuehr DJ. Release of reactive nitrogen intermediates and reactive oxygen intermediates from mouse peritoneal macrophages. Comparison of activating cytokines and evidence for independent production. J Immunol 1988; 141:2407–2412.
84. Leist M, Gantner F, Jilg S, Wendel A. Activation of the 55 kDa TNF receptor is necessary and sufficient for TNF-induced liver failure, hepatocyte apoptosis, and nitrite release. J Immunol 1995; 154:1307–1316.
85. Wang JH, Redmond HP, Watson RW, Bouchier-Hayes D. Role of lipopolysaccharide and tumor necrosis factor-α in induction of hepatocyte necrosis. Am J Physiol 1995; 269:G297–G304.
86. Leist M, Gantner F, Bohlinger I, Germann PG, Tiegs G, Wendel A. Murine hepatocyte apoptosis induced in vitro and in vivo by TNF-α requires transcriptional arrest. J Immunol 1994; 153:1778–1788.
87. Cerami A. Inflammatory cytokines. Clin Immunol Immunopathol 1992; 62:S3–S10.
88. Tracey KJ, Beutler B, Lowry SF, Merryweather J, Wolpe S, Milsark IW, Hariri RJ, Fahey TJ 3rd, Zentella A, Albert JD. Shock and tissue injury induced by recombinant human cachectin. Science 1986; 234:470–474.
89. Gardner CR, Barton D, Goller N, Durham S, Laskin DL. Role of tumor necrosis factor-α (TNF-α) and nitric oxide (NO) in acetaminophen hepatotoxicity (abstr). Toxicologist 1995; 15:136.
90. Gardner CR, Morio LA, Laskin DL. Inhibition of acetaminophen hepatotoxicity by gadolinium chloride ($GdCl_3$) is associated with decreased tumor necrosis factor-α (TNF-α) and interleukin-1 (IL-1) production (abstr). Toxicologist 1994; 14:201.
91. Blazka ME, Elwell MR, Holladay SD, Wilson RE, Luster MI. Histopathology of acetaminophen-induced liver changes: role of interleukin 1α and tumor necrosis factor α. Toxicol Pathol 1996; 24:181–189.
92. Gardner CR, Heck DE, Chiu H, Laskin JD, Durham SK, Laskin DL. Decreased hepatotoxicity of acetaminophen in mice lacking inducible nitric oxide synthase (abstr). Toxicol Sci 1998; 42:363.
93. Curran BD, Ferrari FK, Kispert PH, Stadler J, Stuehr DJ, Simmons RL, Billiar TR.

Nitric oxide and nitric-oxide-generation compounds inhibit hepatocyte protein synthesis. FASEB J 1991; 5:2085–2092.
94. Shinagawa T, Yoshioka K, Kakumu S, Wakita T, Ishikawa T, Itoh Y, Takayangi M. Apoptosis in cultured rat hepatocytes: the effects of tumour necrosis factor-α and interferon-γ. J Pathol 1991; 165:247–253.
95. Dalmau M, Joaquin M, Nakamura T, Bartrons R, Gil J. Nitric oxide inhibits DNA synthesis and induces activation of poly(ADP-ribose) polymerase in cultured rat hepatocytes. Exp Cell Res 1996; 228:14–18.
96. Stadler J, Curran RD, Ochoa JB, Harbrecht BG, Hoffman RA, Simmons RL, Billiar TR. Effect of endogenous nitric oxide on mitochondrial respiration of rat hepatocytes in vitro and in vivo. Arch Surg 1991; 126:186–191.
97. Stadler J, Schmalix WA, Doehmer J. Inhibition of biotransformation by nitric oxide (NO) over production and toxic consequences. Toxicol Lett 1995; 82/83:215–219.
98. Hodgson PD, Renton KW. The role of nitric oxide generation in interferon-evoked cytochrome P450 down-regulation. Int J Immunopharmacol 1995; 17:995–1000.
99. Müller CM, Scierka A, Stiller RL, Kim Y-M, Cook DR, Lancaster JR, Buffington CW. Nitric oxide mediates hepatic cytochrome P450 dysfunction induced by endotoxin. Anesthesiology 1996; 84:35–42.
100. Stadler J, Trockfeld J, Schmalix WA, Brill T, Siewert JR, Greim H, Doehmer J. Inhibition of cytochromes P4501A by nitric oxide. Proc Natl Acad Sci USA 1994; 91:3559–3563.
101. Gergel D, Misik V, Riesz P, Cederbaum AI. Inhibition of rat and human cytochrome P4502E1 catalytic activity and reactive oxygen radical formation by nitric oxide. Arch Biochem Biophys 1997; 337:239–250.
102. Khatsenko OG, Gross SS, Rifkind AB, Vane JR. Nitric oxide is a mediator of the decrease in cytochrome P450-dependent metabolism caused by immunostimulants. Proc Natl Acad Sci USA 1993; 90:11147–11151.
103. Khatsenko OG, Kikkawa Y. Nitric oxide differentially affects constitutive cytochrome P450 isoforms in rat liver. J Pharmacol Exp Ther 1997; 280:1463–1470.
104. Halliwell B, Gutteridge JMC. Oxygen toxicity, oxygen radicals, transition metals and disease. Biochem J 1984; 219:1–14.
105. Stadler J, Bergonia HA, Di Silvio M, Sweetland MA, Billiar TR, Simmons RL, Lancaster JR Jr. Nonheme iron-nitrosyl complex formation in rat hepatocytes: detection by electron paramagnetic resonance spectroscopy. Arch Biochem Biophys 1993; 302:4–11.
106. Sevanian A, Hochstein P. Mechanisms and consequences of lipid peroxidation in biological systems. Annu Rev Nutr 1985; 5:365–390.
107. Rubbo H, Radi R, Trujillo M, Telleri R, Kalyanaraman B, Barnes S, Kirk M, Freeman BA. Nitric oxide regulation of superoxide and peroxynitrite-dependent lipid peroxidation. Formation of novel nitrogen-containing oxidized lipid derivatives. J Biol Chem 1994; 269:26066–26075.
108. Radi R, Beckman JS, Bush KM, Freeman B. Peroxynitrite oxidation of sulfhydryls: the cytotoxic potential of superoxide and nitric oxide. J Biol Chem 1991; 266:4244–4250.
109. Radi R, Beckman JS, Bush KM, Freeman B. Peroxynitrite-induced membrane lipid

peroxidation: the cytotoxic potential of superoxide and nitric oxide. Arch Biochem Biophys 1991; 288:481–487.
110. Niu X-F, Smith CW, Kubes P. Intracellular oxidative stress induced by nitric oxide synthesis inhibition increases endothelial cell adhesion to neutrophils. Circ Res 1994; 74:1133–1140.
111. Sibelle Y, Reynolds NY. Macrophages and polymorphonuclear neutrophils in lung defense. Am Rev Respir Dis 1990; 141:471–501.
112. Wizemann TM, Laskin DL. Enhanced phagocytosis, chemotaxis, and production of reactive oxygen intermediates by interstitial lung macrophages following acute endotoxemia. Am J Respir Cell Mol Biol 1994; 11:358–365.
113. Pendino KJ, Laskin JD, Shuler RL, Punjabi CJ, Laskin DL. Enhanced production of nitric oxide by rat alveolar macrophages after inhalation of a pulmonary irritant is associated with increased expression of nitric oxide synthase. J Immunol 1993; 151:7196–7205.
114. Prokhorova S, Lavnikova N, Laskin DL. Functional characterization of interstitial macrophages and subpopulations of alveolar macrophages from rat lung. J Leukoc Biol 1994; 55:141–146.
115. Wizemann TM, Gardner CR, Laskin JD, Quinones S, Durham SK, Goller NL, Ohnishi ST, Laskin DL. Production of nitric oxide and peroxynitrite in the lung during acute endotoxemia. J Leukoc Biol 1994; 56:759–768.
116. Hirano S. Nitric oxide-mediated cytotoxic effects of alveolar macrophages on transformed lung epithelial cells are independent of the β2 integrin-mediated intercellular adhesion. Immunology 1998; 93:102–108.
117. Lavnikova N, Drapier J-C, Laskin, DL. A single exogenous stimulus activates rat macrophages for nitric oxide production and cytotoxicity. J Leukoc Biol 1993; 54:322–328.
118. Jorens PG, Van Overveld FJ, Bult H, Vermeire PA, Herman AG. L-arginine-dependent production of nitrogen oxides by rat pulmonary macrophages. Eur J Pharmacol 1991; 200:205–209.
119. Chou TC, Li CY, Wu CC, Yen MH, Ding YA. The inhibition by dantrolene of L-arginine transport and nitric oxide synthase in rat alveolar macrophages. Anesth Analg 1998; 86:1065–1069.
120. Mason GL, Yang Z, Olchowy TW, Jian Z, Bochsler PN. Nitric oxide production and expression of inducible nitric oxide synthase by bovine alveolar macrophages. Vet Immunol Immunopathol 1996; 53:15–27.
121. Rylander R, Fogelmark B, Sjostrand M. Free lung cell phagocytosis and lysosomal enzyme activity after inhalation of lipopolysaccharide in guinea-pigs. Agents Actions 1985; 16:353–358.
122. Snella MC, Rylander R. Lung cell reactions after inhalation of bacterial lipopolysaccharides. Eur J Respir Dis 1982; 63:550–557.
123. Hsieh V, Amoruso-Marchat B, Rylander R, Polla BS. Oxygen metabolites from lavage and interstitial lung cells after inhalation of endotoxin in guinea pigs. Int Arch Allergy Immunol 1994; 104:42–47.
124. de Rochemonteix-Galve B, Marchat-Amoruso B, Dayer J-M, Rylander R. Tumor necrosis factor and interleukin-1 activities in free lung cells after single and repeated inhalation of bacterial endotoxin. Infect Immun 1991; 59:3646–3650.

125. Schneeberger EE. Alveolar type I cells. In: Crystal RG, West JB, Barnes PJ, Cherniack NS, Weibel ER, eds. The Lung. Vol. 1. New York: Raven Press, 1991; pp. 229–234.
126. Plopper CG, Dungworth DL, Tyler WS. Pulmonary lesions in rats exposed to ozone. Am J Pathol 1973; 71:375–394.
127. van Klaveren RJ, Demedts M, Nemery B. Cellular glutathione turnover in vitro, with emphasis on type II pneumocytes. Eur Respir J 1997; 10:1392–1400.
128. Kinnula VL, Chang L, Everitt JI, Crapo JD. Oxidant and antioxidants in alveolar epithelial type II cells: in situ, freshly isolated, and cultured cells. Am J Physiol 1992; 262:L69–L77.
129. Mason RJ, Williams MC. Alveolar type II cells. In: Crystal RG, West JB, Barnes PJ, Cherniack NS, Weibel ER, eds. The Lung. Vol. 1. New York: Raven Press, 1991; pp. 235–246.
130. Paine R III, Rolfe MW, Standiford TJ, Burdick MD, Rollins BJ, Strieter RM. MCP-1 expression by rat type II alveolar epithelial cells in primary culture. J Immunol 1993; 150:4561–4570.
131. Punjabi CJ, Laskin JD, Pendino KJ, Goller NL, Durham SK, Laskin DL. Production of nitric oxide by rat type II pneumocytes: increased expression of inducible nitric oxide synthase following inhalation of a pulmonary irritant. Am J Respir Cell Mol Biol 1994; 11:165–172.
132. Kinnula VL, Chang LY, Ho YS, Crapo, JD. Hydrogen peroxide release from alveolar macrophages and alveolar type II cells during adaptation to hyperoxia in vivo. Exp Lung Res 1992; 18:655–673.
133. Hahon N, Castranova V. Interferon production in rat type II pneumocytes and alveolar macrophages. Exp Lung Res 1989; 15:429–445.
134. Peters-Golden M, Feyssa A. Transcellular eicosanoid synthesis in cocultures of alveolar epithelial cells and macrophages. Am J Physiol 1993; 264:L438–L447.
135. Vittori E, Sciacca F, Colotta F, Mantovani A, Mattoli S. Protective effect of nedocromil sodium on the interleukin-1-induced production of interleukin-8 in human bronchial epithelial cells. J Allergy Clin Immunol 1992; 90:76–84.
136. van Overveld FJ, Jorens PG, De Backer WA, Rampart M, Bossaert L, Vermeire PA. Release of arachidonic acid metabolites from isolated human alveolar type II cells. Prostaglandins 1992; 44:101–110.
137. Strunk RC, Eidlen DM, Mason RJ. Pulmonary alveolar type II epithelial cells synthesize and secrete proteins of the classical and alternative complement pathways. J Clin Invest 1988; 81:1419–1426.
138. Schreck R, Meier B, Mannel DN, Droge W, Baeuerle PA. Dithiocarbamates as potent inhibitors of nuclear factor κB activation in intact cells. J Exp Med 1992; 175:1181–1194.
139. Laskin DL, Heck DE, Guo Y, Gardner CR, Laskin JD. Acute endotoxemia is associated with increased nitric oxide production by type II cells and activation of NfκB. Am J Respir Crit Care Med 1997; 155:A616.
140. Xie Q-W, Kashiwabara Y, Nathan C. Role of transcription factor NF-κB/Rel in induction of nitric oxide synthase. J Biol Chem 1994; 269:4705–4708.
141. Moncada S, Palmer RM, Higgs EA. Nitric oxide: physiology, pathophysiology, and pharmacology. Pharmacol Rev 1991; 43:109–142 .

142. Nijkamp FP, van der Linde HJ, Folkerts G. Nitric oxide synthesis inhibitors induce airway hyperresponsiveness in the guinea pig in vivo and in vitro. Role of the epithelium. Am Rev Respir Dis 1993; 148:727–734 .
143. Nathan C. Nitric oxide as a secretory product of mammalian cells. FASEB J 1992; 6:3051–3064.
144. Hibbs J, Taintor R, Vavrin Z, Rachlin E. Nitric oxide: a cytotoxic activated macrophage effector molecule. Biochem Biophys Res Commun 1988; 157:87–94.
145. Heck DE, Laskin DL, Gardner CR, Laskin JD. Epidermal growth factor suppresses nitric oxide and hydrogen peroxide production by keratinocytes. J Biol Chem 1992; 267:21277–21280.
146. Punjabi CJ, Laskin DL, Heck DE, Laskin JD. Production of nitric oxide by murine bone marrow cells. Inverse correlation with cellular proliferation. J Immunol 1992; 149:2179–2184.
147. Stuehr DJ, Nathan CF. Nitric oxide: a macrophage product responsible for cytostatic and respiratory inhibition in tumor target cells. J Exp Med 1989; 168:1543–1555.
148. Pendino KJ, Meidhof TM, Heck DE, Laskin JD, Laskin DL. Inhibition of macrophages with gadolinium chloride abrogates ozone-induced pulmonary injury and inflammatory mediator production. Am J Respir Cell Mol Biol 1995; 13:125–132.
149. Lavnikova N, Burdelya L, Lakhotia A, Patel N, Prokhorova S, Laskin DL. Macrophage and interleukin-1 induced nitric oxide production and cytostasis in hamster tumor cells. J Leukoc Biol 1997; 61:425–458.
150. Pendino KJ, Gardner CR, Quinones S, Laskin DL. Stimulation of nitric oxide production in rat lung lavage cells by anti-Mac-1β antibody: effects of ozone inhalation. Am J Respir Cell Mol Biol 1996; 14:327–333.
151. Pendino KJ, Gardner CR, Shuler RL, Laskin JD, Durham SK, Barton D, Ohnishi ST, Ohnishi T, Laskin DL. Inhibition of ozone-induced nitric oxide synthase expression in the lung by endotoxin. Am J Respir Cell Mol Biol 1996; 14:516–525.
152. Gow AJ, Thom SR, Ischiropoulos H. Nitric oxide and peroxynitrite-mediated pulmonary cell death. Am J Physiol 1998; 274:L112–L118.
153. Ialenti A, Ianaro A, Moncada S, Di Rosa M. Modulation of acute inflammation by endogenous nitric oxide. Eur J Pharmacol 1992; 211:177–182.
154. Ishizaki T, Shigemori K, Yamamura Y, Matsukawa S, Nakai T, Miyabo S, Hayakawa M, Ozawa T, Voelkel NF. Increased nitric oxide biosynthesis in leukotoxin,9,10-epoxy-12-octadecenoate injured lung. Biochem Biophys Res Commun 1995; 210:133–137.
155. Mulligan MS, Warren JS, Smith CW, Anderson DC, Yeh CG, Rudolph AR, Ward PA. Lung injury after deposition of IgA immune complexes: requirements for CD18 and L-arginine. J Immunol 1992; 148:3086–3092.
156. Mulligan MS, Hevel, JM, Marletta MA, Ward PA. Tissue injury caused by deposition of immune complexes is L-arginine dependent. Proc Natl Acad Sci USA 1991; 88:6338–6342.
157. Berisha HI, Pakbaz H, Absood A, Said SI. Nitric oxide as a mediator of oxidant lung injury due to paraquat. Proc Natl Acad Sci USA 1994; 91:7445–7449.
158. Putensen C, Rasanen J, Downs JB. Effect of endogenous and inhaled nitric oxide

on the ventilation-perfusion relationships in oleic-acid lung injury. Am J Respir Crit Care Med 1994; 150:330–336.

159. Sun D, Zhao M, Ma D, Liao S, Di C. Protective effect of interleukin-1 receptor antagonist on oleic acid-induced lung injury. Chin Med J 1996; 109:522–526.

160. Leeman M, De Beyl VZ, Gilbert E, Melot C, Naeije R. Is nitric oxide released in oleic acid lung injury? J Appl Physiol 1993; 74:650–654.

161. Laskin DL, Pendino KJ, Punjabi CJ, Rodriguez del Valle M, Laskin JD. Pulmonary and hepatic effects of inhaled ozone in rats. Environ Health Perspect 1994; 102:61–64.

162. Peavy DL, Fairchild EJ. Toxicologic interactions between ozone and bacterial endotoxin. Environ Res 1987; 42:63–71.

163. Frank L, Yam J, Roberts RJ. The role of endotoxin in protection of adult rats from high oxygen lung toxicity. J Clin Invest 1978; 61:269–275.

164. Ruetten H, Thiemermann C. Prevention of the expression of inducible nitric oxide synthase by aminoguanidine or aminoethyl-isothiourea in macrophages and in the rat. Biochem Biophys Res Commun 1996; 225:525–530.

165. Ruetten H, Southan GJ, Abate A, Thiemermann C. Attenuation of endotoxin-induced multiple organ dysfunction by 1-amino-2-hydroxy-guanidine, a potent inhibitor of inducible nitric oxide synthase. Br J Pharmacol 1996; 118:261–270.

166. Blackford JA Jr, Antonini JM, Castranova V, Dey RD. Intratracheal instillation of silica up-regulates inducible nitric oxide synthase gene expression and increases nitric oxide production in alveolar macrophages and neutrophils. Am J Respir Cell Mol Biol 1994; 11:426–431.

167. Huffman LJ, Judy DJ, Castranova V. Regulation of nitric oxide production by rat alveolar macrophages in response to silica exposure. J Toxicol Environ Health 1998; 53:29–46.

168. Warner RL, Paine R 3rd, Christensen PJ, Marletta MA, Richards MK, Wilcoxen SE, Ward PA. Lung sources and cytokine requirements for in vivo expression of inducible nitric oxide synthase. Am J Respir Cell Mol Biol 1995; 12:649–661.

169. Ischiropoulos H, Mendiguren I, Fisher D, Fisher AB, Thom SR. Role of neutrophils and nitric oxide in lung alveolar injury from smoke inhalation. Am J Respir Crit Care Med 1994; 150:337–341.

170. Weinberger B, Fakhrzadeh L, Heck DE, Laskin JD, Gardner CR, Laskin DL. Inhaled nitric oxide primes lung macrophages to produce reactive oxygen and nitrogen intermediates. Am J Respir Crit Care Med 1998; 158:931–938.

171. Numata M, Suzuki S, Miyazawa N, Miyashita A, Nagashima Y, Inoue S, Kaneko T, Okubo T. Inhibition of inducible nitric oxide synthase prevents LPS-induced acute lung injury in dogs. J Immunol 1998; 160:3031–3037.

172. Putensen C, Rasanen J, Downs JB. Effect of endogenous and inhaled nitric oxide on the ventilation-perfusion relationships in oleic-acid lung injury. Am J Respir Crit Care Med 1994; 150:330–336.

173. Mikawa K, Nishina K, Tamada M, Takao Y, Maekawa N, Obara N. Aminoguanidine attenuates endotoxin-induced acute lung injury in rabbits. Crit Care Med 1998; 26:605–911.

174. Cuzzocrea S, Zingarelli B, Gilad E, Hake P, Salzman AL, Szabo C. Protective ef-

fects of 3-aminobenzamide, an inhibitor of poly(ADP-ribose) synthase in a carrageenan-induced model of local inflammation. Eur J Pharmacol 1998; 342:67–76.

175. Cuzzocrea S, Zingarelli B, Hake P, Salzman AL, Szabo C. Antiinflammatory effects of mercaptoethylguanidine, a combined inhibitor of nitric oxide synthase and peroxynitrite scavenger, in carrageenan-induced models of inflammation. Free Radic Biol Med 1998; 24:450–459.

176. Saleh D, Barnes PJ, Giaid A. Increased production of the potent oxidant peroxynitrite in the lungs of patients with idiopathic pulmonary fibrosis. Am J Respir Crit Care Med 1997; 155:1763–1769.

177. Ischiropoulos H, al-Mehdi AB, Fisher AB. Reactive species in ischemic rat lung injury: contribution of peroxynitrite. Am J Physiol 1995; 269:L158–L164.

178. Kooy NW, Royall JA, Ye YZ, Kelly DR, Beckman JS. Evidence for in vivo peroxynitrite production in human acute lung injury. Am J Respir Crit Care Med 1995; 151:1250–1254.

179. Royall JA, Kooy NW, Beckman JS. Nitric oxide-related oxidants in acute lung injury. New Horiz 1995; 3:113–122.

180. Haddad IY, Pataki G, Hu P, Galliani C, Beckman JS, Matalon S. Quantitation of nitrotyrosine levels in lung sections of patients and animals with acute lung injury. J Clin Invest 1994; 94:2407–2413.

181. Haddad IY, Crow JP, Hu P, Ye Y, Beckman J, Matalon S. Concurrent generation of nitric oxide and superoxide damages surfactant protein A. Am J Physiol 1994; 267:L242–L249.

182. Haddad IY, Ischiropoulos H, Holm BA, Beckman JS, Baker JR, Matalon S. Mechanisms of peroxynitrite-induced injury to pulmonary surfactants. Am J Physiol 1993; 265:L555–L564.

183. McDermott CD, Gavita SM, Shennib H, Giaid A. Immunohistochemical localization of nitric oxide synthase and the oxidant peroxynitrite in lung transplant recipients with obliterative bronchiolitis. Transplantation 1997; 64:270–274.

184. Eiserich JP, van der Vliet A, Handelman GJ, Halliwell B, Cross CE. Dietary antioxidants and cigarette smoke-induced biomolecular damage: a complex interaction. Am J Clin Nutr 1995; 62:1490S–1500S.

185. Gaston B, Drazen JM, Loscalzo J, Stamler JS. The biology of nitrogen oxides in the airways. Am J Respir Crit Care Med 1994; 149:538–551.

186. Meng QH, Bishop AE, Jacoub MH, Polak JM. Expression of inducible nitric oxide synthase and formation of peroxynitrite in alveolar type II cells in emphysema (abstr). Am J Respir Crit Care Med 1998; 157:A233.

10

Nitric Oxide and Cell Death in Inflammation

Mark J. S. Miller and Manuel Sandoval
Albany Medical College, Albany, New York

Karen A. Ribbons
Progen Industries, Ltd., Darra, Queensland, Australia

Elizabeth E. Mannick
University of Rochester, Rochester, New York

I. SOURCE OF NITRIC OXIDE IN INFLAMMATION

One can reasonably predict that NO will promote cellular dysfunction and tissue injury only when it is produced in large amounts, and this invariably requires the expression of the inducible form of nitric oxide synthase (iNOS). If iNOS is absent, then NO will primarily exhibit anti-inflammatory effects (1). Conversely, the presence of iNOS is an excellent indicator that NO is directly contributing to cellular injury (2). The best exception for this "rule" is pregnancy, and the possible explanations for this exception are discussed at the end of the chapter. Infection, where the high rate of NO production is necessary for the clearance of the infection, is also a state where the contribution of NO to tissue injury must be tempered by the pathology of the infection itself. Under these conditions NOS inhibition may exacerbate or hinder the pathology depending on the relative strengths of the host defense response versus the aggressiveness of the invading micro-organism (3).

The sustained and robust calcium-independent production of nitric oxide is thought to be responsible for the association of cytostasis and cytotoxicity with iNOS expression. However, the correlease of other free radicals, cytokines, and adhesion molecules and their interplay may be required for an effective toxic cocktail. The regulation of these processes by the transcription factor NF-κB (see below) is a further example of diverse pathways cooperating in states of inflammation, with iNOS being one component as are cytokines, like TNFα, or IL-1β. The relative importance of individual components, in this case iNOS-derived NO, can be difficult to decipher.

Expression of iNOS has been confirmed in a variety of experimental forms of chronic inflammation, including rat and guinea pig ileitis/colitis induced by the hapten TNBS (4–6), adjuvant ileitis or arthritis (7–10), HLA-B27 transgenic rat colitis (11), streptococcal cell wall arthritis and colitis (12), autoimmune glomerulonephritis (13), and encephalitis (14).

In these studies iNOS expression has been confirmed by RT-PCR, Northern or Western analysis. Immunohistochemical analysis reveals a wide range of cellular source of NO, well beyond macrophages, the original site of iNOS discovery. There remains to this day some misconception that iNOS has little role in humans because of the difficulty of expressing iNOS in peripheral blood neutrophils or monocytes. Nevertheless, in human inflammation NO levels are equivalent to animal models, and iNOS can be detected in inflamed tissue, including neutrophils and macrophages (Fig. 1) (15). The dilemma appears to be

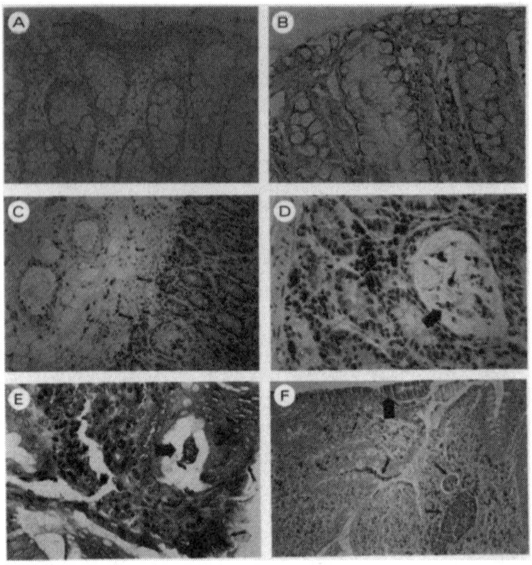

Figure 1 Immunohistochemical localization of iNOS in spontaneous colitis in monkeys and humans. Panels A and B display results from noninflamed human and monkey colon, respectively. Panels C and D show positive iNOS staining colitis in rhesus monkeys, as indicated by the brown color reaction and highlighted by the arrows. Positive staining was localized to the submucosa, mucosa, and crypt abscesses as seen on higher magnification (Panel D). Examples of human colitis are shown in Panels E and F. Positive iNOS staining was observed in epithelia and crypt abscesses in both Crohn's colitis (Panel E) and ulcerative colitis (Panel F).

due to the requirement of human leukocytes for a cocktail of cytokines (16), as in the inflammatory milieu, in order to activate iNOS gene expression. Numerous cell types have been found to express iNOS with the appropriate provocation or in chronic inflammation. Primary sources appear to be leukocytes, epithelia, plasma cells, smooth muscle, and even neurons (4–6,16,17).

II. NITRIC OXIDE AND SIGNAL TRANSDUCTION IN INFLAMMATION

Nitric oxide's ability to act upon intracellular signal transduction pathways contributes to both its physiologic (smooth muscle relaxant, neutrotransmitter) and pathophysiologic (inflammation and cell death-related) actions. One of the primary second messengers for nitric oxide is cyclic guanosine 3′,5′-monophosphate (cGMP) (18,19). By nitrosating the heme moiety of soluble guanyl cyclase and thereby altering its three-dimensional configuration, the production of cGMP is facilitated and a wide range of cellular responses is initiated.

One intracellular target of cGMP stimulated by NO is p_{21}ras (20,21) (Table 1), a proto-oncogene which feeds into a signal cascade involving raf kinase and MAP kinase. Although p_{21}ras plays a role in both cell death and inflammation (22,23), the precise relationship among nitric oxide, ras activation, and these events has not been established.

Activation of cGMP may account for some of the anti-inflammatory effects of NO. By activating cGMP, NO lowers intracellular calcium levels and can attenuate platelet and neutrophil aggregation (1,24,25). Nitric oxide may also inhibit growth factor receptor signaling by inositol 1,4,5-triphosphate (IP_3) by a similar mechanism (26,27). In contrast, however, cGMP induced by NO has the

Table 1 Cell Cycle Effects of Nitric Oxide

Intracellular target	Effects	Consequences	References
cGMP	Stimulation	↓Intracellular Ca^{2+}	18
		↓Platelet aggregation	24,25
		↓Neutrophil activation	1
		↑p21 ras	20,23
		↑JNK and ERK kinase	
		↓Growth factor signaling	26,27,29
ADP-ribose	↑Ribosylation	↓GAPDH	31,32
		PARS	33
Protein-thiols	Nitrosothiol/formation	↓PKC	35,36
		↑p21 ras	23
		↓NF-κB	38–40
		↓Adhesion molecules	38

potential to mobilize intracellular calcium by stimulating cyclic ADP-ribose synthesis (28). cGMP may mediate some of the effects of NO on cell death. Although nitric oxide frequently induces apoptosis, in certain cell types it prevents cell death (Table 2). A study in PC-12 cells and sympathetic neurons revealed that nitric oxide donors prevented cell death due to growth factor withdrawal (29). This effect could be mimicked by cGMP analogs, suggesting that the protective effects of NO were mediated by the cGMP signal transduction pathway. Similarly, Genero et al. (30) reported that both nitric oxide and cGMP analogs prevents antigen-induced apoptosis in B cells, possibly by preventing the fall in Bcl-2, which accompanies apoptosis.

In addition to its effects on cGMP, nitric oxide also stimulates the ADP-ribosylation of a number of intracellular proteins. The best-described of these is glyceraldehyde-3-phosphate dehydrogenase (GAPDH), an enzyme involved in the generation of adenosine triphosphate (ATP) during glycolysis (31,32). By inhibiting GADPH, NO may also contribute to cellular energy depletion and, possibly, apoptosis. Nitric oxide also activates poly (ADP-ribose) synthetase, an enzyme involved in DNA repair and apoptosis (33).

NO-mediated signal transduction can be mediated by the nitrosation of protein thiol groups (34). A prime example is nitric oxide's ability to inhibit protein kinase C (PKC) via S-nitrosylation of a thiol on the enzyme (35). Since PKC is known to upregulate the production of NO by iNOS, nitric oxide's ability to inhibit PKC may represent one possible autoregulatory, negative-feedback mechanism (36). This inhibition of PKC by NO may also contribute to the ability of NO to induce apoptosis, depending on the cell type. Indirect evidence sup-

Table 2 Apoptosis Induced by Nitric Oxide

Cell type	Comments	Reference
Macrophages	Endogenous (autocrine) and exogenous sources	92,93
	LDL-activated macrophages (foam cells)	94
	Prevented by PKC or PKA activation Bcl-2	37
	Correlated with p53 accumulation	95
	NO prevents LPS-induced cell death	96
Epithelia	Low dose = apoptosis; high dose = necrosis	97
Vascular smooth muscle	Associated with upregulation of Fas	98
	Cytokine-induced apoptosis partly via NO	99
Leukemia cells	Vascular NO-induced apoptosis in leukemia cells	100
	Inhibitors of poly(ADP-ribose) polymerase protective	101
Hepatocytes	NO donors protect against TGFβ-induced apoptosis	102
Myoblasts	Effects analogous to reactive oxygen species	103
Lymphocytes	NO protects B cells from apoptosis	104

porting this hypothesis is the observation that PKC agonists protect macrophages from cell death (37).

Finally, nitric oxide may mediate inflammation and contribute to cell death by acting directly on transcriptional factors. The effects of NO on the transcription factor NF-κB have been examined in a number of in vitro studies (38–40). NF-κB binds to the promoter region of a variety of inflammation-related genes including tumor necrosis factor-α, interleukins, 1, 6, and 8, and a variety of adhesion molecules. Three studies have shown that NO can downregulate NF-κB activity, suggesting an anti-inflammatory role for NO (38,39). One study documented NO's ability to upregulate NF-κB activity, however (40). Based on these results, it appears that NO may play a biphasic role in the activation of NF-κB. However, the circumstances determining its stimulatory or inhibitory roles in the activation of NF-κB have yet to be defined, with one possibility being the influence of NO-derived reactive nitrogen species. We have evaluated the effects of peroxynitrite on NF-κB activation in RAW 264.7 macrophages and found that ONOO– can either inhibit or activate NF-κB depending on the dose of ONOO– and the agent used to stimulate NF-κB (41). Both NO and peroxynitrite's biphasic effects may explain why nitric oxide has been described to possess both anti- and pro-inflammatory effects (1,2).

III. ACTIONS OF NITRIC OXIDE IN INFLAMMATION

Pharmacological probes that inhibit the formation of nitric oxide have been demonstrated to attenuate or even prevent chronic inflammation (2). These beneficial effects have been demonstrated in a variety of species, organs, and forms of inflammation (Table 3). To a large extent these studies have used inhibitors that are not specific for iNOS but, rather, nonselective inhibitors of iNOS. While only slightly selective for iNOS, aminoguanidine has proven to be effective in models of inflammatory bowel disease-TNBS ileitis and colitis, PG-PS colitis, and HLA-B27 colitis (5,11,12,42), as well as adjuvant arthritis (7), where the iNOS inhibitor L-NIL was also proven to be effective. It should be noted that it remains to be proven that inhibitors of NOS production are beneficial in chronic inflammation in humans, largely because of the lack of suitable compounds (potent and selective for iNOS) for clinical evaluation.

In contrast to the regulation of physiological responses, the effects of NO or NO-derived products on cellular function in inflammation are not mediated by cGMP. The deleterious effects of NO are the result of its free radical nature its avidity for transition metals, and as for NO^+ for reactions with amines and sulfhydryl groups. These three reactions are the basis for NO-mediated biological actions; other effects may be derived from reactive nitrogen species generated from NO, as NO itself is limited in its reactivity. Three mitochondrial en-

Table 3 Conditions Where NOS Inhibitors Have Displayed Anti-Inflammatory Actions

Organ	Model	Reference
Stomach	H. pylori gastritis	15
Ileum	TNBS	1,5
Colon	TNBS, PG-PS, HLA-B27	6,11,12,42
	Iodoacetamide	111
Joints	Adjuvant	7,9,10
Brain	Immune encephalomyelitis	14
	Pneumococcal meningitis	
Kidney	Spontaneous glomerulonephritis	13
Pancreas	Diabetes in BB/E rats	112
Lungs	Carrageenan-induced pleurisy	113,114
Eyes	LPS-induced uveitis	115
Skin	Air pouch granulomas	116

zymes are susceptible to inhibition by NO (43–45)—aconitase (part of the tricarboxylic acid cycle), NADPH-ubiquinone oxidoreductase, and succinate-ubiquinone oxidoreductase (Complex I and Complex II—the mitochondrial respiratory chain). This inhibition is the result of the degradation of the iron-sulfur clusters and may result in the liberation of nonheme iron from the cells affected (46). The net result of this effect is compromised cellular respiration—in effect, metabolic poisoning. This will result in the attenuation/cessation of cellular functions including replication and may also be involved in the evocation of programmed cell death. Inhibition of ribonucleotide reductase may also contribute to the inhibition of DNA synthesis.

Nitric oxide may promote and cancel lipid peroxidation. Initiation of lipid peroxidation is likely to be mediated by compounds other than NO, e.g., peroxynitrite (47). Nitric may actually negate the propagation of established lipid peroxidation by a radical-radical annihilation reaction (48). This may explain why nitric oxide synthase inhibitors promote mortality and morbidity in neonatal rats exposed to hyperoxic oxygen (49). Under these circumstances NO is beneficial, because as a free radical it is less toxic than the lipid peroxides it annihilates.

IV. REACTIVE NITROGEN SPECIES AND DNA DAMAGE

The ability of NO and related reactive nitrogen species to react with DNA and cause adducts and consequently mutations, provides a potential link between chronic inflammation and the development of cancer. Thus, sustained overproduction of these compounds may be a critical component of the precancerous process, as described for gastric cancer (15,50). If mutations initiated by these agents are not repaired before cellular replication occurs, then they may remain a permanent component of the genome. The types of alterations caused by reactive oxygen and nitrogen species is primarily point mutations, particularly G:C → A:T mutations (2) (Table 4). Importantly, these are a common type of transformation in cancer. Of the 12 potential sites for alkylation of DNA including the ring nitrogen positions in adenine, guanine, cytosine, and thymine; the oxygen atoms of guanine, thymine, and cytosine; and the phosphate groups, experimental data confirm that the O^6-alkylation of guanine represents the most important reaction. Mutations at this site bear a stronger link to carcinogenicity.

NO-derived products promote deamination of DNA. The process involves the N-nitrosation of aromatic amines (Ar-NH2) to form the nitrosamine (Ar–NH–NO) intermediate, which then decomposes and rearranges to produce the diazonium ion (Ar–N2+), which in turn decomposes to the hydroxyl derivative of the aromatic amine (Ar-OH). These reactions have been proposed for the oxidative deamination of cytosine, methylcytosine, guanine, and adenine (see Table 4) and the observed increased formation of oxidative DNA adducts being formed following exposure to oxidative or nitrosative species. It has recently been described that nitroguanine adducts can be detected in response to peroxynitrite exposure (51,52) expanding the chemical processes and potential endogenous mediators for mutations. Peroxynitrite has recently been described to not result in a buildup of 8-oxodeoxyguanosine (53), either because peroxynitrite does not produce 8-oxodeoxyguanosine or because it inactivates an intermediate in its formation. Thus, the formation of these adducts can be a useful tool for examining oxidative stress, but there remains a lot to be understood about the basic chemistry and biology of these events. Nitric oxide has also been shown to compromise the ability of natural repair processes, e.g., PARS, to repair DNA damage. Thus, in addition to causing DNA damage, these endogenous compounds may actually limit the mechanisms to prevent these mutations from being permanently incorporated into the genome.

Table 4 Types of Alterations and Mutations that May Arise from the Deamination of DNA Bases by NOx

Conversion	Type of mutation
Cytosine + NOx →	G:C → A:T
mCytosine + NOx → thymine	G:C → A:T
Guanine + NOx → xanthine	G:C → A:T
Adenine + NOx → hypoxanthine	A:T → G:C

An interesting aspect of recent research is the ability of reactive nitrogen species to form nitroso (NO) and nitro (NO_2) adducts. A wide range of nitrosated compounds have been linked to the development of gastrointestinal cancer (54–57), and now it appears that they may operate through mechanisms that appear similar, if not identical, to the effects of endogenous compounds (58,59). For example, exogenous nitrosamines bear a resemblance to the intermediate products formed during nitrosative reactions produced endogenously. Thus, the ability of nitrosamines and related compounds to promote gastrointestinal cancer may not be the same spurious toxicological phenomenon, but rather a means of duplicating the reactions elicited in the natural precancerous process.

Cells have several choices in the way they respond to genotoxic species. Repair of DNA damage is an ongoing process, but one that may be compromised by NO (60,61) (Fig. 2). If the extent of DNA damage is extensive and perhaps beyond the capacity of repair processes, then the cell may initiate programmed cell death or apoptosis (shrinkage necrosis). The advantage of this process is that damaged DNA will not be replicated, and thus the risk for disrupted cellular replication is reduced. The role of reactive oxygen and nitrogen species in this process can be highlighted by the recognition that the tumor suppressor gene *Bcl2* is an antioxidant (62,63). Endogenous Bcl-2 has been demonstrated to prevent NO-mediated apoptosis (60). Nitric oxide either produced endogenously or added to culture media has been demonstrated to cause apoptosis of a variety of cell types (Table 2), including many of which that are transformed and "immortalized." Presumably, cells that have a more appropriate regulation of their cell cycle are more susceptible than transformed cells to the induction of apoptosis by reactive nitrogen intermediates.

The evidence that nitric oxide or related species initiates apoptosis in inflammation is based on in situ indices of apoptosis like the TUNEL or ISEL immunohistochemical techniques for DNA fragmentation. In severe inflammation there is little evidence for apoptosis; rather, the primary form of cell death is necrosis (Fig. 2). Reactive nitrogen species may well participate in this cellular toxicity because of the therapeutic effects of NOS inhibitors; however, necrotic cell death is likely to represent an effect of multiple influences, including cytokines, proteases, and free radicals. In milder forms of inflammation there appears to be significant apoptosis, for example in Helicobacter pylori gastritis (Fig. 3) (15). The colocalization of iNOS and nitrotyrosine with apoptosis in the gastric mucosa hints at cause-and-effect but remains an association (15). However, either clearance of the infection with antimicrobial agents or antioxidants (ascorbate or β-carotene) reduced the degree of apoptosis. Importantly, the expression of iNOS and the degree of tyrosine nitration were reduced by the same agents. This suggests strong links between the expression of iNOS, formation

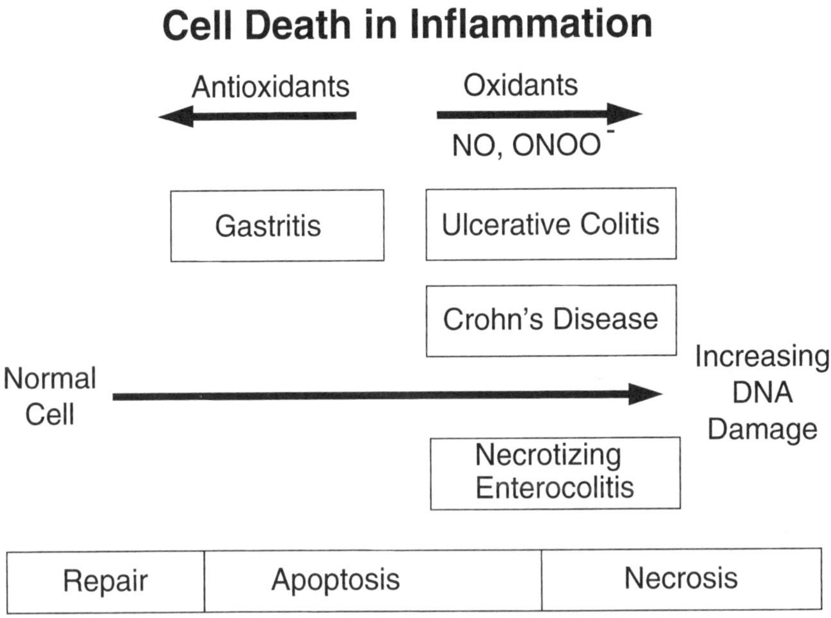

Figure 2 Schematic diagram of the effects of genotoxic species on the pathways of cell death. In response to a DNA damage cells may either arrest their cell cycle and repair the damage or, alternatively, undergo apoptosis. If the burden of genotoxic species is excessive, the cell will die by necrosis. With this stimulus, antioxidant defenses may limit the degree of DNA damage, but it may not lead to the prevention of cell death unless the damage is reduced beyond the point where a cell commits to apoptosis; otherwise, the cell will still die but in a manner delayed in respect to necrosis.

of NO and nitrating species, and the induction of apoptosis in gastric mucosa. Peroxynitrite, linked to the formation of nitrotyrosine, has also been demonstrated to induce apoptosis in a variety of cell types (Table 5). To date we cannot discern if nitrotyrosine is simply a marker of peroxynitrite formation or the generation of other nitrating species, or alternatively, if the levels of intracellular nitrotyrosine influences cellular function, e.g., inhibits tyrosine kinase activity. Genistein, a tyrosine kinase inhibitors, is anti-inflammatory and reduces NO release in gut inflamation (64), a finding that things at a possible cause-and-effect. Other reports have confirmed that genistein inhibits iNOS expression, but inhibition of tyrosine kinase may not be the mechanism of action; for example, this plant-derived product also has substantive estrogenic properties and effects on the cell cycle regulation (65–67).

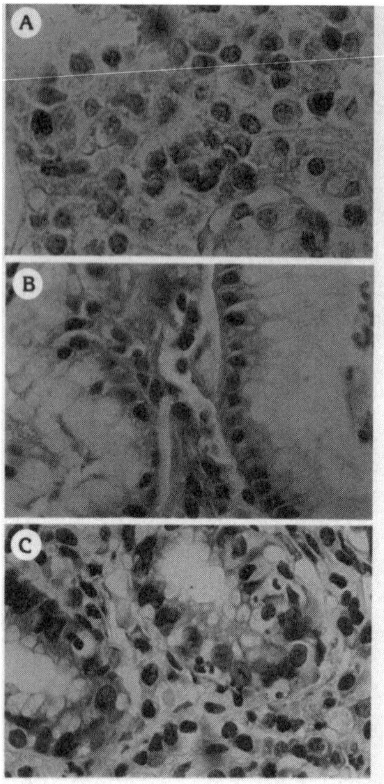

Figure 3 Colocalization of immunohistochemical evidence for iNOS, nitrotyrosine, and apoptosis in human H. pylori gastritis. From biopsies of patients with H. pylori gastritis it can be noted that iNOS (A), evident in neutrophils and macrophages, colocalizes with nitrotyrosine (B) and cell death by apoptosis (C) using the TUNEL histochemical approach. As NO and peroxynitrite induce apoptosis in gut epithelia, the excessive formation of reactive nitrogen species associated with gut inflammation may contribute to cell death mechanisms in these disease states.

V. ANTIOXIDANT DEFENSES AND NOx-MEDIATED TOXICITY

Unlike with reactive oxygen species we have yet to identify a system of enzyme catalysts for the inactivation of reactive nitrogen species. Superoxide is the prototypical reactive oxygen metabolite, and it is dismutated to hydrogen peroxide by superoxide dismutase. The product, hydrogen peroxide, is then inactivated by a variety of peroxidases (catalase, glutathione peroxidase) to yield water and oxygen. No such corollary exists for NO. However, the paradigm of reactive

Table 5 Apoptosis Induced by Peroxynitrite

Cell type	Comments	Reference
Neurons	Low dose = apoptosis; high dose = necrosis	105
PC-12 cells	Inhibits DOPA synthesis	106
	Protected by NGF or insulin	107
	Induced by downregulation of SOD	108
Thymocytes	Demonstrated in fresh cell suspensions	109
Macrophages and colonic epithelia	Prevented by ascorbic acid	110

oxygen species gaining potency after the initial production of superoxide is relevant for NOx because NO may be converted to a host of nitrogen species of greater reactivity than the parent NO, e.g., NO_2, N_2O_3, N_2O_4, $ONOO^-$. These parallels are a useful for understanding the biological consequences of NO formation. Nevertheless, how is NO inactivated biologically, and are their different mechanisms for the different reactive nitrogen species?

Transition metals catalyze the transformation of NO to the relatively inactive forms nitrite/nitrate, although the nitrosating agent NO^+ may be formed under some circumstances. Thus metals are an important source of NO degradation. Alternative substrates, amines, and thiols can act as sequestrants for NO, where it may be rereleased or precipitate a new biological activity. Being a free radical, NO can be inactivated by radical-radical annihilation; for example, in the case of lipid peroxidation NO may display antioxidant properties by preventing the propagation of lipid peroxidation (48). Alternatively, if the alternate free radical is superoxide, the product is $ONOO^-$, which is a powerful oxidant and nitrating agent (68,69), and the potential for cellular injury is enhanced, not reduced. The consequences of these reactions depend on whether the radical that NO is reacting with is more or less reactive than the parent or the resultant product.

Endogenous antioxidants may become depleted during inflammation predisposing the cells/tissue for injury. The thiol, glutathione, is reduced in chronic inflammation as well as the enzymes responsible for glutathione regeneration (70,71). However, this not likely to be due to a direct effect of NO because of its relative poor reactivity, but rather more powerful oxidants, e.g., OCl^-.

We have recently evaluated the ability of endogenous and clinically used antioxidants to negate the toxic effects of peroxynitrite. A single exposure to peroxynitrite, despite its short half-life at physiological pH, results in apoptosis or necrosis, depending on the dose of epithelial cells and macrophages (72). Coadministration of ascorbic acid greatly diminishes the propensity of peroxynitrite to promote cell death, probably because of its ability to degrade

$ONOO^-$ as has also been described for bicarbonate (73). Clinically we have demonstrated the presence of iNOS, nitrotyrosine, and apoptosis in close approximation in the gastric mucosa of patients with Helicobacter pylori-induced gastritis. Supplemental ascorbic acid reduced the nitrotyrosine staining and the degree of apoptosis in these patients. These findings suggest that the effects of ascorbic acid observed in vitro may have direct clinical applications. Interestingly, patients that present with atrophic gastritis, the form of gastritis that precedes the development of gastric cancer, have an impaired mechanism for the secretion of ascorbic acid into the gastric lumen (74), raising the possibility that the imbalance associated with enhanced formation of genotoxic pro-oxidants and the reduced levels of antioxidants may be critical in predisposing these patients for gastric cancer (75).

Another antioxidant, 5-aminosalicylic acid (5-ASA), used in various formulations for the treatment of clinical inflammatory bowel disease (76), also negates the ability of peroxynitrite to promote apoptosis and necrosis of intestinal epithelial cells. However, 5-ASA does not react with NO directly, so its therapeutic efficacy in vivo does not reflect a direct interaction with NO, but rather may result from reactions with NO-derived species like peroxynitrite. Nitrotyrosine, a marker for peroxynitrite formation, is colocalized with iNOS in models of gut inflammation; nitrotyrosine formation can be prevented by inhibiting the NO synthesis (5,8). Thus antioxidant defenses may limit cellular toxicity by inactivating NO, sequestering it, or terminating the effects of reactive nitrogen intermediates derived from NO.

An additional emerging consideration is the role of oxidants in the activation of NF-κB (see above). While the direct effects of NO and $ONOO^-$ on NF-κB activity remains to be determined, antioxidants may limit the expression of iNOS by preventing proteosome-dependent activation of NF-κB. In our trial of H. pylori-infected gastritis patients, long-term administration of antioxidants (ascorbate or β-carotene) reduced iNOS expression, independent of the H. pylori infection (15). As NF-κB activation has been described in H. pylori infection, it is possible that antioxidants could attenuate NO formation through effects at the transcriptional level.

VI. CELLULAR INJURY AND DEATH: ARE THEY DUE TO NO OR NOx?

The best estimates for NO production under iNOS-dependent conditions is in the micromolar range, 1 to 10 μM. Basal conditions where iNOS is not expressed is <50 nM, for comparison (77). Bearing this in mind, how can NO, which is inherently a weakly reactive free radical, initiate cellular injury and death? One possibility is that the cellular reserves of antioxidant defenses, thiols, etc., and intracellular Fe levels required for mitochondrial function, are depleted with sustained exposure to NO. Another is that in inflammation NO is only part of the host of noxious mediators released and insulting the integrity of tissue function. However, under this scenario NO is a small component of the pathophysiological process that is not in total agreement with the remarkable beneficial effects of NOS inhibitors in inflammation (2).

A more likely explanation is that NO itself is not the cause of cellular injury and dysfunction, but rather that NO-derived products mediate these actions. Discovering the nature of these products is a difficult task based on their marked reactivity, ephemeral nature, and difficulty in assaying. For example, there is little evidence that products like NO_2, N_2O_4, and N_2O_3 are produced from endogenous NO. Nitrosothiols are generated and therefore the NO^- species must be generated. To highlight our limitations in this are, we still do not know what intermediates are formed in the auto-oxidation of NO to nitrite in aqueous solutions. What is clear is that NO is far less stable in biological solutions (or just water) than it is in air, and the route of oxidation in water is distinct from gaseous NO.

One product that has received considerable attention is peroxynitrite. This is largely because the kinetics for its formation from NO and O_2^- are so favorable, exceeding that for superoxide dismutase (54). With the knowledge that both NO and O_2^- are produced in abundance in inflammation, it seems reasonable that peroxynitrite is formed in vivo in these states, although definitive proof is lacking. What has been used to track peroxynitrite formation is peroxynitrite because of its ability to nitrate tyrosine residues (5,8,79,80). It should also be recognized that this reaction is not exclusive for $ONOO^-$, and at best is circumstantial evidence for the formation of peroxynitrite. Furthermore, Grisham and colleagues have demonstrated that unless NO and O_2^- are present in stoichemetric amounts, there is a cancelation reaction (81). Nevertheless, with these caveats there is evidence that nitrotyrosine formation in a variety of inflammatory conditions, e.g., models of inflammatory bowel disease (5,8), gastritis (15), arthritis (82), pulmonary injury in sepsis (80), and atherosclerosis (59). Peroxynitrite has also been described to inhibit alpha-1-proteinase inhibitor (84) and directly cause colitis in rats with luminal application (85). Peroxynitrite has numerous toxic effects on mitochondrial function (86–89) and is genotoxic (27,28,90), actions reported for nitric oxide, raising the possibility that many of the deleterious effects of NO in biological settings are dictated by the formation of other more potent reactive nitrogen species. This concept is consistent with the ubiquitous role of NO in nature as a signal transduction agent regulating a host of organ functions, a role that seems paradoxical if NO is a potent mediator of cell injury.

The expression of iNOS, normally a hallmark of inflammation and NO-related cytotoxicity, is not always associated with tissue injury. The best example of this exception is pregnancy. The nonpregnant uterus does not express iNOS, and NO release is minimal. However, the gravid uterus and placenta are excellent sources of iNOS and produce micromolar amounts of NO (77). Coincidentally, the uteroplacental unit does not display evidence for nitrotyrosine despite the presence of iNOS and high levels of NO, except in models of fetal growth retardation where nitrotyrosine is colocalized with apoptosis (77). In addition, it is now clear that the fetus expresses iNOS on a continual basis, at least from the second trimester, in variety of organs, and in particular uses iNOS-derived NO to regulate fetal vascular tone as indicated by the ability of iNOS inhibitors to constrict the fetal vasculature (91). Thus it is clear that the expression of iNOS and the subsequent release of NO in large amounts is not necessarily associated with cytostasis or cytotoxicity, and in the fetus it is associated with marked cellular proliferation. From these exceptions we need to readdress the role of NO as a mediator of inflammation, perhaps by focusing on the transformation of NO to other, more toxic species.

REFERENCES

1. Kubes P, Wallace JL. Nitric oxide at a mediator of gastrointestinal mucosal injury? Say it ain't so. Mediat Inflamm 1995; 4:397–405.
2. Miller MJS, Grisham MB. Nitric oxide as a mediator of inflammation? You had better believe it. Mediat Inflamm 1995; 4:387–396.
3. Koedel U., Beinatowicz A, Paul R, Frei K, Fontana A, Pfister H-W. Experimental pneumococcal meningitis: cerebrovascular alterations, brain edema, and meningeal inflammation are linked to the production of nitric oxide. Ann Neurol 1995; 37:313–323.
4. Miller MJS, Sadowska-Krowicka H, Chotinaruemol S, Kakkis JL, Clark DA. Amelioration of chronic ileitis by nitric oxide synthase inhibition. J Pharmacol Exp Ther 1993; 264:11–16.
5. Miller MJS, Thompson JH, Zhang X-J, Sadowska-Krowicka H, Kakkis JL, Munshi UK, Sandoval M, Rossi JL, Eloby-Childress S, Bechman YS, Ye YZ, Rodi CP, Manning PT, Currie MG, Clark DA. Role of inducible nitric oxide synthase expression and peroxynitrite formation in guinea pig ileitis. Gastroenterology 1995; 109:1475–1483.
6. Hogaboam CM, Jacobson K, Collins SM, Blennerhassett MG. The selective beneficial effects of nitric oxide inhibition in experimental colitis. Am J Physiol 1995; 268:G673–G684.
7. Connor JR, Manning PT, Settle SL, Moore WM, Jerome GM, Webber RK, Tjoeng FS, Currie MG. Suppression of adjuvant-induced arthritis by selective inhibitor of inducible nitric oxide synthase. Eur J Pharmacol 1995; 273:15–24.
8. Seago ND, Thompson JH, Zhang X-J, Clark DA, Miller MJS. Inducible nitric oxide synthase and guinea-pig ileitis induced by adjuvant. Mediat Inflamm 1995; 4:19–24.
9. Stefanovic-Radic M, Stadler J, Evans CH. Nitric oxide and arthritis. Arthritis Rheumatol 1993; 36:1036–1069.

10. Ialenti A, Moncada S, Di Rosa M. Modulation of adjuvant arthritis by endogenous nitric oxide. Br J Pharmacol 1993; 110:701–706.
11. Aiko S, Grisham MB. Spontaneous intestinal inflammation and nitric oxide metabolism in HLA-B27 transgenic rats. Gastroenterology 1995; 109:142–150.
12. Grisham MG, Specian RD, Zimmerman TE. Effects of chronic nitric oxide synthase inhibition on the pathophysiology observed in a model of chronic granulomatous colitis. J Pharmacol Exp Ther 1994; 271:1114–1121.
13. Weinberg JB, Granger DL, Pisetsky DS, Seldin MF, Misukonis MA, Mason SN, Pippin AM, Ruiz P, Wood CR, Gilkeson GS. The role of nitric oxide in the pathogenesis of spontaneous murine autoimmune disease: increased nitric oxide production and nitric oxide synthase expression in MRI-1 pr/1pr mice, and reduction of spontaneous glomerulonephritis and arthritis by orally administered N^G-monomethyl-L-arginine. J Exp Med 1994; 179:651–659.
14. Cross AH, Misko TP, Liu RF, Hickey WF, Trotter JL, Tilton RG. Aminoguanidine, an inhibitor of inducible nitric oxide synthase, ameliorates experimental autoimmune encephalomyelitis in SLJ mice. J Clin Invest 1994; 93:2684–2690.
15. Mannick EE, Bravo LE, Zarama G, Realpe JL, Zhang X-J, Ruiz B, Fontham ETH, Mera R, Miller MJS, Correa P. Inducible nitric oxide synthase, nitrotyrosine, and apoptosis in *Helicobacter pylori* gastritis: effect of antibiotics and antioxidants. Cancer Res 1996; 56:3238–3243.
16. Salzman AL, Danenberg AG, Ueta I, O'Connor M, Linn SC, Szabo C. Induction and activity of nitric oxide synthase in cultured human intestinal epithelial monolayers. Am J Physiol 1996; 270:G565–G573.
17. Ribbons KA, Zhang X-J, Thompson JH, Greenberg SS, Moore WM, Kornmeier CM, Currie MG, Lerche N, Blanchard J, Clark DA, Miller MJS. Potential role of nitric oxide in a model of chronic colitis in rhesus macaques. Gastroenterology 1995; 108:705–711.
18. Arnold WP, Mittal CK, Katsuki S, Murad. F. Nitric oxide activates guanylate cyclase and increases guanosine 3′,5′-cyclic monophosphate levels in various tissue preparations. Proc Natl Acad Sci USA 1977; 74:3203–3207.
19. Schultz K-D, Schultz K, Schultz G. Sodium nitroprusside and other smooth muscle relaxants increase cyclic GMP levels in rat ductus deferens. Nature 1977; 265:750–751.
20. Lander HM, Ogiste JS, Pearce SF, Len R, Novogrodsky A. Nitric oxide-stimulated guanine nucleotide exchange on p21 ras. J Biol Chem 1995; 270:7017–7020.
21. Lander HM, Jacovina AT, Davis RJ, Taurus JM. Differential activation of mitogen-activated protein kinases by nitric oxide-related species. J Biol Chem 1996; 271:19705–19709.
22. Ferrari G, Greene LA. Proliferative inhibition by dominant-negative Ras rescues naive and neuronally differentiated PC12 cells from apoptotic death. EMBO J 1994; 13:5922–5928.
23. Lander HM, Ogiste JS, Teng KK, Novogrodsky A. P21 ras as a common signaling target of reactive free radicals and cellular redox stress. J Biol Chem 1995; 270:21195-21198.
24. Mellion BT, Ignarro LJ, Ohlstein EH, Pontecovro EG, Hyman AL, Kadowitz PJ.

N′-nitro-N-nitrosoguanidine through a stomach tube. Top Chem Carcinog 1972; 121–132.
57. Mirvish S. The etiology of gastric cancer: intragastric nitrosamide formation and other theories. J Natl Cancer Inst 1993; 71:629–647.
58. Grisham MB, Ware K, Gilleland HE Jr, Gilleland LB, Abell CL, Yamada T. Neutrophil-mediated nitrosamine formation: role of nitric oxide in rats. Gastroenterology 1992; 103:1260–1266.
59. Ohshima H, Bandaletora TY, Brouet I, Bartsch H, Kirby G, Ogunbiyi F, Vatansapt V, Pipitgool V. Increased nitrosamine and nitrate biosynthesis mediated by nitric oxide synthase induced in hamsters infected with liver fluke (Opisthorchis viverrini). Carcinogenesis (Lond) 1994; 15:271–275.
60. Messmer UK, Reimer DM, Brune B. Nitric oxide induced poly(ADP-ribose) polymerase cleavage in RAW 264.7 macrophage apoptosis is blocked by Bcl-2. FEBS Lett 1996; 384:162–166.
61. Wink DA, Laval J. The Fpg protein, a DNA repair enzyme, is inhibited by the biomediator nitric oxide in vitro and in vivo. Carcinogenesis 1994; 15:2125–2129.
62. Kane DJ, Sarafian TA, Anton R, Hahn H, Gralla EB, Valentine JS, Ord T, Bredesem DE. Bcl-2 inhibition of neural death: decreased generation of reactive oxygen species. Science 1993; 262:1276–1277.
63. Hockenbery DM, Oltval ZN, Yin X-M, Milliman CL, Korsmayer SJ. Bcl-2 functions in an antioxidant pathway to prevent apoptosis. Cell 1993; 75:241–251.
64. Sadowska-Krowicka H, Mannick EE, Oliver PD, Sandoval M, Zhang X J, Elohy-Childress S, Clark DA, Miller MJS. Genistein and gut inflammation: role of nitric oxide. Proc Soc Exp Biol Med 1998; 217:351–357.
65. Miksicek RJ. Interaction of naturally occurring non-steroidal estrogens with expressed recombinant human estrogen receptor. J Steroid Biochem Mol Biol 1994; 49:153–160.
66. Yanagihara K, Ito A, Toge T, Numoto M. Antiproliferative effects of isoflavones on human cancer cell lines established from the gastrointestinal tract. Cancer Res 1993; 53:5815–5821.
67. Lee K-M, Toscas K, Villareal ML. Inhibition of bradykinin- and tharpsigargin-induced Ca^{2+} entry by tyrosine kinase inhibitors. J Biol Chem 1993; 268:9945–9948.
68. Pryor WA, Squadrito GL. The chemistry of peroxynitrite: a product from the reaction of nitric oxide with superoxide. Am J Physiol 1995; 268:L699–L722.
69. Radi R, Beckman JS, Bush KM, Freeman BA. Peroxynitrite-induced membrane lipid peroxidation: the cytotoxic potential of superoxide and nitric oxide. Arch Biochem Biophys 1991; 288:481–487.
70. Buffinton GD, Doe WF. Altered ascorbic acid status in the mucosa from inflammatory bowel disease patients. Free Radic Res 1995; 22:131–143.
71. McKenzie SJ, Baker MS, Buffinton GD, Doe WF. Evidence of oxidant-induced injury to epithelial cells during inflammatory bowel disease. J Clin Invest 1996; 98:136–141.
72. Sandoval M, Zhang X-J, Liu X, Mannick EE, Clark DA, Miller MJS. Peroxynitrite-induced apoptosis in T84 and RAW 264.7 cells: attenuation by L-ascorbic acid. Free Radic Biol Med 1997; 22:489–495.

73. Uppu RM, Squadrito GL, Pryor WA. Acceleration of peroxynitrite oxidations by carbon dioxide. Arch Biochem Biophys 1996; 327:335–343.
74. Ruiz B, Rood JC, Fontham ETH, Malcom GT, Hunter FM, Sobhan M, Johnson WD, Correa P. Vitamin C concentration in gastric juice before and after anti-Helicobacter pylori treatment. Am J Gastroenterol 1994; 89:533–539.
75. Correa P. Helicobacter pylori and gastric carcinogenesis. Am J Surg Pathol 1995; 19:S37–S43.
76. Pearson DC, Jourd'heil D, Meddings JB. The antioxidant properties of 5-amino-salicydic acid. Free Radic Biol Med 1996; 21:367–373.
77. Miller MJS, Voelker CA, Olister SA, Thompson JH, Zhang X-J, Rivera D, Eloby-Childress S, Liu X, Clark DA, Pierce MR. Fetal growth retardation in rats may result from apoptosis: role of peroxynitrite. Free Radic Biol Med. In press.
78. Beckman JS, Beckman TW, Chen J, Marshall PA, Freeman BA. Apparent hydroxyl radical production by peroxynitrite: implications for endothelial injury from nitric oxide and superoxide. Proc Natl Acad Sci USA 1990; 87:1620–1624.
79. Ischiropoulos H, Zhu L, Beckman JS. Peroxynitrite formation from macrophage-derived nitric oxide. Arch Biochem Biophys 1992; 298:446–451.
80. Kooy NW, Royall JA, Ye YZ, Kelly DR, Beckman JS. Evidence for in vivo peroxynitrite production in human acute lung injury. Am J Respir Crit Care Med 1995; 151:1250–1254.
81. Miles AM, Bohle DS, Glassbrenner PA, Hansert B, Wink DA, Grisham MB. Modulation of superoxide-dependent oxidation and hydroxylation reactions by nitric oxide. J Biol Chem 1996; 271:40–47.
82. Kaur H, Halliwell B. Evidence of nitric oxide-mediated oxidative damage in chronic inflammation. Nitrotyrosine in serum and synovial fluid from rheumatoid patients. FEBS *Lett* 1994; 350:9–12.
83. Beckman JS, Ye YZ, Anderson PG, Chen J, Accavitti MA, Tarpey MM, White CR. Extensive nitration of protein tyrosines in human atherosclerosis detected in immunohistochemistry. Biol Chem Hoppe Seyler 1994; 375:81–88.
84. Moreno JJ, Pyror WA. Inactivation of α1-proteinase inhibitor of peroxynitrite. Chem Res Toxicol 1992; 5:425–431.
85. Rachmilewitz D, Stamler JS, Karmeli F, Mullins ME, Singel DJ, Loscalzo J, Xavier RJ, Padolsky DK. Peroxynitrite-induced rat colitis—a new model of colonic inflammation. Gastroenterology 1993; 105:1681–1686.
86. Radi R, Beckman JS, Bush KM, Freeman BA. Peroxynitrite oxidation of sulfhydryls. J Biol Chem 1991; 266:4244–4250.
87. Rubbo H, Dennicola A, Radi R. Peroxynitrite inactivates thiol-containing enzymes of Trypanosoma cruzi energetic metabolism and inhibits cell respiration. Arch Biochem Biophys 1994; 308:96–102.
88. Zhu L, Gunn C, Beckman JS. Bactericidal activity of peroxynitrite. Arch Biochem Biophys 1992; 298:452–457.
89. Radi R, Rodriguez M, Castro L, Telleri R. Inhibition of mitochondrial electron transport by peroxynitrite. Arch Biochem Biophys 1993; 308:89–95.
90. Zingarelli B, O'Connor M, Wong H, Salzman AL, Szabó C. Peroxynitrite-mediated DNA strand breakage activates poly-adenosine diphosphate ribosyl synthetase

and causes cellular energy depletion in macrophages stimulated with bacterial lipopolysaccharide. J Immunol 1996; 156:350–358.
91. Bustamante SA, Pang Y, Romero S, Pierce MR, Voelker CA, Thompson JH, Sandoval M, Liu X, Miller MJS. Inducible nitric oxide synthase and the regulation of central vessels caliber in the fetal rat. Circulation 1996; 94:1948–1953.
92. Albina JE, Cui S, Mateo RB, Reichner JS. Nitric oxide-mediated apoptosis in murine penitoneal macrophages. J Immunol 1993; 150:5080–5085.
93. Cui S, Reichner JS, Mateo RB, Albina JE. Activated murine macrophages induce apoptosis in tumor cells through nitric oxide dependent or independent mechanisms. Cancer Res 1994; 54:2462–2467.
94. Yang X, Galeano NF, Szabolcs M, Sciacca RR, Cannon PJ. Oxidized low density lipoproteins alter macrophage lipid uptake, apoptosis, viability and nitric oxide synthesis. J Nutr 1996; 126:10725–10755.
95. Messmer UK, Brune B. Nitric oxide (NO) in apoptotic versus necrotic RAW 264.7 macrophage cell death: the role of the NO^- donor, NAD^+ content, and p53 accumulation. Arch Biochem Biophys 1996; 327:1–10.
96. Mannick EE, Oliver PD, Sadowska-Krowicka H, Miller MJS. Inhibition of inducible nitric oxide synthase reverses endotoxin tolerance in cultured macrophages. In: Moncada A, Stamler J, Gross S, Higgs EA, eds. *The Biology of Nitric Oxide*. 1996; Vol. 5, p 164.
97. Sandoval M, Liu X, Oliver PD, Zhang X-J, Clark DA, Miller MJS. Nitric oxide induces apoptosis in a human colonic epithelial cell line, T84. Mediat Inflamm 1995; 4:248–250.
98. Fukuo K, Hata S, Suhara T, Nakahashi T, Shinto Y, Tsujimoto Y, Morimoto S, Ogihara T. Nitric oxide induces upregulation of Fas and apoptosis in vascular smooth muscle. Hypertension 1996; 27:823–826.
99. Geng Y, Wu Q, Musynski M, Hansson GK, Libby P. Apoptosis of vascular smooth muscle cells induced by in vitro stimulation with interferon-gamma, tumor necrosis factor-alpha, and interleukin-1-beta. Arteroscler Thromb Vasc Biol 1996; 16:19–27.
100. Geng Y, Hellstrand K, Wennmalm A, Hansson GK. Apoptotic death of leukemic cells induced by vascular cells expressing nitric oxide synthase in response to gamma-interferon and tumor necrosis factor-alpha. Cancer Res 1996; 56:866–874.
101. Keu ML, Chau YP, Wang JH, Shiah SG. Inhibitors of poly(ADP-ribose) polymerase block nitric oxide-induced apoptosis but not differentiation in human leukemia HL-60 cells. Biochem Biophys Res Commun 1996; 219:502–508.
102. Martin-Sanz P, Diaz-Guerra MJ, Casada M, Bosca L. Bacterial lipopolysaccharide antagonizes transforming growth factor beta 1-induced apoptosis in primary cultures of hepatocytes. Hepatology 1996; 23:1200–1207.
103. Stangel M, Zetts UK, Mix E, Zielasek J, Toyka KV, Hartung HP, Gold R. H_2O_2 and nitric oxide-mediated oxidative stress induce apoptosis in rat skeletal muscle myoblasts. J Neuropathol Exp Neurol 1996; 55:36–43.
104. Mannick JB, Asano K, Izumi K, Kieff E, Stamler JS. Nitric oxide produced by human B lymphocytes inhibits apoptosis and Epstein-Barr virus reactivation. Cell 1994; 79:1137–1146.

105. Bonfoco E, Krainc D, Ankarcrona M, Nicotera P, Lipton SA. Apoptosis and necrosis: two distinct events induced, respectively, by mild and intense insults with N-methyl-D-aspartate or nitric oxide/superoxide in cortical cell cultures. Proc Natl Acad Sci USA 1995; 92:7162–7166.
106. Ischiropoulos H, Duran D, Horwitz J. Peroxynitrite-mediated inhibition of DOPA synthesis in PC12 cells. J Neurochem 1995; 65:2366–2372.
107. Estevez AG, Radi R, Barbeito L, Shin JT, Thompson JA, Beckman JS. Peroxynitrite-induced cytotoxicity in PC12 cells: evidence for an apoptotic mechanism differentially modulates by neurotrophic factors. J Neurochem 1995; 65:1543–1550.
108. Troy CM, Defossi D, Prochiantz A, Greene LA, Shelanski MC. Down regulation of Cu/Zn superoxide dismutase leads to cell death via the nitric oxide-peroxynitrite pathway. J Neurosci 1996; 16:253–261.
109. Salgo WA, Squadrito GL, Pryor WA. Peroxynitrite induces apoptosis in thymocytes. Biochem Biophys Res Commun 1995; 215:1111–1118.
110. Sandoval M, Zhang X-J, Liu X, Mannick EE, Clark DA, Miller MJS. Peroxynitrite-induced apoptosis in T84 and RAW 264.7 cells lines: effect of plant antioxidants. Nitric Oxide: Biology and Chemistry. 1997; 1:476–483.
111. Rachmilewitz D, Karmeli F, Okon F. Sulfhydryl blocker-induced rat colonic inflammation is ameliorated by inhibition of nitric oxide synthase. Gastroenterology 1995; 109:98–106.
112. Lindsay RM, Smith W, Rossiter WP, McIntyre MA, Williams BC, Baird JD. N^{ω}-Nitro-arginine methylester reduces the incidence of IDDM in BB/E rats. Diabetes 1995; 44:365–368.
113. Ianaro A, O'Donnell CA, Di Rosa M, Liew FY. A nitric oxide synthase inhibitor reduces inflammation, down-regulates inflammatory cytokines and enhances interteukin-10 production in carrageenin-induced edema in mice. Immunology 1994; 82:370–375.
114. Tracey WR, Nakame M, Kulk J, Budzik G, Klinghoffer V, Harris R, Carter CR. The nitric oxide synthase inhibitor, L-N^{G}-monometh-L-arginine, reduces carrageenin-induced pleurisy in the rat. J Pharmacol Exp Ther 1995; 273:1285–1299.
115. Parks DJ, Cheung MK, Chan CFC, Rosberge FC. The role of nitric oxide in uveitis. Arch Ophthalmol 1995; 112:544–546.
116. Vane JR, Mitchell JA, Appleton I, Tomlinson A, Bishop-Bailey D, Croxtail J, Willoughby DA. Inducible iso forms of cyclooxygenase and nitric oxide synthase in inflammation. Proc Natl Acad Sci USA 1994; 91:2046–2050.

11
Nitric Oxide as a Cellular Mediator of Apoptosis

Jeffrey D. Laskin and Thomas M. Mariano
University of Medicine and Dentistry of New Jersey–Robert Wood Johnson Medical School, Piscataway, New Jersey

I. NITRIC OXIDE AND APOPTOSIS

Nitric oxide is a free radical gas and perhaps the smallest known biological messenger and cytotoxic mediator. Since its discovery more than a decade ago as an endothelium-derived relaxing factor (EDRF) (1–3), nitric oxide has been implicated in many physiological and pathophysiological processes (4–10). Recently, considerable attention has focused on the role of nitric oxide in the regulation of cell growth. In earlier reports, nitric oxide was shown to inhibit cell growth, a process that was associated with suppression of DNA synthesis and, eventually, cell death (11). It is now well established that in mammalian cells, death can occur via multiple pathways including necrosis and apoptosis, and an emerging literature indicates that nitric oxide can induce cell death by both pathways. The present review will focus on the regulation of apoptosis by nitric oxide and its oxidation products.

Apoptosis, or gene-directed (programmed) cell death, is an important and highly regulated mechanism used for the clearance of cells that are virus infected, injured beyond the capabilities of repair mechanisms, or no longer needed by the host (24,25). The elimination of cells occurs without the induction of an immune inflammatory response. Apoptosis occurs during normal cellular development and differentiation of multicellular organisms, tumor regression, and a number of pathological disorders (12–16). Morphologically, apoptosis is characterized by a series of structural changes which include a decrease in cell volume, condensation of nuclear chromatin, nuclear fragmentation, plasma membrane blebbing, lack of mitochondrial swelling, and the sequestration of cellular contents into membrane-bound structures called apoptotic bodies (17). The apoptotic bodies are then extruded into the extracellular environment, where they are phagocytized by neighboring cells (18–21).

Over the last several years it has been established that proteases, in particular, cysteine proteases such as interleukin-1β-converting enzyme, are integral components of the apoptotic cell death metabolic pathway (22). A large number of homologous proteases, now referred to as caspases (cysteine proteases that cleave at aspartic acid residues), have been identified (23,24). The caspases themselves are expressed as inactive proenzymes which become activated by cleavage, usually by another caspase. The activated caspases then cleave and inactivate targets such as poly(ADP-ribose) polymerase (PARP), which may act as a molecular nick sensor and plays a role in rejoining DNA strand breaks (25,26), DNA-dependent protein kinases, and nuclear laminins (27). One of the hallmark characteristics of apoptosis is activation of endogenous endonucleases which result initially in the production of large DNA fragments (50 to 300 kb) and ultimately small oligonucleosomal-size products (28) that, when displayed on agarose gels, are commonly referred to as DNA ladders.

A variety of cell types have been reported to undergo apoptosis when stimulated to produce nitric oxide by agents such as endotoxin, cytokines, and/or growth factors. For example, nitric oxide-induced apoptosis occurs in both mouse and human macrophages (29–31), in tumor cells (33–35,110), primary cultures of rat pancreatic islet cells (35) and pancreatic beta cell lines (35–38), rat lung fibroblasts (39), renal mesangial cells (40), vascular smooth muscle cells (41), and adrenal vascular endothelial cells (42). Primary cultures of cerebeller granule cells (43–45), cortical cultures (46), embryonic motor neurons (47), and astrocytes (48) as well as neuronal cell lines (44) can also undergo apoptosis in direct response to nitric oxide. Interestingly, the presynaptic actions of nitric oxide in neuronal tissue can also lead to toxicity (43).

A number of studies have also demonstrated that tumor cells co-cultured with nitric oxide-producing cells such as macrophages can undergo apoptosis (49–52), a process that is likely to be important in macrophage-mediated cytotoxicity. Similar results have been reported with primary cultures of neurons co-cultured with nitric oxide-producing glial cells (53), in rat Kupffer cells co-cultured with hepatoma cells (54), in mixed cultures of mouse osteoblasts and bone marrow cells (55), in type II pneumocytes cocultured with alveolar macrophages (56), and in co-cultures of thymocytes and endothelial cells stimulated to express nitric oxide with endotoxin and cytokines (57). This phenomenon is important since it suggests that it is not just the nitric oxide-producing cells that are susceptible to apoptosis. Thus, nitric oxide can freely diffuse across cell membranes (58) and affect neighboring cells. In this regard, a number of studies have shown that addition of exogenous nitric oxide in the form of saturated solutions of nitric oxide or nitric oxide-donating compounds can induce apoptosis in neuronal cells (53), macrophages (30,59), smooth muscle cells (41,60,61), fibroblasts (33,62), melanoma cells (34), glomerular mesangial (63,64), epithelial and endothelial cells (63), pancreatic beta cells (38), chondrocytes (65), thymocytes (29,57), and neutrophils (66). The fact that nitric oxide released by one cell can affect nearby cells may be important in tissue injury where apoptosis is observed (67,68).

For example, in a cardiac model of allograft rejection, increased numbers of cells producing the enzyme responsible for synthesizing nitric oxide, i.e., nitric oxide synthase (NOS), and apoptotic cardiac myocytes, macrophages, and endothelial cells were observed in close proximity (69,70). In the thymus of endotoxin-treated mice, NOS was evident in areas adjacent to apoptotic foci (57); in human lupus nephritis, there is a direct correlation between NOS-positive glomeruli and apoptosis (71), while in a rabbit model of myocardial infarction, induction of NOS correlated with areas of cellular apoptosis (72). In these model systems, apoptosis can be observed in those cells producing nitric oxide, as well as in surrounding cells. It is important to note that in the case of immune cells such as macrophages and granulocytes, apoptosis may not necessarily be due to their ability to produce nitric oxide as these cells normally undergo apoptosis during the resolution of inflammation (73).

II. SIGNALING MECHANISMS BY WHICH NITRIC OXIDE AFFECTS APOPTOSIS

At the present time, little is known about the signaling mechanisms by which nitric oxide triggers apoptosis. Approaches that have been used to begin to understand this problem include identifying the precise targets of nitric oxide in cells and determining if these targets are integral components of the apoptotic signaling cascade. Because of its ability to extract electrons from electron-rich substrates, it appears that nitric oxide has the a ability to react with a large number of cellular components including membranes, enzymes, and DNA. When key structural elements of cells or enzymes are subject to electrophilic attack, altered cell functioning may result, and this may be critical for initiating apoptosis. Important targets include cellular sulfhydryls as well as heme- and iron-sulfur-containing proteins (74). Inhibition of enzymes important in DNA synthesis such as ribonucleotide reductase or intermediary metabolism such as *cis*-aconitase, glyceraldehyde-3-dehydrogenase (75), or inhibition of mitochondrial respiration (76) may lead to cell growth arrest, a prerequisite for apoptosis (77). Nitric oxide inhibits cytochrome C oxidase and complexes I–III and this may cause the release of mitrochondrial-derived factors regulating the apoptotic process (78–82).

It is also well recognized that nitric oxide stimulates the production of cGMP in target cells by virtue of its ability to activate guanylyl cyclase (83). Indeed, activation of cGMP-dependent protein kinase (protein kinase G) may be a key step by which nitric oxide regulates apoptosis. A number of studies have implicated nitric oxide-induced cGMP in apoptosis. For example, using cGMP analogs and soluble guanylyl cyclase inhibitors, Loweth et al. (38) provided evidence that cGMP and protein kinase G were involved in nitric oxide-mediated apoptosis in pancreatic beta-cells. Similar results were reported with embryonic neurons cultured with brain-derived neurotrophic factor (84).

A key enzyme in the protein kinase signaling cascade that has been demonstrated to be a component of the apoptotic process is protein kinase C (85,86). Membrane localization of protein kinase C is thought to be critical in the activation of the enzyme (87). Both the presence and absence of membrane-associated protein kinase C isoforms have been reported to regulate the activity of nitric oxide. For example, in human smooth muscle cells, activation of protein kinase C by growth factors is inhibited by sodium nitroprusside, and this has been suggested as a possible mechanism for the antiproliferative effects of nitric oxide (88). In contrast, S-nitroso-N-acetylpenicillamine-induced apoptosis in rabbit vascular smooth muscle cells was associated with prolonged membrane translocation of protein kinase C-epsilon (89). That protein kinase C is important in regulating apoptosis in these studies was further supported by the findings that inhibitors of the enzyme prevented, while activators of the enzyme enhanced, nitric oxide donor-induced apoptosis in these cells (89). Interestingly, protein kinase A inhibitors also suppressed apoptosis in these cells, providing support for the idea that this enzyme is also involved in regulating the actions of nitric oxide. The substrates phosphorylated by protein kinases and their role in the apoptotic process have not yet been elucidated.

It is important to note that two key events in the apoptotic process, activation of caspases and proteolytic cleavage of poly(ADP-ribose) polymerase, have been observed in cells during nitric oxide- and peroxynitrite-induced apoptosis (90–93). At the present time the mechanisms leading to caspase activation are unclear. Since both nitric oxide and peroxynitrite modify and inactivate caspases (94–96), it appears unlikely that apoptosis is initiated directly via these proteases. Presumably, the nitric oxide-induced signaling cascade, for example, protein kinase activation, leads to activation of the caspases.

Nitric oxide and peroxynitrite are known to induce DNA damage, and the protein product of the p53 tumor suppressor gene plays an important role in this cellular response. The p53 protein is a transcription factor that increases the rate of transcription of several genes shown to contain p53-dependent, *cis*-acting, DNA-response elements (97). The p53 concentration in a cell is normally very low, but upon exposure to the appropriate signals, the levels increase and the protein is activated. In earlier work it was demonstrated that apoptosis induced by nitric oxide was associated with increased expression of p53 and apoptosis (98–102). For example, Messmer and co-workers (98) reported that nitric oxide produced endogenously by cytokines or exogenously by nitric oxide donors resulted in the stimulation of p53 and apoptosis of mouse RAW 264.7 macrophages and pancreatic RINm5F cells. NOS inhibitors prevented the inducible nitric oxide production as well as p53 expression and apoptosis. The fact that nitric oxide readily induces p53 suggests that this may be an important upstream event in the signaling pathways leading to apoptosis.

III. NITROSOTHIOLS AND PEROXYNITRITE AS MEDIATORS OF APOPTOSIS?

A number of laboratories have raised the possibility that a metabolite or oxidation product of nitric oxide is responsible for mediating its signaling, potentially, by acting as a sink for nitric oxide or by directly mediating cellular responses. For example, nitrosothiols, formed by the facile reaction of nitric oxide with sulfhydryl groups in proteins and/or low-molecular-weight peptides such as glutathione, may be key reactive nitrogen intermediates (103). Metabolites such as these can potentially act directly in cellular signaling and enhance the biological effects of nitric oxide, or are metabolized further to initiate cellular events such as apoptosis. The nitrosothiols can also release nitric oxide, thus acting as a cellular store for this mediator or transnitrosate critical thiols in proteins and possibly modulating their biological activity (104). The ability of nitric oxide to form S-nitrosothiols and inhibit cysteine-containing proteins may be a critical determinant in the regulation of the apoptotic cell death (78) (see below).

One important oxidation product of nitric oxide is peroxynitrite (105,106). Formed by the reaction of nitric oxide with superoxide anion, peroxynitrite is a potent oxidizing agent that is formed in a large number of cell types and is thought to mediate many of the toxic actions of nitric oxide (105,107,108). In the presence of transition metals, peroxynitrite readily nitrates tyrosine residues in proteins (109), and the idea that peroxynitrite plays a role in the action of nitric oxide in vivo is largely based on the detection of nitrate-modified proteins in tissues (107). Peroxynitrite is also highly reactive with iron-sulfur centers, thiols, thiol ethers, and zinc fingers (105,107). In addition to modifying sulfhydryl-containing proteins, peroxynitrite oxidizes membrane phospholipids as well as DNA (110,111). However, with a half-life of less than 1 sec at physiological pH, the activity of peroxynitrite will be dependent on its concentration and accessibility to critical cellular targets before it decomposes (107). Cytotoxicity may be due directly to peroxynitrite, activated forms of this reactive nitrogen intermediate (107), and breakdown products including hydroxyl radicals (112). Almost all cells that can produce nitric oxide form peroxynitrite, since superoxide anion is a rather ubiquitous cellular metabolite formed during mitochondrial respiration and as a byproduct of numerous oxidoreductase reactions (113). Elevated levels of superoxide are formed in cells deficient in the detoxifying enzyme superoxide dismutase (114,115), in phagocytic cells undergoing a respiratory burst (116), and in immunologically activated cells (11). These last cells are often exposed to proinflammatory cytokines that can induce nitric oxide (11).

Although peroxynitrite is formed endogenously in many cell types, whether or not it plays a role in mediating the actions of nitric oxide is uncertain. In rat motor neurons, endogenous peroxynitrite has been implicated in apoptosis because both nitric oxide and superoxide anion appear to be necessary for the process and nitrotyrosine-containing proteins are found in the apoptotic cells (47). Direct support for a role for this compound in modulating apoptosis is based on studies demonstrating that in some cell types, exogenous administration of peroxynitrite causes apoptosis. For example, using rat AR4-2J pancreatic acinar cells, Sata et al. (117) found that low concentrations of peroxynitrite (2 to 5 μM) induce DNA fragmentation, an indication of apoptosis. Similarly, Salgo and colleagues (118–120) demonstrated that near physiological concentrations of peroxynitrite initiate apoptosis in rat thymocytes. These same concentrations of peroxynitrite were found to readily nick DNA and oxidize cellular thiols in the cells.

These investigators also reported that peroxynitrite induces oxidative stress, as evidenced by the formation of thiobarbituric acid-reactive material and DNA protein crosslinks in the cells (118). Peroxynitrite-induced oxidative stress, as well as apoptosis, was found to be inhibited by Trolox, an antioxidant that is a water-soluble vitamin E analog, and several additional phenolic antioxidants including 3-*tert*-butyl-4-hydroxyanisole, butylated hydroxytoluene and 2,6-diisopropylphenol (118). Similarly, Lin et al. (121,122) found that peroxynitrite-induced apoptosis in the human HL60 promyelocytic cell line was associated with increased production of superoxide anion and hydrogen peroxide. Production of these reactive oxygen intermediates as well as apoptosis was inhibited by N-acetyl-L-cysteine, a free radical scavenger. In several other cell lines, antioxidants have also been reported to inhibit peroxynitrite-induced apoptosis, including RAW 264.7 macrophages, human T84 colon tumor cells (123), and bovine retinal pigmented epithelial cells (124). Taken together, these data suggest that reactive oxygen intermediates may be important components of the peroxynitrite-induced apoptotic signaling pathway.

In rat pheochromocytoma PC12 cells, Estevez et al. (125) also found that peroxynitrite induces apoptosis. However, in these cells, nerve growth factor and insulin protected against apoptosis, while acidic and basic fibroblast growth factors enhanced peroxynitrite-induced apoptosis. Presumably, these growth factors modulate peroxynitrite-induced apoptosis by altering the redox balance in the cells. In this regard, Heck et al. (11) reported that epidermal growth factor suppressed hydrogen peroxide as well as nitric oxide production by keratinocytes.

Mechanisms by which epidermal growth factor can alter the production of reactive oxygen intermediates in keratinocytes include the modulation of mitochondrial respiration and/or the activity of a variety of cellular oxidoreductases (11). In PC12 cells, Spear et al. (126) found that phosphatidylinositol 3-kinase inhibitors reduced the ability of nerve growth factor to protect against peroxynitrite-induced apoptosis, suggesting that this enzyme mediates the effects of the growth factor. These investigators proposed a mechanism whereby peroxynitrite inhibits the interaction of phosphatidylinositol 3-kinase with phosphotyrosine-containing proteins (126). Presumably, these phosphotyrosine-containing proteins are critical in regulating the apoptotic process.

It should be noted that in all studies using exogenous peroxynitrite to induce apoptosis, either a nitric oxide-donating compound or chemically synthesized material was used. In some cases, the chemically synthesized product was contaminated with hydrogen peroxide as well as potentially toxic oxidation products (127) and thus, the precise nature of the chemical species responsible for modulating apoptosis is uncertain. In this regard, many of the cytokines that stimulate nitric oxide production in cells also increase intracellular hydrogen peroxide (11). Hydrogen peroxide is known to modulate a variety of growth factor-mediated signaling cascades including those leading to apoptosis (81,128,129), as well as the activity of transcription factors such as nuclear factor-κB (130) and thus, it may be important in signal transduction. In this regard, hydrogen peroxide induces apoptosis in a variety of cell types including sinusoidal endothelial cells (131,132). Appropriate control experiments for peroxynitrite include hydrogen peroxide, peroxynitrite incubated with MnO_2, a treatment which removes excess hydrogen peroxide, and decomposed peroxynitrite (122) or peroxynitrite prepared without hydrogen peroxide (133).

It is important to note that detection of oxidized products such as 3-nitrotyrosine in cells and tissues producing nitric oxide does not necessarily mean that the reaction is mediated by peroxynitrite. For example, it has been demonstrated that the reaction of hypochlorous acid or myeloperoxidase with the nitric oxide metabolite nitrite also promotes tyrosine nitration via nitrogen dioxide and nitryl chloride (124). These latter reactive nitrogen intermediates are readily formed in neutrophils, a rich source of myeloperoxidase (134), suggesting that nitrite is a potentially important regulator of the inflammatory process.

IV. INHIBITION OF APOPTOSIS BY NITRIC OXIDE

It is important to note that in some instances nitric oxide has been shown to inhibit apoptosis. For example, using rat hepatocytes, Kim et al. (95) reported that both exogenously applied nitric oxide and nitric oxide generated endogenously can prevent growth factor deprivation as well as tumor necrosis factor- or anti-

FAS antibody-induced apoptosis. It was hypothesized that caspase-3-like enzyme activity may be the target for nitric oxide in hepatocytes. In this regard, these investigators found that nitric oxide readily inhibited this enzyme, presumably via protein S-nitrosylation. Inhibition of apoptosis via S-nitrosylation of caspases has also been described in human umbilical endothelial cells (94,135), in sheep arterial endothelial cells transfected with iNOS (136), in human leukemia cells (78,137), in an embryonic kidney cell line (138), and in primary cerebrocortical neurons (138). S-nitrosylation of recombinant reconstituted caspase-3 in vitro also suppresses enzyme activity (94–96). In human umbilical venous endothelial cells, Dimmeler et al. (139) showed that nitric oxide-induced inhibition of apoptosis is associated with S-nitrosylation of Cys163 in caspase CPP-32, a functionally essential amino acid, thus providing a precise site of action for nitric oxide.

In hepatocytes, Kim et al. (95) further demonstrated that the effects of nitric oxide on apoptosis were reduced by suppressing the formation of cGMP, suggesting that inhibition of caspase-3 by nitric oxide may also occur via a cGMP-dependent mechanism. This dependence has also been described in rat dorsal root ganglion neurons (140) and in rat ovarian follicles (141). As indicated above, nitric oxide-induced apoptosis has also been associated with activation of caspases; however, it should be pointed out that several different members of the caspase family appear to be involved in the apoptotic process (142), and whether or not nitric oxide induces or prevents apoptosis may depend on which proteases are activated in a particular cell type (143). Interestingly, the inhibitory effects of nitric oxide on rat hepatocyte apoptosis is in contrast to earlier work demonstrating that nitric oxide can promote apoptosis in hepatocytes isolated from endotoxin-treated rats (144). In these latter studies, nitric oxide was found to contribute to endotoxin-induced tissue injury. Furthermore, endotoxin treatment enhanced cytokine- and lipopolysaccharide-induced nitric oxide production by hepatocytes, an effect that was suppressed by nitric oxide synthase inhibitors. Differences in the results between the above studies may be due to differences in the strains of animals or protocols used prior to isolation of the hepatocytes.

The ability of nitric oxide to suppress apoptosis may also be due to direct inhibition of signaling cascades necessary to induce cell death. For example, nitric oxide is known to inhibit the Janus kinases (JAK), an important protein kinase family mediating the action of a host of cytokines, including those known to induce nitric oxide and initiate apoptosis (145). Nitric oxide-donating compounds have also been reported to inhibit apoptosis in trophic factor-deprived rat pheochromocytoma PC12 cells (146). In these cells, c-Jun kinase appears to be necessary for apoptosis, and nitric oxide inhibits activation of this important regulatory protein (146). Whether or not nitric oxide regulates the apoptotic process may also depend on the its concentration. Using rat thymocytes, Sandau and Brune (147) demonstrated that high concentrations of nitric oxide derived from a nitric oxide donor inhibited apoptosis while low concentrations promoted apoptosis. However, it should be noted that in most systems, the concentrations of nitric oxide that regulate apoptosis are not known. In fact, in several different systems it has been reported that high concentrations of nitric oxide can promote apoptosis (29,57). Inhibition of apoptosis by nitric oxide may also be dependent on the growth state of the cells. Thus, in bovine endothelial cells, nitric oxide has been found to inhibit apoptosis in confluent, but not proliferating cells (148).

V. ENZYMES RESPONSIBLE FOR PRODUCING THE NITRIC OXIDE THAT CAN LEAD TO APOPTOSIS

Nitric oxide is produced by the NOS family of enzymes which catalyze the oxidation of the amino acid L-arginine to nitric oxide and L-citrulline. The active form of the enzyme is a dimer which utilizes FMN, FAD, heme, NADPH, and tetrahydrobiopterin as cofactors. There is also a site on the enzyme for Ca^{2+}-calmodulin binding. Three isoforms of NOS have been characterized. Although the first two, neuronal (nNOS, NOS1) NOS and endothelial (eNOS, NOS3) NOS, were originally thought to be constitutive isoforms, it has been shown that they can be induced under certain circumstances (149–154). The third isoform of NOS (iNOS, NOS2) is often called the inducible form because of its ability to be induced by endotoxins and a variety of cytokines and growth factors (155–160). High-output production of nitric oxide by iNOS is thought to be a significant factor in the body's response to infection and disease, whereas nNOS and eNOS produce relatively smaller amounts of nitric oxide which are involved in physiological functions such as the regulation of vascular tone and nerve transmission.

In many studies, relatively large quantities of nitric oxide are required to induce apoptosis, suggesting that the high-output iNOS isoform of NOS mediates this process. Indeed, in cells such as macrophages (54,161), cardiac myocytes (69), embryonic motor neurons (47), and astrocytes (48), there is a direct correlation among cellular expression of iNOS, nitric oxide production, and apoptosis. As indicated above, expression of iNOS is also observed in in vivo models of tissue apoptosis, including endotoxin-induced thymic injury (57), human lupus nephritis (71), and myocardial infarction (72). However, at the present time, the possibility that nNOS or eNOS contributes to the regulation of apoptosis cannot be excluded. For example, excessive neurotransmitter release in NMDA receptor activation induced apoptosis is thought to be mediated by nNOS (44). In rat embryonic motor neurons, trophic factor deprivation-induced apoptosis is associated with induction of nNOS (47). These cells also constitutively express eNOS, which may also support cell survival (84). It should be noted that cell death is not obligatory following exposure to high concentrations of nitric oxide. In fact, many cells can withstand fairly high concentrations of nitric oxide without cell death possibly due to the activation of protective mechanisms (see below).

VI. MECHANISMS BY WHICH CELLS PROTECT THEMSELVES AGAINST NITRIC OXIDE

During wounding and inflammation, excessive production of nitric oxide can lead to tissue injury, and thus it is not unexpected that cells possess protective mechanisms against this damage. Of particular importance is the status of cellular antioxidant defenses. By controlling oxidants such as superoxide anion, cells can detoxify intracellular nitric oxide. For example, in human articular chondrocytes, reactive oxygen intermediates appear to directly protect against nitric oxide-induced apoptosis (162). Thus, levels of enzymes that metabolize reactive oxygen intermediates, including catalase and superoxide dismutase, may be critical determinants of nitric oxide-induced toxicity. Nitric oxide is also known to modulate the cellular thiol redox state of cells, and this can regulate apoptosis. For example, the reaction of nitric oxide or peroxynitrite with cellular glutathione can reduce intracellular reactive nitrogen intermediates. In vascular smooth muscle cells, reduced glutathione has been reported to protect against nitric oxide-induced apoptosis (112), while N-acetylcysteine, a glutathione pro-drug, suppresses nitric oxide-induced apoptosis in COLO 205 human colon carcinoma cells (163).

Proteins such as Bcl-2, p53, and heat shock proteins also afford protection against nitric oxide-induced apoptosis. Bcl-2 is the prototypic member of a family of proteins that regulate the apoptotic process. Family members such as Bcl-2 and Bcl-xL are apoptosis suppressors; others, such as Bax and Bak, are apoptosis promotors (164–166). The proteins interact with one another and cell survival is determined by the ratio of the antiapoptotic and pro-apoptotic proteins. Genaro and co-workers (167) have demonstrated that nitric oxide enhances Bcl-2 expression in B lymphocytes and prevents apoptosis. In a subsequent study, Xie and co-workers (168) showed that IL-1 and interferon-gamma induced high levels of endogenous nitric oxide and apoptosis in murine L929 cells, but not in NIH3T3 cells. NIH3T3 cells were, however, more sensitive to nitric oxide from exogenous sources. It has also been shown that transfection of the cell lines with the human bcl-2 gene led to increased resistance to both endogenous and exogenous nitric oxide-mediated apoptosis. These data again suggest that nitric oxide-mediated apoptosis can be suppressed by expression of bcl-2.

In general, activation of p53 leads to the initiation of the apoptotic pathway. However, several years ago, it was reported that human cells treated with a nitric oxide donor or cells overexpressing iNOS accumulate p53 in the nucleus but do not undergo apoptosis (169). Transfection of DLD-1 human colon carcinoma cells with wild-type p53 resulted in a decrease in iNOS protein synthesis and enzymatic activity after the cells were incubated with cytokines, a treatment that usually leads to a substantial increase in iNOS synthesis and activity. Thus it was hypothesized that a negative feedback loop may exist in which endogenous nitric oxide-induced DNA damage results in the accumulation of wild-type p53, which then transrepresses iNOS expression, thereby reducing the potential for further nitric oxide-induced DNA damage.

Organisms respond to harmful environmental agents by the induction of a family of proteins termed "heat shock" or "stress" proteins. The heat shock proteins (hsp) are further divided into subfamilies based on the molecular weights of the proteins encoded. There are subfamilies that contain proteins with molecular weights of 90, 70, and 60 kDa as well as a family of much smaller proteins. The heat shock proteins are highly conserved and play essential roles as molecular chaperones under both normal and stressed conditions. They facilitate the folding of nascent proteins, aid in the resolubilizing of protein aggregates, and assist in refolding denatured proteins (170–173). Recently, roles for the stress proteins in immune phenomena and apoptosis have been suggested (173). Studies during the past several years have indicated that induction of the heat shock response, in particular, the induction of hsp70, can protect cells against injury mediated by iNOS (174–179). For example, Feinstein and co-workers (180) showed that hyperthermic treatment of brain glial cells led to a decrease in the iNOS mRNA and enzymatic activity normally induced by cytokine treatment. Rat fibroblasts transfected with the human hsp70 cDNA did not express iNOS after treatment with cytokines and lipopolysaccharide. They also showed that the transfected cells, as well as heat-shocked glial cells, exhibited decreased NF-κB p65 subunit nuclear accumulation. These results suggest that iNOS induction can be modulated by the stress response of cells and that hsp70 may play a role in this process. Several recent reports have even suggested that nitric oxide itself can induce hsp70 expression and prevent apoptosis (180,181).

SUMMARY

An important mechanism by which nitric oxide mediates toxicity is by inducing apoptosis. This process has been observed in cells stimulated to produce nitric oxide endogenously, for example, in response to endotoxin, cytokines, and growth factors, as well as in cells treated with nitric oxide-donating compounds. Oxidation products of nitric oxide including peroxynitritc may mediate the action of nitric oxide. The mechanisms by which nitric oxide or its oxidation products regulate apoptosis are not clear at the present time. Nitric oxide can cause nonspecific injury to cell membranes and DNA as well as inhibit many different kinds of enzymes. As cells attempt to repair injury, any one of these targets may trigger the apoptotic process. Nitric oxide is also known to activate guanylyl cyclase, increasing intracellular cGMP, an important regulator of cellular protein kinases. Nitric oxide may induce apoptosis by both cGMP-dependent and -independent process. In some instances, nitric oxide also inhibits apoptosis. Under these circumstances, nitric oxide may act to inhibit critical enzymes in the apoptotic signaling pathway, such as the caspases. Alternatively, where reactive oxygen intermediates mediate apoptosis, nitric oxide, by virtue of its ability to react with radicals such as superoxide anion, may suppress the apoptotic process. Thus, the redox state of the cell may be crucial for determining whether or not nitric oxide either induces or protects against apoptosis. Identification of the precise cellular targets for nitric oxide during cellular injury and inflammation may provide important clues as to the mechanisms by which nitric oxide regulates apoptosis.

REFERENCES

1. Furchgott RF, Zawadzki JV. The obligatory role of endothelial cells in the relaxation of arterial smooth muscle by acetylcholine. Nature 1980; 288:373–376.
2. Palmer RM, Ferrige AG, Moncada S. Nitric oxide release accounts for the biological activity of endothelium-derived relaxing factor. Nature 1987; 327:524–526.
3. Ignarro LJ, Buga GM, Wood KS, Byrns RE, Chaudhauri G. Endothelium-derived relaxing factor produced and released from artery and vein is nitric oxide. Proc Natl Acad Sci USA 1987; 84:9265–9269.
4. Laskin JD, Heck DE, Laskin DL. Multifunctional role of nitric oxide in inflammation. Trends Endocrinol Metab 1994; 5:377–382.
5. Schmidt HH, Walter U. NO at work. Cell 1994; 78:919–925.
6. Nathan C, Xie QW. Nitric oxide synthase: roles, tolls, and controls. Cell 1994; 78:915–918.
7. Cifone MG, Cironi L, Meccia MA, Roncaioli P, Festuccia C, De Nuntiis G, D'Alo S, Santoni A. Role of nitric oxide in cell-mediated tumor cytotoxicity. Adv Neuroimmunol 1995; 5:443–461.
8. Jenkins DC, Charles IG, Thomsen LL, Moss DW, Holmes LS, Baylis SA, Rhodes P, Westmore K, Emson PC, Moncada S. Roles of nitric oxide in tumor growth. Proc Natl Acad Sci USA 1995; 92:4392–4396.
9. Nathan C. Inducible nitric oxide synthase: what difference does it make? J Clin Invest 1997; 100:2417–2423.
10. Albina JE, Reichner JS. Role of nitric oxide in mediation of macrophage cytotoxicity and apoptosis. Cancer Metastasis Rev 1998; 17:39–53.
11. Heck DE, Laskin DL, Gardner CR, Laskin JD. Epidermal growth factor suppresses nitric oxide and hydrogen peroxide production by keratinocytes. Potential role for nitric oxide in the regulation of wound healing. J Biol Chem 1992; 267:21277–21280.

12. Sen S. Programmed cell death: concept, mechanism and control. Biol Rev 1992; 67:287–319.
13. White E. Life, death, and the pursuit of apoptosis. Genes Dev 1996; 10:1–15.
14. Fotedar R, Diederich L, Fotedar A. Apoptosis and the cell cycle. Prog Cell Cycle Res 1996; 2:147–163.
15. Savitz SI, Rosenbaum DM. Apoptosis in neurological disease. Neurosurgery 1998; 42:555–572.
16. Leist M, Nicotera P. Apoptosis, excitotoxicity, and neuropathology. Exp Cell Res 1998; 239:183–201.
17. Ormerod MG, Apoptosis: flow cytometric studies. Micros Ana 1993; 3:25–27.
18. Bellone M, Iezzi G, Rovere P, Galati G, Ronchetti A, Protti MP, Davoust J, Rugarli C, Manfredi AA. Processing of engulfed apoptotic bodies yields T cell epitopes. J Immunol 1997; 159:5391–5399.
19. Nguyen KB, Pender MP. Phagocytosis of apoptotic lymphocytes by oligodendrocytes in experimental autoimmune encephalomyelitis. Acta Neuropathol (Berl) 1998; 95:40–46.
20. Rubartelli A, Poggi A, Zocchi MR. The selective engulfment of apoptotic bodies by dendritic cells is mediated by the alpha(v)beta3 integrin and requires intracellular and extracellular calcium. Eur J Immunol 1997; 27:1893–1900.
21. Nishikawa S, Sasaki F. Phagocytotic processing of apoptotic bodies of transitional ameloblasts by MHC class II-expressing macrophages in rat incisor. J Histochem Cytochem 1996; 44:1459–1467.
22. Yuan J, Shaham S, Ledoux S, Ellis HM, Horvitz HR. The *C. elegans* cell death gene ced-3 encodes a protein similar to mammalian interleukin-1β-converting enzyme. Cell 1993; 75:641–652.
23. Lincz LF. Deciphering the apoptotic pathway: all roads lead to death. Immunol Cell Biol 1998; 76:1–19.
24. Kidd VJ. Proteolytic activities that mediate apoptosis. Annu Rev Physiol 1998; 60:533–573.
25. de Murcia G, de Murcia JM. Poly(ADP-ribose) polymerase: a molecular nick sensor. Trends Biochem Sci 1994; 19:172–176.
26. Lindahl T, Satoh MS, Poirier GG, Klugland A. Post-translational modification of poly(ADP-ribose) polymerase induced by DNA strand breaks. Trends Biochem Sci 1995; 20:405–411.
27. Shimizu T, Cao CX, Shao RG, Pommier Y. Lamin B phosphorylation by protein kinase C alpha and proteolysis during apoptosis in human leukemia HL60 cells. J Biol Chem 1998; 273: 8669–8674.
28. Wyllie AH. Glucocorticoid-induced thymocyte apoptosis is associated with endogenous endonuclease activation. Nature 1980; 284:555–556.
29. Albina JE, Cui S, Mateo RB, Reichner JS. Nitric oxide-mediated apoptosis in murine peritoneal macrophages. J Immunol 1993; 150:5080–5083.
30. Messmer UK, Reed UK, Brune B. Bcl-2 protects macrophages from nitric oxide-induced apoptosis. J Biol Chem 1996; 271:20192–20197.
31. Brune B, Golkel C, von Knethen A. Cytokine and low-level nitric oxide

prestimulation block p53 accumulation and apoptosis of RAW 264.7 macrophages. Biochem Biophys Res Commun 1996; 229:396–401.

32. Ramirez R, Carracedo J, Castedo M, Zamzami N, Kroemer G. CD60-induced monocyte apoptosis involves multiple nonredundant signaling pathways. Cell Immunol 1996; 172:192–199.
33. Xie K, Wang Y, Huang S, Xu L, Bielenberg D, Salas T, McConkey DJ, Jiang W, Fidler IJ. Nitric oxide-mediated apoptosis of K-1735 melanoma cells is associated with downregulation of Bcl-2. Oncogene 1997; 15:771–779.
34. Xie K, Huang S, Wang Y, Beltran PJ, Juang SH, Dong Z, Reed JC, McDonnell TJ, McConkey DJ, Fidler IJ. Bcl-2 protects cells from cytokine-induced nitric-oxide-dependent apoptosis. Cancer Immunol Immunother 1996; 43:109–115.
35. Kaneto H, Fujii J, Seo HG, Suzuki K, Matsuoka T, Nakamura M, Tatsumi M,Yamasaki Y, Kamada T,Taniguichi N. Apoptotic cell death triggered by nitric oxide in pancreatic beta-cells. Diabetes 1995; 44:733–738.
36. Iwahashi H, Hanafusa T, Eguchi Y, Nakajima H, Miyagawa J, Itoh N, Tomita K, Namba M, Kuwajima M, Noguchi T, Tsujimoto Y, Matsuzawa Y. Cytokine-induced apoptotic cell death in a mouse pancreatic beta-cell line: inhibition by Bcl-2. Diabetologia 1996; 39:530–536.
37. Ankarcrona M, Dypbukt JM, Brune B, Nicotera P. Interleukin-1 beta-induced nitric oxide production activates apoptosis in pancreatic RINm5F cells. Exp Cell Res 1994; 213:172–177.
38. Loweth AC, Williams GT, Scarpello JH, Morgan NG. Evidence for the involvement of cGMP and protein kinase G in nitric oxide-induced apoptosis in the pancreatic B-cell line, HIT-T15. FEBS Lett 1997; 400:285–288.
39. Zhang HY, Gharaee-Kermani M, Phan SH. Regulation of lung fibroblast alpha-smooth muscle actin expression, contractile phenotype, and apoptosis by IL-1beta. J Immunol 1997; 158:1392–1399.
40. Kunz D, Walker G, Eberhardt W, Messmer UK, Huwiler A, Pfeilschifter J. Platelet-derived growth factor and fibroblast growth factor differentially regulate interleukin 1beta- and cAMP-induced nitric oxide synthase expression in rat renal mesangial cells. J Clin Invest 1997; 100:2800–2809.
41. Fukuo K, Hata S, Suhara T, Nakahashi T, Shinto Y, Tsujimoto Y, Morimoto S, Ogihara T. Nitric oxide induces upregulation of Fas and apoptosis in vascular smooth muscle cells. Hypertension 1996; 27:823–826.
42. Lopez-Collazo E, Mateo J, Miras-Portugal MT, Bosca L. Requirement of nitric oxide and calcium mobilization for the induction of apoptosis in adrenal vascular endothelial cells. FEBS Lett 1997; 413:124–128.
43. Bonfoco E, Leist M, Zhivotovsky B, Orrenius S, Lipton SA, Nicotera P. Cytoskeletal breakdown and apoptosis elicited by NO donors in cerebellar granule cells require NMDA receptor activation. J Neurochem 1996; 67:2484–2493.
44. Nicotera P, Bonfoco E, Brune B. Mechanisms for nitric oxide-induced cell death: involvement of apoptosis. Adv Neuroimmunol 1995; 5:411–420.
45. Hu S, Peterson PK, Chao CC. Cytokine-mediated neuronal apoptosis. Neurochem Int 1997; 30:427–431.

46. Palluy O, Rigaud M. Nitric oxide induces cultured cortical neuron apoptosis. Neurosci Lett 1996; 208:1–4.
47. Estevez AG, Spear N, Manuel SM, Radi R, Henderson CE, Barbeito L, Beckman JE. Nitric oxide and superoxide contribute to motor neuron apoptosis induced by trophic factor deprivation. J Neurosci 1998; 18:923–931.
48. Hu J, Van Eldik LJ. S100 beta induces apoptotic cell death in cultured astrocytes via a nitric oxide-dependent pathway. Biochim Biophys Acta 1996; 1313:239–245.
49. Sveinbjornsson B, Olsen R, Seternes OM, Seljelid R. Macrophage cytotoxicity against murine meth A sarcoma involves nitric oxide-mediated apoptosis. Biochem Biophys Res Commun 1996; 223:643–649.
50. Albina JE, Martin BA, Henry WL Jr, Louis CA, Reichner JS. B cell lymphoma-2 transfected P815 cells resist reactive nitrogen intermediate-mediated macrophage-dependent cytotoxicity. J Immunol 1996; 157:279–283.
51. Albina JE, Reichner JS. Role of nitric oxide in mediation of macrophage cytotoxicity and apoptosis. Cancer Metastasis Rev 1998; 17:39–53.
52. Cui S, Reichner JS, Mateo RB, Albina JE. Activated murine macrophages induce apoptosis through nitric oxide-dependent or -independent mechanisms. Cancer Res 1994; 54:2462–2467.
53. Nomura Y, Uehara T, Nakazawa M. Neuronal apoptosis by glial NO: involvement of inhibition of glyceraldehyde-3-phosphate dehydrogenase. Hum Cell 1996; 9:205–214.
54. Kurose I, Higucho H, Yonei Y, Ebinuma H, Watanabe N, Hokari R, Fukumura D, Miura S, Takaishi M, Saito H, Nakatsumi RC, Ishii H. Rat Kupffer cell-derived nitric oxide suppresses proliferation and induces apoptosis of syngeneic hepatoma cells. Gastroenterology 1996; 111:1058–1070.
55. Van't Hof RJ, Ralston SH. Cytokine-induced nitric oxide inhibits bone resorption by inducing apoptosis of osteoclast progenitors and suppressing osteoclast activity. J Bone Miner Res 1997; 12:1797–1804.
56. Hirano S. Nitric oxide-mediated cytotoxic effects of alveolar macrophages on transformed lung epithelial cells are independent of the beta 2 integrin-mediated intercellular adhesion. Immunology 1998; 93:102–108.
57. Fehsel K, Kroncke KD, Meyer KL, Huber H, Wahn V, Kolb-Bachofer V. Nitric oxide induces apoptosis in mouse thymocytes. J Immunol 1995; 155:2858–2865.
58. Lancaster JR. Diffusion of free nitric oxide. Methods Enzymol 1996; 268:31–50.
59. Messmer UK, Brune B. Nitric oxide (NO) in apoptotic versus necrotic RAW 264.7 macrophage cell death: the role of NO-donor exposure, NAD+ content and p53 accumulation. Arch Biochem Biophys 1996; 327:1–10.
60. Pollman MJ, Yamada T, Horiuchi M, Gibbons GH. Vasoactive substances regulate vascular smooth muscle cell apoptosis. Countervailing influences of nitric oxide and angiotensin II. Circ Res 1996; 79:748–756.
61. Nishio E, Fukushima K, Shiozaki M, Wantanabe Y. Nitric oxide donor SNAP induces apoptosis through cGMP-independent mechanism. Biochem Biophys Res Commun 1996; 221:163–168.
62. Khan S, Kayahara M, Joashi U, Mazarakis N, Sarraf C, Edwards A, Hughes M,

Mehmet H. Differential induction of apoptosis in Swiss 3T3 cells by nitric oxide and the nitrosonium cation. J Cell Sci 1997; 110:2315–2322.
63. Muhl H, Sandau K, Brune N, Briner VA, Pfeilschifter J. Nitric oxide donors induce apoptosis in glomerular mesangial cells, epithelial cells and endothelial cells. Eur J Pharmacol 1996; 317:137–149.
64. Sandau K, Pfeilschifter J, Brune B. The balance between nitric oxide and superoxide determines apoptotic and necrotic death of rat mesangial cells. J Immunol 1997; 158:4938–4946.
65. Blanco FJ, Ocha RL, Schwartz H, Lotz M. Chondrocyte apoptosis induced by nitric oxide. Am J Pathol 1995; 146:75–85.
66. Fortenberry JD, Owens ML, Brown MR, Atkinson D, Brown LAS. Exogenous nitric oxide enhances neutrophil cell death and DNA fragmentation. Am J Respir Cell Mol Biol 1998; 18:421–428.
67. Laskin DL. Sinusoidal lining cells and hepatotoxicity. Toxicol Pathol 1996; 24:112–118.
68. Laskin DL, Pendino KJ. Macrophages and inflammatory mediators in tissue injury. Annu Rev Pharmacol Toxicol 1995; 35:655–677.
69. Szabolcs N, Michler RE, Yang X, Aji W, Roy D, Athan E, Sciacca RR, Minanov OP, Cannon PJ. Apoptosis of cardiac myocytes during cardiac allograft rejection. Relation to induction of nitric oxide synthase. Circulation 1996; 94:1665–1673.
70. Szaboles MJ, Ravalli S, Minanov O, Sciacca RR, Michler RE, Cannon PJ. Apoptosis and increased expression of inducible nitric oxide synthase in human allograft rejection. Transplantation 1998; 65:804–812.
71. Wang JS, Tseng HH, Shih DF, Jou HS, Ger LP. Expression of inducible nitric oxide synthase and apoptosis in human lupus nephritis. Nephron 1977; 77:404–411.
72. Suzuki H, Wildhirt SM, Dudek RR, Narayan KS, Bailey AH, Bind RJ. Induction of apoptosis in myocardial infarction and its possible relationship to nitric oxide synthase in macrophages. Tissue Cell 1996; 28:89–97.
73. Savill J. Apoptosis in resolution of inflammation. J Leukoc Biol 1997; 61:375–380.
74. MacMicking J, Xie QW, Nathan C. Nitric oxide and macrophage function. Annu Rev Immunol 1997; 15:323–350.
75. Nathan C. Nitric oxide as a secretory product of mammalian cells. FASEB J 1992; 6:3051–3064.
76. Bolanos JP, Heales SJR, Land JM, Clark JB. Effects of peroxynitrite on the mitochondrial respiratory chain: differential susceptibility of neurons and astrocytes in primary culture. J Neurosci 1995; 64:1965–1972.
77. Kidd VJ. Proteolytic activities that mediate apoptosis. Annu Rev Physiol 1998; 60:533–573.
78. Melino G, Bernassola F, Knight RA, Corasaniti MT, Nistico G, Finazzi-Agro A. S- nitrosylation regulates apoptosis. Nature 1997; 388:432–433.
79. Ankarcrona M, Dypbukt JM, Bonfoco E, Zhivotovsky B, Orrenius S, Lipton SA, Nicotera P. Glutamate-induced neuronal death: a succession of necrosis or apoptosis depending on mitochondrial function. Neuron 1995; 15:961–973.

80. Lizasoain I, Moro MA, Knowles RG, Darley-Usmar V, Moncada S. Nitric oxide and peroxynitrite exert distinct effects on mitochondrial respiration which are differentially blocked by glutathione or glucose. Biochem J 1996; 314:877–880.
81. Kroemcr G, Petit P, Zamzami N, Vayssiere JL, Mignotte B. The biochemistry of programmed cell death. FASEB J 1995; 9:1277–1287.
82. Kluck RM, Bossy-Wetzel E, Green DR, Newmeyer DD. The release of cytochrome c from mitochondria: a primary site for Bcl-2 regulation of apoptosis. Science 1997; 275:1132–1136.
83. McDonald LJ, Murad F. Nitric oxide and cyclic GMP signaling. Proc Soc Exp Biol Med 1996; 211:1–6.
84. Estevez AG, Spear N, Thompson JA, Cornwell TL, Radi R, Barbeito L, Beckman JS. Nitric oxide-dependent production of cGMP supports the survival of rat embryonic motor neurons cultured with brain-derived neurotrophic factor. J Neurosci 1998; 18:3708–3714.
85. Shao RG, Cao CX, Pommoer Y. Activation of PKC alpha downstream from caspases during apoptosis induced by 7-hydroxystaurosporine or the topoisomerase inhibitors, camptothecin and etoposide, in human myeloid leukemia HL60 cells. J Biol Chem 1997; 272:31321–31325.
86. Mizuno K, Noda K, Araki T, Imaoka T, Kobayashi Y, Akita Y, Shimonaka M, Kishi S, Ohno S. The proteolytic cleavage of protein kinase C isotypes, which generates kinase and regulatory fragments, correlates with Fas-mediated and 12-O-tetradecanoyl-phorbol-13-acetate-induced apoptosis. Eur J Biochem 1997; 250:7–18.
87. Deacon EM, Pongracz J, Griffiths G, Lord JM. Isoenzymes of protein kinase C: differential involvement in apoptosis and pathogenesis. Mol Pathol 1997; 50:124–131.
88. Guh JH, Hwang TL, Ko FN, Chueh SC, Lai M.K, Teng CM. Antiproliferative effect in human prostatic smooth muscle cells by nitric oxide donor. Mol Pharm 1998; 53:467–474.
89. Nishio E, Watanabe Y. Nitric oxide donor-induced apoptosis in smooth muscle cells is modulated by protein kinase C and protein kinase A. Eur J Pharmacol 1997; 339:245–251.
90. Lin KT, Xue JY, Lin MC, Spokas EG, Sun FF, Wong PY. Peroxynitrite induces apoptosis of HL-60 cells by activation of a caspase-3 family protease. Am J Physiol 1998; 274:C855-C860.
91. Wang H, Keiser JA. Molecular characterization of rabbit CPP32 and its function in vascular smooth muscle cell apoptosis. Am J Physiol 1998; 274:H1132–H1140.
92. Nishio E, Watanabe Y. NO induced apoptosis accompanying the change of oncoprotein expression and the activation of CPP32 protease. Life Sci 1998; 62:239–245.
93. Brockhaus F, Brune B. U937 apoptotic cell death by nitric oxide: Bcl-2 downregulation and caspase activation. Exp Cell Res 1998; 238:33–41.
94. Haendeler J, Weiland U, Zeiher AM, Dimmeler S. Effects of redox-related congeners of NO on apoptosis and caspase-3 activity. Nitric Oxide 1997; 1:282–293.

95. Kim Y-M, Talanian RV, Billiar TR. Nitric oxide inhibits apoptosis by preventing increases in caspase-3-like activity via two distinct mechanisms. J Biol Chem 1997: 272:31138–31148.
96. Li J, Billiar TR, Talanian RV, Kim YM. Nitric oxide reversibly inhibits seven members of the caspase family via S-nitrosylation. Biochem Biophys Res Commun 1997; 240:419–424.
97. Levine AJ. p53, the cellular gatekeeper for growth and division. Cell 1997; 88:323–331.
98. Messmer UK, Ankarcrona M, Nicotera P, Brune B. p53 Expression in nitric oxide-induced apoptosis. FEBS Lett 1994; 355:23-26.
99. Fehsel K, Kroncke KD, Meyer KL, Huber H, Wahn V, Kolb-Bachofen V. Nitric oxide induces apoptosis in mouse thymocytes. J Immunol 1995; 155:2858–2865.
100. Ho YS, Wang YJ, Lin JK. Induction of p53 and p21/WAF1/CIP1 expression by nitric oxide and their association with apoptosis in human cancer cells. Mol Carcinog 1996; 16:20–31.
101. Sandau K, Pfeilschifter J, Brune B. Nitric oxide and superoxide induced p53 and Bax accumulation during mesangial cell apoptosis. Kidney Int 1997; 52:378–386.
102. Kitamura Y, Kamoshima W, Shimohama S, Nomura Y, Taniguchi T. Nitric oxide donor-induced p53-sensitive cell death is enhanced by Bcl-2 reduction in human neuroblastoma cells. Neurochem Int 1998; 32:93–102.
103. Stamler JS, Hausladen A. Oxidative modifications in nitrosative stress. Nat Struct Biol 1998; 5:247–249.
104. Liu Rudd MA, Freedman JE, Loscalzo J. S-Transnitrosation reactions are involved in the metabolic fate and biological actions of nitric oxide. J Pharmacol Exp Ther 1998; 284:526–534.
105. Hippeli S, Eistner EF. OH-radical-type reactive oxygen species: a short review on the mechanisms of OH-radical and peroxynitrite toxicity. Z Naturforsch 1997; 52c:555–563.
106. Pryor WA, Squadrito GL. The chemistry of peroxynitrite: a product from the reaction of nitric oxide with superoxide. Am J Physiol 1995; 268:L699–L722.
107. Beckman JS, Koppenol WH. Nitric oxide, superoxide, and peroxynitrite: the good, the bad, and ugly. Am J Physiol 1996; 271:C1424–C1437.
108. Muijsers RB, Folkerts G, Henricks PA, Sadeghi-Hashjin G, Nijkamp FP. Peroxynitrite: a two-faced metabolite of nitric oxide. Life Sci 1997; 60:1833–1845.
109. Beckman JS, Chen J, Ischiropoulos H, Crow JP. Oxidative chemistry of peroxynitrite. Methods Enzymol 1995; 233:229–240.
110. Szabo C, Ohshima H. DNA damage induced by peroxynitrite: subsequent biological effects. Nitric Oxide 1997; 1:373–385.
111. Szabo C. DNA strand breakage and activation of poly-ADP ribosyltransferase: a cytotoxic pathway triggered by peroxynitrite. Free Radic Biol Med 1996; 21:855–869.
112. Hippeli S, Elstner EF. OH-radical-type reactive oxygen species: a short review on the mechanisms of OH-radical and peroxynitrite toxicity. Z Naturforsch 1997; 52c:555–563.

113. Fridovich I. Superoxide radical and superoxide dismutases. Annu Rev Biochem 1995; 64: 97–112.
114. Troy CM, Derossi D, Prochiantz A, Greene LA, Shelanski ML. Downregulation of Cu/Zn superoxide dismutase leads to cell death via the nitric oxide-peroxynitrite pathway. J Neurosci 1996; 16:253–261.
115. Fridovich I. Superoxide dismutases. Adv Enzymol 1986; 58:61–97.
116. Babior BM. The respiratory burst oxidase. Curr Opin Hematol 1995; 2:55–60.
117. Sata N, Klonowski-Stumpe H, Han B, Haussinger D, Niederau C. Cytotoxicity of peroxynitrite in rat pancreatic acinal AR4-2J cells. Pancreas 1997; 15:278–284.
118. Salgo MG, Pryor WA. Trolox inhibits peroxynitrite-mediated oxidative stress and apoptosis in rat thymocytes. Arch Biochem Biophys 1996; 333:482–488.
119. Salgo MG, Squadrito GL, Pryor WA. Peroxynitrite causes apoptosis in rat thymocytes. Biochem Biophys Res Commun 1995; 215:1111–1118.
120. Salgo MG, Bermudez E, Squadrito GL, Pryor WA. Peroxynitrite causes DNA damage and oxidation of thiols in rat thymocytes. Arch Biochem Biophys 1995; 322:500–505.
121. Lin KT, Xue JY, Sun FF, Wong PY. Reactive oxygen species participate in peroxynitrite-induced apoptosis in HK-60 cells. Biochem Biophys Res Commun 1997; 230:115–119.
122. Lin K-T, Xue H-Y, Nomen M, Spur B, Wong P. Peroxynitrite-induced apoptosis in HL-60 cells. J Biol Chem 1995; 270:16487–16490.
123. Sandoval M, Zhang XJ, Liu X, Mannick EE, Clark DA, Miller MJ. Peroxynitrite-induced apoptosis in T84 and RAW 264.7 cells: attenuation by L-ascorbic acid. Free Radic Biol Med 22:489–495.
124. Behar-Cohen FF, Hetdolf S, Faure V, Droy-Lefaix MT, Courtois Y, Goureau O. Peroxynitrite cytotoxicity on bovine retinal pigmented epithelial cells in culture. Biochem Biophys Res Commun 1996; 226:842–849.
125. Estevez AG, Radi R, Barbeito L, Shin JT, Thompson JA, Beckman JS. Peroxynitrite-induced cytotoxicity in PC12 cells: evidence for an apoptotic mechanism differentially modulated by neurotrophic factors. J Neurochem 1995; 65:1543–1550.
126. Spear N, Estevez AG, Barbeito L, Beckman JS, Johnson GV. Nerve growth factor-induced apoptosis via a mechanism dependent on phosphatidylinositol 3-kinase. J Neurochem 1997; 69:53–59.
127. Szabo C. The pathophysiological role of peroxynitrite in shock, inflammation, and ischemia-reperfusion injury. Shock 1996; 6:79–88.
128. Jacobson MD. Reactive oxygen species and programmed cell death. Trends Biochem Sci 1996; 21:83–86.
129. Buttke TM, Sandstrom PA. Oxidative stress as a mediator of apoptosis. Immunol Today 1994; 15:7–10.
130. Kaul N, Gopalakrishna R, Gundieda U, Choi J, Forman HJ. Role of protein kinase C in basal and hydrogen peroxide-stimulated NF-kappa B activation in the murine macrophage J774A.1 cell line. Arch Biochem Biophys 1998; 350:79–86.
131. Hermann C, Zeiher AM, Dimmeler S. Shear stress inhibits H_2O_2-induced apoptosis

of human endothelial cells by modulation of the glutathione redox cycle and nitric oxide synthase. Arterioscler Thromb Vasc Biol 1997; 17:3588–3592.

132. Riou C, Remy C, Rabilloud R, Rousset B, Fonlupt P. H_2O_2 induces apoptosis of pig thyrocytes in culture. J Endocrinol 156:315–322.
133. Uppu RM, Squadrito GL, Cueto R, Pryor WA. Synthesis of peroxynitrite by azide-ozone reaction. Methods Enzymol 1996; 269:311–321.
134. Eiserich J, Hristova M, Cross CE, Jones AD, Freeman BA, Halliwell B, van der Vliet A. Formation of nitric oxide-derived inflammatory oxidants by myeloperoxidase in neutrophils. Nature 1998; 391:393–397.
135. Dimmeler S, Rippmann V, Weiland U, Haendeler J, Zeiher AM. Angiotensin II induces apoptosis of human endothelial cells. Protective effect of nitric oxide. Circ Res 1997; 81:970–976.
136. Tzeng E, Kim YM, Pitt BR, Lizonova A, Kovesdi I, Billiar TR. Adenoviral transfer of the inducible nitric oxide synthase gene blocks endothelial cell apoptosis. Surgery 1997; 122:255–263.
137. Mohr S, Zech B, Lapetina EG, Brune B. Inhibition of caspase-3 by S-nitrosation and oxidation caused by nitric oxide. Biochem Biophys Res Commun 1997; 238:387–391.
138. Tenneti L, D'Emilia DM, Lipton SA. Suppression of neuronal apoptosis by S-nitrosylation of caspases. Neurosci Lett 1997; 236:139–142.
139. Dimmeler S, Haendeler J, Nehls M, Zeiher AM. Suppression of apoptosis by nitric oxide via inhibition of interleukin-1-beta-converting enzyme (ICE)-like and cysteine protease protein (CPP)-32-like proteases. J Exp Med 1997; 185:601–607.
140. Thippeswamy T, Morris R. Cyclic guanosine 3′,5′-monophosphate-mediated neuroprotection by nitric oxide in dissociated cultures of rat dorsal root ganglion neurons. Brain Res 1997; 774:116–122.
141. Chun SY, Eisenhauer KM, Kubo M, Hsueh HJ. Interleukin-1 beta suppresses apoptosis in rat ovarian follicles by increasing nitric oxide production. Endocrinology 1995; 136:3120–3127.
142. Thornberry NA. Caspases: key mediators of apoptosis. Chem Biol 1998; 5:R97–R103.
143. Brockhous F, Brune B. U937 apoptotic cell death by nitric oxide: Bcl-2 down-regulation and caspase activation. Exp Cell Res 1998; 238:33–41.
144. Laskin DL, Rodriguez del Valle M, Heck DE, Hwang SM, Ohnishi ST, Durham SK, Goller NL, Laskin DL. Hepatic nitric oxide production following acute endotoxemia in rats is mediated by increased inducible nitric oxide synthase gene expression. Hepatology 1995; 22:223–234.
145. Duhe RJ, Evans GA, Erwin RA, Kirken RA, Cox GW, Farrar WL. Nitric oxide and thiol redox regulation of Janus kinase activity. Proc Natl Acad Sci USA 1998; 95:126–131.
146. Park DS, Stefanis L, Yan CYI, Farinelli SE, Greene LA. Ordering the cell death pathway. Differential effects of BCL-2, an interleukin-1-converting enzyme family protease inhibitor, and other survival agents on JUNK activation in serum/nerve growth factor-deprived PC12 cells. J Biol Chem 1996; 271:21898–21905.

147. Sandau K, Brune B. The dual role of S-nitrosoglutathione (GSNO) during thymocyte apoptosis. Cell Signal 1996; 8:173–177.
148. Lopez-Farre A, Sancheze Miguel L, Caramelo C, Gomez-Macias J, Garcia R, Mosquera JR, de Frutos T, Millas I, Rivas F, Eschezarreta G, Casado S. Role of nitric oxide in autocrine control of growth and apoptosis of endothelial cells. Am J Physiol 1997; 272:H760–H768.
149. Wu W, Liuzzi FJ, Schinco FP, Depto AS, Mong JA, Dawson TM, Snyder SH. Neuronal nitric oxide synthase is induced in spinal neurons by traumatic injury. Neuroscience 1994; 61:719–726.
150. Cahill PA, Redmond EM, Hodges R, Zhang S, Sitzmann JV. Increased endothelial nitric oxide synthase activity in the hyperemic vessels of portal hypertensive rats. J Hepatol 1996; 25:370–378.
151. Sheehy AM, Phung YT, Riemer RK, Black SM. Growth factor induction of nitric oxide synthase in rat pheochromocytoma cells. Brain Res Mol Brain Res 1997; 52:71–77.
152. Chen K, Inoue M, Wasa M, Fukuzawa M, Kamata S, Okada A. Expression of endothelial constitutive nitric oxide synthase mRNA in gastrointestinal mucosa and its downregulation by endotoxin. Life Sci 1997; 61:1323–1329
153. Neugarten J, Ding Q, Friedman A, Lei J, Silbiger S. Sex hormones and renal nitric oxide synthases. J Am Soc Nephrol 1997; 8:1240–1246.
154. Hennington BS, Zhang H, Miller MT, Granger JP, Reckelhoff JF. Angiotensin II stimulates synthesis of endothelial nitric oxide synthase. Hypertension 1998; 31:283–288.
155. Hewett SJ, Corbett JA, McDaniel, ML, Choi DW. Interferon-gamma and interleukin-1 beta induce nitric oxide formation from primary mouse astrocytes. Neurosci Lett 1993; 164:229–232.
156. Gilbert RS, Herschman HR. Transforming growth factor beta differentially modulates the inducible nitric oxide synthase gene in distinct cell types. Biochem Biophys Res Commun 1993; 195:380–384.
157. Koide M, Kawahara Y, Tsuda T, Nakayama I, Yokoyama M. Expression of nitric oxide synthase by cytokines in vascular smooth muscle cells. Hypertension 1994; 23:I45–I48.
158. Williams G, Brown T, Becker L, Prager M, Giroir BP. Cytokine-induced expression of nitric oxide synthase in C2C12 skeletal muscle myocytes. Am J Physiol 1994; 267:R1020–R1025.
159. Sparrow JR, Nathan C, Vodovotz Y. Cytokine regulation of nitric oxide synthase in mouse retinal pigment epithelial cells in culture. Exp Eye Res 1994; 59:129–139.
160. Chesrown SE, Monnier J, Visner G, Nick HS. Regulation of inducible nitric oxide synthase mRNA levels by LPS, INF-gamma, TGF-beta, and IL-10 in murine macrophage cell lines and rat peritoneal macrophages. Biochem Biophys Res Commun 1994; 200:126–134.
161. Nitsch DD, Ghilardi N, Muhl H, Nitsch C, Brune B, Pfeilschifter J. Apoptosis and expression of inducible nitric oxide synthase are mutually exclusive in renal mesangial cells. Am J Pathol 1997; 150:889–900.

162. Blanco FJ, Ochs RL, Schwaraz H, Lotz M. Chondrocytes apoptosis induced by nitric oxide. Am J Pathol 1995; 146:75–85.
163. Ho YS, Lee HM, Mou TC, Wang YJ, Ling JK. Suppression of nitric oxide-induced apoptosis by N-acetyl-L-cysteine through modulation of glutathione, bcl-2, and bax protein levels. Mol Carcinog 1997; 19:101–113.
164. Rao L, White E. Bcl-2 and ICE family of apoptotic regulators: making a connection. Curr Opin Genet Dev 1997; 7:52–58.
165. Yang J, Liu X, Bhalla K, Kim C N, Ibrado A M, Cai J, Peng T-I, Jones D P, Wang X. Prevention of apoptosis by Bcl-2: release of cytochrome c from mitochondria blocked. Science 1997; 275:1129–1132.
166. Boise LH, Gonzalez-Garcia M, Postema CE, Ding L, Lindsten T, Turka LA, Mao X, Nuñez G, Thompson C. bcl-x, a bcl-2-related gene that functions as a dominant regulator of apoptotic death. Cell 1993; 74:597–608.
167. Genaro AM, Hortelano S, Alvarez A, Martinez C, Bosca L. Splenic B lymphocyte programmed cell death is prevented by nitric oxide release through mechanisms involving sustained Bcl-2 levels. J Clin Invest 1995; 95:1884–1890.
168. Xie K, Huang S, Beltran PJ, Juang SH, Dong Z, Reed JC, McDonnell TJ, McConkey DJ, Fidler IJ. Bcl-2 protects cells from cytokine-induced nitric-oxide-dependent apoptosis. Cancer Immunol Immunother 1996; 43:109–115.
169. Forrester K, Ambs S, Lupold SE, Kapust RB, Spillare EA, Weinberg WC, Felley-Bosco E, Wang XW, Geller DA, Tzeng E, Billiar TR, Harris CC. Nitric oxide-induced p53 accumulation and regulation of inducible nitric oxide synthase expression by wild-type p53. Proc Natl Acad Sci USA 1996; 93:2442–2447.
170. Lindquist S, Craig EA. The heat-shock proteins. Annu Rev Genet 1988; 22:631–677.
171. Morimoto RI, Tissieres A, Georgopoulous C. The stress response, function of the proteins, and perspectives. In: Morimoto RI, Tissieres A, Georgopoulous C, eds. Stress Proteins in Biology and Medicine, Cold Spring Harbor, NY: Cold Spring Harbor Laboratory Press, 1990; pp. 1–13.
172. Schlesinger MJ. Heat shock proteins. J Biol Chem 1990; 265:12111–12114.
173. Liossis S-N, Via CS, Tsokos GC. The alter ego of heat shock proteins. Clin Immunol Immunopathol 1998; 86:235–236.
174. Wong HR, Finder JD, Wasserloos K, Pitt BR. Expression of iNOS in cultured rat pulmonary artery smooth muscle cells is inhibited by the heat shock response. Am J Physiol 1995; 269:L843–L848.
175. Hirvonen MR, Brune B, Lapetina EG. Heat shock proteins and macrophage resistance to the toxic effects of nitric oxide. Biochem J 1996; 315:845–849.
176. Feinstein DL, Galea E, Aquino DA, Li GC, Xu H, Reis DJ. Heat shock protein 70 suppresses astroglial-inducible nitric-oxide synthase expression by decreasing NfkappaB activation. J Biol Chem 1996; 271:17724-17732.
177. de Vera ME, Kim YM, Wong HR, Wang Q, Billiar TR, Geller DA. Heat shock response inhibits cytokine-inducible nitric oxide synthase expression in rat hepatocytes. Hepatology1996; 24:1238–1245.
178. de Vera ME, Wong JM, Zhou JY, Tzeng E, Wong HR, Billiar TR, Geller DA. Cytokine-induced nitric oxide synthase gene transcription is blocked by the heat shock response in human liver cells. Surgery 1996; 120:144–149.

179. Wong HR, Ryan M, Menendez IY, Denenberg A, Wispe JR. Heat shock protein induction protects human respiratory epithelium against nitric oxide-mediated cytotoxicity. Shock 1997; 8:213–218.
180. Xu Q, Hu Y, Kleindienst R, Wick G. Nitric oxide induces heat-shock protein 70 expression in vascular smooth muscle cells via activation of heat shock factor 1. J Clin Invest 1997; 100:1089–1097.
181. Kim YM, de Vera ME, Watkins SC, Billiar TR. Nitric oxide protects cultured rat hepatocytes from tumor necrosis factor-alpha-induced apoptosis by inducing heat shock protein 70 expression. J Biol Chem 1997; 272:1402–1411.

12

Nitric Oxide Biosynthesis and Metabolism in Keratinocytes:

The Functions and Consequences of Nitric Oxide Production in the Skin

Diane E. Heck
Rutgers University, Piscataway, New Jersey

I. INTRODUCTION

In mammals a specialized squamous epithelium, skin, comprises the outermost tissue and functions as the initial site of interaction with the environment. This continually renewing organ forms a protective yet plastic barrier, relaying information from, and responding to, stimuli from the environment. (1,2; for review see 3,4) . The skin employs an array of strategies to defend the organism from damage. Tissue defense mechanisms range from remodeling its architecture in response to external stimuli, which occurs during chemical or ultraviolet light - induced hyperplasia, to reversibly establishing an inflammatory microenvironment as a defense from microbial invasion (1,2,5,6). Results from numerous recent studies indicate that nitric oxide plays critical roles in most, if not all, dermal responses to stimuli from the external environment (5–9). Mammalian skin consists of diverse cell types in the outermost layer of the skin, the epidermis; keratinocytes comprise the predominant cell type populating the tissue (1,2). This review will discuss nitric oxide production by keratinocytes, focusing largely on the biological functions of nitric oxide in the skin. Unlike the majority of immune and neural cells, epithelial cells express all three isoforms of nitric oxide synthase found in mammals (10–12). In this review we will also address the physiological roles unique to each isoform and the potential significance of the

cesses (7). For example, both basilar levels and cytokine-stimulated production of nitric oxide and hydrogen peroxide are inhibited by EGF in cultured keratinocytes. Furthermore, expression of type II major histocompatibility antigen (murine Ia antigen) in response to g-IFN is downregulated by EGF (7). It is recognized that EGF, a peptide growth factor that stimulates tyrosine as well as serine and threonine phosphorylation of its receptor, inhibits nitric oxide production by blocking induction of nitric oxide synthase in keratinocytes (18,47–49), however, the biochemical mechanism by which this occurs remains unclear. It is likely that EGF may act in a similar manner to regulate Ia expression, through altering the functioning of intermediates in the g-IFN signaling pathways.

To better understand mechanisms by which cellular factors such as EGF and g-IFN regulate wound healing, it is important to examine the signaling pathways (and the potential for their interaction) which are activated by these mediators. Growth stimulatory polypeptide hormones of the epidermal growth factor family induce cellular effects through specific membrane-bound receptors. Ligand binding stimulates intrinsic tyrosine kinase activity of these receptors. Activation of these kinases initiates a cascade of phosphorylation events. Ultimately, phosphorylation of cellular proteins is required for initiation of biological responses by this family of growth factors (for review see 50–54).

Alternatively, many cytokine receptors, such as those for the interferon family of polypeptide hormones, do not have intrinsic tyrosine kinase activity. These multimeric receptors act through assembly of membrane-bound subunits and accessory proteins. Once assembled, active receptor complexes then recruit cytosolic kinases (55–58). For example, association ofa-, b-, or g-interferon with its specific cellular receptor initiates the formation of active multimeric receptor complexes. Following ligand-induced multimerization, cytosolic tyrosine kinases associate with the active membrane-bound receptor complexes. This process, in turn, results in phosphorylation of specific latent cytosolic transcription factors, referred to as signal transduction and activator of transcription (STAT) proteins. Phosphorylated STAT proteins translocate to the nucleus, where they form active transcription regulatory complexes (59–61).

EGF may function upstream of this process to alter the activity and/or expression of critical members of the signaling cascade, ultimately leading to the inhibition of nitric oxide production. For example, we have demonstrated that EGF differentially modulates the activity of STAT 1-a during differentiation, leading to altered cellular responses (7,18). We have also demonstrated that EGF downregulates hydrogen peroxide and nitric oxide production in response to g-IFN (18). In addition, EGF inhibits g-IFN-induced expression of type II major histocompatability antigen in keratinocytes (18). Potentially, regulation of the STAT signal transduction system may be an important function of EGF as well as other growth factors on a variety of cytokine-induced responses in the skin.

III. THE CONSTITUTIVE NITRIC OXIDE SYNTHASES: OR XX FUTURE OR REMEMBRANCES OF THINGS PAST

In contrast to its role in inflammation, the functions of nitric oxide produced in the skin by the constitutive nitric oxide synthases are, as of yet, poorly defined. Although little is understood about low-level nitric oxide production by keratinocytes, recent findings indicate that nitric oxide may be a critical factor in a sequella of responses initiated by external stimuli which ultimately regulate the composition of the skin (62–65). Within this continually renewing organ, the rate of keratinocyte proliferation, differentiation, and programmed cell death are largely responsible for determining tissue structure. Intriguingly, nitric oxide has been implicated in the regulation of all of these processes, cellular proliferation, differentiation, and even cell death (7,18,66–68), potentially assigning this reactive free radical a critical role in dermal homeostasis. It is well established that in keratinocytes intracellular calcium concentration is central to the regulation of differentiation (for review see 69–71). Although many of the molecular events mediating cell differentiation remain poorly understood, it is clear that calcium acts as a singularly critical regulatory molecule central to the coordination this process. In this regard it is also important to recognize that intracellular calcium levels are critical regulators of nitric oxide production by the constitutive nitric oxide synthases (27,29,39,72–74). Potentially, nitric oxide may act downstream of calcium signaling, amplifying, and directing the initial signal, ultimately altering the functioning of numerous cellular proteins.

From the earliest investigations it has been apparent that calcium-dependent nitric oxide production is associated with both proliferation and differentiation of keratinocytes (7,8,18,45); however, the significance and mechanisms regulating this remained unclear. Recently, results from studies on nitric oxide production by epithelial and other cell types have begun to shed light on the role(s) of nitric oxide in differentiation. It has been demonstrated that nitric oxide derived from the constitutively expressed nitric oxide synthases can directly interact with potassium and ATP-sensitive calcium channels (75–77). Potentially, following activation of the enzymes, nitric oxide may then itself alter intracellular calcium levels in keratinocytes and, in this manner, mediate a host of calcium-dependent events. It has also been suggested that ultraviolet light directly activates NOS1 and subsequently initiates a sequella of events some of which result in the typically observed hyperplastic response to sunlight (78,79). Although each of these scenarios may not preclude the other, the biochemical events mediating them remain poorly defined.

Keratinocytes express two forms of NOS that can be isolated from the particulate section of cell extracts. Immunohistochemistry and sequence analysis has revealed these proteins to be brain-type NOS1 and endothelial-type NOS3

(62,63,80–83), both of which are regulated by intracellular calcium and calmodulin levels. It is interesting to note that similar to endothelial cells, keratinocytes respond to stimulation with the neurotransmitter acetylcholine by producing nitric oxide. In this instance, interaction of agonists with specific subsets of nicotinic acetylcholine receptors initiates transient calcium fluxes activating nitric oxide synthase (84,85). As found in endothelial cells, nitric oxide synthase derived from the particulate fraction of broken cell preparations from keratinocytes produced relatively low concentrations of nitric oxide (80–83).

For some time it has been recognized that keratinocytes produce, recycle, and degrade acetylcholine (83–86). Muscarinic acetylcholine receptors have been identified as important in cell migration and adherence; however, the cellular functions regulated by dermal nicotinic acetylcholine receptors were not well understood (83–87). Studies of pig heart synaptosomes have shed light on this problem, revealing that release of nitric oxide stimulates synthesis and recycling of norepinephrine. Similarly numerous reports indicate that nitric oxide stimulates the release of norepinephrine from the vascular endothelium, microvascular beds, hippocampal slices, cardiac myocytes, and cutaneous afferent neurons (89–94). In a similar manner acetylcholine-stimulated nitric oxide may induce keratinocyte norepinephrine release. Whether released from surrounding keratinocytes, cutaneous nerve fibers, or dendritic or endothelial cells, norepinephrine has a profound effects on keratinocytes.

Norepinephrine acts as a potent antimitotic agent inducing a block in the G2 phase of the cell cycle in undifferentiated keratinocytes (95). In addition, through interaction with B2 adrenergic receptors, norepinephrine stimulates an enhanced influx of extracellular calcium as well as release of calcium from intracellular stores (95–97). As noted above, keratinocyte differentiation is exquisitely sensitive to calcium concentration, enhanced calcium levels stimulating cell differentiation. Furthermore, poorly differentiated keratinocytes found in the lower regions of the epidermis express abundant quantities of enzymes required for the synthesis and recycling of norepinephrine. In the highly differentiated keratinocytes found in the upper regions of the epidermis, expression of these enzymes is reduced. In this latter subset of cells, synthesis of catecholamines is diminished and significantly lower levels of norepinephrine and other catecholamines are expressed (95–97). Depressed levels of norepinephrine result in decreased expression of B2-type adrenergic receptors, and the cells become unresponsive to adrenergic stimulation (98,99). In this manner terminally differentiating keratinocytes are protected from further adrenergic stimulation.

One of the best-characterized biochemical actions induced by nitric oxide is the stimulation of soluble guanylate cyclase and resultant increases in intracellular cGMP (27–30). In many cell types, including keratinocytes, acetylcholine stimulates calcium-dependent cGMP synthesis in a nitric oxide-dependent manner, ultimately resulting in decreased levels of intracellular calcium, a feedback mechanism mediated by diminished transport of calcium through membrane channels. Because increases in intracellular calcium levels are critical for keratinocyte differentiation, this dichotomy has been difficult to reconcile with the observation that nitric oxide, which ultimately lowers intracellular calcium levels, mediates differentiation. Modulation of catecholamine release by nitric oxide during this process provides a tantalizing explanation for stimulation of differentiation in the presence of enhanced cGMP levels, and may help to further elucidate the complex role of nitric oxide in keratinocyte differentiation. In this manner acetylcholine-induced nitric oxide production by the calcium-dependent nitric oxide synthases may stimulate a carefully orchestrated, temporally organized series of biochemical events resulting in keratinocyte differentiation. It is interesting to note that in psoriasis, a condition characterized by deregulated keratinocyte differentiation, continuous aberrant release of low levels of nitric oxide by dermal Langerhans cells has been identified as a major factor in the pathology of the disease (25).

An important focus of current investigation into the role(s) of nitric oxide in the skin is centered on its importance as a regulator of cellular morphogenesis and migration. For some time it has been apparent that nitric oxide is a critical mediator of endothelial cell migration during neovascularization (99–101). Nitric oxide production is also believed to be an important factor facilitating the migration of tumor cells during metastasases (102–106). Interestingly, we have demonstrated that inhibition of nitric oxide production is an important mechanism by which 5-fluorouracil acts to inhibit the progression of carcinoma derived from the epithelium of the colon in cancer chemotherapy. Nitric oxide has been found to dilate tight junctions and alter permeability in the epithelium (102–109). Nicotinic cholinergic receptors expressed in suprabasal and differentiating keratinocytes regulate keratinocyte adhesion and lateral migration, presumably through a nitric oxide-mediated mechanism (110,111). Additionally, recent reports indicate that nitric oxide activates a specific subset of ADP-ribosyl transferases located at the cell membrane. When active, these enzymes ribosylate f-actin, dramatically altering the organization of the cytoskeleton (103–112). Therefore, nitric oxide acts to regulate the dynamic architecture of the skin via diverse pathways including stimulating the production of cGMP, altering the activity of ion channels, or functionally altering cytoskeletal elements.

IV. NOTES FROM THE UNDERGROUND: NITRIC OXIDE AND PROGRAMMED CELL DEATH

In epithelial tissues differentiation culminates in cell death. In continually renewing tissues such as the skin, tissue functioning is dependent on a critical balance between proliferation and apoptosis, programmed cell death. In the squamous epithelium apoptosis comprises the terminal stage of keratinocyte differentiation. This is followed by desquamation, the sloughing of the dead cells (1–3,112–116). In this continually renewing tissue the process of human keratinocyte differentiation and migration upward through the epidermis is believed to take approximately 12 days (1). It is clearly demonstrated by the variant pathophysiology of diseases such as psoriasis, which alter this process, that precise spatial and temporal organization of the course of events culminating in apoptosis is also critical to tissue homeostasis.

Recent reports indicate that low levels of nitric oxide produced by the calcium-dependent nitric oxide synthases inhibit apoptosis induced in response to signals initiated by the tumor necrosis factor/FAS family of cellular mediators (117–119). In this regard, it has been suggested that low levels of nitric oxide produced by endothelial nitric oxide in response to acetylcholine or by brain-type nitric oxide synthase in response to ultraviolet light, may act to protect differentiating keratinocytes from apoptosis. In the terminal stages of differentiation, diminished nitric oxide production resulting from norepinephrine-mediated inhibition of calcium fluxes may serve to remove this protective mechanism, allowing the induction of apoptosis. In contrast, a high concentration of nitric oxide produced by the macrophagelike inducible isoform of nitric oxide synthase in response to inflammatory stimuli has been found to induce apoptosis (120–124). During the initial stages of microbial-induced inflammation, nitric oxide-mediated apoptosis may act to limit hyperplasia. However, persistent or deregulated stimulation of the enzyme may be important in impaired tissue regeneration, as observed in microbial and chemical-induced chronic skin ulceration (125–127).

Clearly, the work of numerous investigators has revealed that nitric oxide is integral to virtually every aspect of skin physiology. Nitric oxide is an important signal regulating keratinocyte proliferation, differentiation, activation, and even cell death. The challenge for the future lays largely in defining regulatory mechanisms by which this free radical gas conveys signals between specific stimuli and distinct cellular responses.

REFERENCES

1. Holbrook KA, Wolff K. In: Fitzpatrick TB, Eisen AZ, Wolff K, Freedberg IM, Austin KF, eds. Dermatology in General Medicine. New York: McGraw-Hill, 1993.
2. Mathias CG, Maibach H. Dermatotoxicology monographs I. Cutaneous irritation: factors influencing the response to irritants. Clin Toxicol. 1978; 13(3):333–346.
3. Fitzpatrick TB, Dermatology 1945–95: the golden age of treatment. J Dermatol 1996; 23(11):728–734.
4. Wolff K. The epidermis as an immune organ. Hautarzt 1988; 39(8):534–543.

5. Kahari VM, Saarialho-Kere U. Matrix metalloproteinases in skin. Exp Dermatol 1997; 6(5):199–213.
6. Misery L. Skin, immunity and the nervous system. Br J Dermatol 1997; 137(6):843–850.
7. Heck DE. Molecular mechanisms regulating nitric oxide biosynthesis: role of xenobiotics in epithelial inflammation. Adv Exp Med Biol 1996; 387:171–176.
8. Bruch-Gerharz D, Ruzicka T, Kolb-Bachofen V. Nitric oxide in human skin: current status and future prospects. J Invest Dermatol 1998; 110(1):1–7.
9. Laskin JD, Heck DE, Laskin DL. Multifunctional Role of Nitric Oxide in Inflammation. Trends Endrocrinol Metab 1994; 5:377–382.
10. Vane JR, Mitchell JA, Appleton I, Tomlinson A, Bishop-Bailey D, Croxtall J, Willoughby DA. Inducible isoforms of cyclooxygenase and nitric-oxide synthase in inflammation. Proc Natl Acad Sci USA 1994; 91(6):2046–2050.
11. Shimizu Y, Sakai M, Umemura Y, Ueda H. Immunohistochemical localization of nitric oxide synthase in normal human skin: expression of endothelial-type and inducible-type nitric oxide synthase in keratinocytes. J Dermatol 1997; 24(2):80–87.
12. Ormerod AD, Weller R, Copeland P, Benjamin N, Ralston SH, Grabowksi P, Herriot R. Detection of nitric oxide and nitric oxide synthases in psoriasis. Arch Dermatol Res 1998; 290(1–2):3–8.
13. Moulin V. Growth factors in skin wound healing. Eur J Cell Biol 1995; 68(1):1–7.
14. Vignola AM. Monocytes. J Allergy Clin Immunol 1998; 101(2 Pt 1):149–152.
15. Baggiolini M. Chemokines and leukocyte traffic. Nature 1998; 392(6676):565–568.
16. Downey GP, Fukushima T, Fialkow L. Signaling mechanisms in human neutrophils. Curr Opin Hematol 1995; 2(1):76–88.
17. Laskin DL, Heck DE, Punjabi CJ, Laskin JD. Role of nitric oxide in hematosuppression and benzene-induced toxicity. Environ Health Perspect 1996; 104(suppl 6):1283–1287.
18. Heck DE, Laskin DL, Gardner CR, Laskin JD. Epidermal growth factor suppresses nitric oxide and hydrogen peroxide production by keratinocytes. Potential role for nitric oxide in the regulation of wound healing. J Biol Chem 1992; 267(30):21277–21280.
19. Norris DA. Cytokine modulation of adhesion molecules in the regulation of immunologic cytotoxicity of epidermal targets. J Invest Dermatol 1990; 95(6 suppl):111S–120S.
20. Luger TA, Schwarz T. Evidence for an epidermal cytokine network. J Invest Dermatol 1990; 95(6 suppl):100S–104S.
21. Boh EE. Role of reactive oxygen species in dermatologic diseases. Clin Dermatol 1996; 14(4):343–352.
22. Deliconstantinos G, Villiotou V, Stavrides JC. Alterations of nitric oxide synthase and xanthine oxidase activities of human keratinocytes by ultraviolet B radiation. Potential role for peroxynitrite in skin inflammation. Biochem Pharmacol 1996; 51(12):1727–1738.

23. Weller R. Nitric oxide—a newly discovered chemical transmitter in human skin. Br J Dermatol 1997; 137(5):665–672.
24. DeGeorge GL, Heck DE, Laskin JD. Arginine metabolism in keratinocytes and macrophages during nitric oxide biosynthesis: multiple modes of action of nitric oxide synthase inhibitors. Biochem Pharmacol 1997; 54(1):103–112.
25. Morhenn VB. Langerhans cells may trigger the psoriatic disease process via production of nitric oxide. Immunol Today 1997; 18(9):433–436.
26. Ross R, Gillitzer C, Kleinz R, Schwing J, Kleinert H, Forstermann U, Reske-Kunz AB. Involvement of NO in contact hypersensitivity. Int Immunol 1998; 10(1):61–69.
27. Moncada S. Nitric oxide in the vasculature: physiology and pathophysiology. Ann N Y Acad Sci 1997; 811:60–67.
28. Nathan C. Inducible nitric oxide synthase: what difference does it make? J Clin Invest 1997; 100(10):2417–2423.
29. Marletta MA. Nitric oxide synthase structure and mechanism. J Biol Chem 1993; 268(17):12231–12234.
30. Beckman JS, Koppenol WH. Nitric oxide, superoxide, and peroxynitrite: the good, the bad, and ugly. Am J Physiol 1996; 271(5 Pt 1):C1424–C1437.
31. Knowles RG. Nitric oxide biochemistry. Biochem Soc Trans 1997; 25(3):895–901.
32. Lin KT, Xue JY, Lin MC, Spokas EG, Sun FF, Wong PY. Peroxynitrite induces apoptosis of HL-60 cells by activation of a caspase-3 family protease. Am J Physiol 1998; 274(4 Pt 1):C855–C860.
33. Albina JE, Reichner JS. Role of nitric oxide in mediation of macrophage cytotoxicity and apoptosis. Cancer Metastasis Rev 1998; 17(1):39–53.
34. Mannick JB, Miao XQ, Stamler JS. Nitric oxide inhibits Fas-induced apoptosis. J Biol Chem 1997 ; 272(39):24125–24128.
35. Gute DC, Ishida T, Yarimizu K, Korthuis RJ. Inflammatory responses to ischemia and reperfusion in skeletal muscle. Mol Cell Biochem 1998; 179(1–2):169–187.
36. Gordge MP. How cytotoxic is nitric oxide? Exp Nephrol 1998; 6(1):12–16.
37. Lala PK, Orucevic A. Role of nitric oxide in tumor progression: lessons from experimental tumors. Cancer Metastasis Rev 1998r; 17(1):91–106.
38. Robertson FM, Long BW, Tober KL, Ross MS, Oberyszyn TM. Gene expression and cellular sources of inducible nitric oxide synthase during tumor promotion. Carcinogenesis 1996; 17(9):2053–2059.
39. Vanheel B, Van de Voorde J. Nitric oxide induced membrane hyperpolarization in the rat aorta is not mediated by glibenclamide-sensitive potassium channels. Can J Physiol Pharmacol 1997; 75(12):1387–1392.
40. Feletou M, Vanhoutte PM. Endothelium-derived hyperpolarizing factor. Clin Exp Pharmacol Physiol 1996; 23(12):1082–1090.
41. Zhang Y, Ramos BF, Jakschik BA. Role of mast cells in plasma permeation due to immune injury of the skin basement membrane. Immunology 1992; 77(3):422–427.
42. Davies KJ. Oxidative stress: the paradox of aerobic life. Biochem Soc Symp 1995; 61:1–31.
43. White MJ, Heckler FR. Oxygen free radicals and wound healing. Clin Plast Surg 1990; 17(3):473–484.

44. Martin A. The use of antioxidants in healing. Dermatol Surg 1996; 22(2):156–160.
45. Shearer JD, Richards JR, Mills CD, Caldwell MD. Differential regulation of macrophage arginine metabolism: a proposed role in wound healing. Am J Physiol 1997; 272(2 Pt 1):E181–E190.
46. Moulin V. Growth factors in skin wound healing. Eur J Cell Biol 1995; 68(1):1–7.
47. Sciorati C, Nistico G, Meldolesi J, Clementi E. Nitric oxide effects on cell growth: GMP-dependent stimulation of the AP-1 transcription complex and cyclic GMP-independent slowing of cell cycling. Br J Pharmacol 1997; 122(4):687–697.
48. Estrada C, Gomez C, Martin-Nieto J, De Frutos T, Jimenez A, Villalobo A. Nitric oxide reversibly inhibits the epidermal growth factor receptor tyrosine kinase. Biochem J 1997; 326(Pt 2):369–376.
49. Goldman R, Moshonov S, Chen X, Berchansky A, Furstenberger G, Zor U. Crosstalk between elevation of $[Ca^{2+}]i$, reactive oxygen species generation and phospholipase A2 stimulation in a human keratinocyte cell line. Adv Exp Med Biol 1997; 433:41–45.
50. Hunter T, Cooper JA. Protein-tyrosine kinases. Annu Rev Biochem 1985; 54:897–930.
51. Riese DJ 2nd, Stern DF. Specificity within the EGF family/ErbB receptor family signaling network. Bioessays 1998; 20(1):41–48.
52. Belsches AP, Haskell MD, Parsons SJ. Role of c-Src tyrosine kinase in EGF-induced mitogenesis. Front Biosci 1997; 2:D501–D518.
53. Castellino AM, Chao MV. Trans-signaling by cytokine and growth factor receptors. Cytokine Growth Factor Rev 1996; 7(4):297–302.
54. Goodlad RA, Wright NA. Epidermal growth factor (EGF). Baillieres Clin Gastroenterol. 1996; 10(1):33–47.
55. Naismith JH, Sprang SR. Modularity in the TNF-receptor family. Trends Biochem Sci 1998; 23(2):74–79.
56. Haque SJ, Williams BR. Signal transduction in the interferon system. Semin Oncol 1998; 25(1 suppl 1):14–22.
57. Pestka S, Kotenko SV, Muthukumaran G, Izotova LS, Cook JR, Garotta G. The interferon gamma (IFN-gamma) receptor: a paradigm for the multichain cytokine receptor. Cytokine Growth Factor Rev 1997; 8(3):189–206.
58. Drachman JG, Kaushansky K. Structure and function of the cytokine receptor superfamily. Curr Opin Hematol 1995; 2(1):22–28.
59. Darnell JE Jr. STATs and gene regulation. Science 1997; 277(5332):1630–1635.
60. Ransohoff RM. Cellular responses to interferons and other cytokines: the JAK-STAT paradigm. N Engl J Med 1998; 338(9):616–618.
61. Silvennoinen O, Saharinen P, Paukku K, Takaluoma K, Kovanen P. Cytokine receptor signal transduction through Jak tyrosinc kinases and Stat transcription factors. APMIS 1997; 105(7):497–509.
62. Deliconstantinos GG, Villiotou V, Stavrides JC. Nitric oxide and peroxynitrite released by ultraviolet B-irradiated human endothelial cells are possibly involved in skin erythema and inflammation. Exp Physiol 1996; 81(6):1021–1033.
63. Warren JB. Nitric oxide and human skin blood flow responses to acetylcholine and ultraviolet light. FASEB J 1994; 8(2):247–251.

64. Kang-Rotondo CH, Major S, Chiang TM, Myers LK, Kang ES. Upregulation of nitric oxide synthase in cultured human keratinocytes after ultraviolet B and bradykinin. Photodermatol Photoimmunol Photomed. 1996; 12(2):57–65.
65. Weller R. Nitric oxide—a newly discovered chemical transmitter in human skin. Br J Dermatol 1997; 137(5):665–672.
66. Henseleit U, Rosenbach T, Kolde G. Induction of apoptosis in human HaCaT keratinocytes. Arch Dermatol Res 1996; 288(11):676–683.
67. Thomsen LL, Miles DW. Role of nitric oxide in tumour progression: lessons from human tumours. Cancer Metastasis Rev 1998; 17(1):107–118.
68. Szab C, Ohshima H. DNA damage induced by peroxynitrite: subsequent biological effects. Nitric Oxide 1997; 1(5):373–385.
69. Kruszewski FH, Hennings H, Yuspa SH, Tucker RW. Regulation of intracellular free calcium in normal murine keratinocytes. Am J Physiol 1991; 261(5 Pt 1):C767–C773.
70. Yuspa SH, Hennings H, Tucker RW, Jaken S, Kilkenny AE, Roop DR. Signal transduction for proliferation and differentiation in keratinocytes. Ann NY Acad Sci 1988; 548:191–196.
71. Fuchs E. Epidermal differentiation and keratin gene expression. J Cell Sci Suppl 1993; 17:197–208.
72. Leist M, Nicotera P. Calcium and neuronal death. Rev Physiol Biochem Pharmacol 1998; 132:79–125.
73. Hurford WE. The biologic basis for inhaled nitric oxide. Respir Care Clin North Am 1997; 3(3):357–369.
74. Yun HY, Dawson VL, Dawson TM. Nitric oxide in health and disease of the nervous system. Mol Psychiatry 1997; 2(4):300–310.
75. Aghdasi B, Reid MB, Hamilton SL. Nitric oxide protects the skeletal muscle Ca^{2+} release channel from oxidation induced activation. J Biol Chem 1997; 272(41):25462–25467.
76. Clementi E, Meldolesi J. The cross-talk between nitric oxide and Ca^{2+}: a story with a complex past and a promising future. Trends Pharmacol Sci 1997; 18(8):266–269.
77. Vanhoutte P. Role of calcium and endothelium in hypertension, cardiovascular disease, and subsequent vascular events. J Cardiovasc Pharmacol 1992; 19(suppl 3):S6–S10.
78. Deliconstantinos G, Villiotou V, Stravrides JC. Release by ultraviolet B (u.v. B) radiation of nitric oxide (NO) from human keratinocytes: a potential role for nitric oxide in erythema production. Br J Pharmacol 1995; 114(6):1257–1265.
79. Warren JB, Loi RK, Coughlan ML. Nitric oxide and human skin blood flow responses to acetylcholine and ultraviolet light. FASEB J 1994; 8(2):247–251.
80. Baudouin JE, Tachon P. Constitutive nitric oxide synthase is present in normal human keratinocytes. J Invest Dermatol 1996; 106(3):428–431.
81. Belmont HM, Levartovsky D, Goel A, Amin A, Giorno R, Rediske J, Skovron ML, Abramson SB. Increased nitric oxide production accompanied by the up-regulation of inducible nitric oxide synthase in vascular endothelium from patients with systemic lupus erythematosus. Arthritis Rheum 1997 ; 40(10):1810–1816.

82. Wang R, Ghahary A, Shen YJ, Scott PG, Tredget EE. Nitric oxide synthase expression and nitric oxide production are reduced in hypertrophic scar tissue and fibroblasts. J Invest Dermatol 1997 ; 108(4):438–444.
83. Romero-Graillet C, Aberdam E, Clement M, Ortonne JP, Ballotti R. Nitric oxide produced by ultraviolet-irradiated keratinocytes stimulates melanogenesis. J Clin Invest 1997; 99(4):635–642.
84. Grando SA. Biological functions of keratinocyte cholinergic receptors. J Invest Dermatol Symp Proc 1997; 2(1):41–48.
85. Grando SA, Horton RM, Mauro TM, Kist DA, Lee TX, Dahl MV. Activation of keratinocyte nicotinic cholinergic receptors stimulates calcium influx and enhances cell differentiation. J Invest Dermatol 1996; 107(3):412–418.
86. Lindstrom J. Nicotinic acetylcholine receptors in health and disease. Mol Neurobiol 1997; 15(2):193–222.
87. Grando SA, Zelickson BD, Kist DA, Weinshenker D, Bigliardi PL, Wendelschafer-Crabb G, Kennedy WR, Dahl MV. Keratinocyte muscarinic acetylcholine receptors: immunolocalization and partial characterization. J Invest Dermatol 1995; 104(1):95–100.
88. Ogata H, Podolsky DK. Trefoil peptide expression and secretion is regulated by neuropeptides and acetylcholine. Am J Physiol 1997; 273(2 Pt 1):G348–G354.
89. Sears CE, Choate JK, Paterson DJ. Inhibition of nitric oxide synthase slows heart rate recovery from cholinergic activation. J Appl Physiol 1998; 84(5):1596–1603.
90. Vanhatalo S, Lumme A, Soinila S. Colchicine differentially induces the expressions of nitric oxide synthases in central and peripheral catecholaminergic neurons. Exp Neurol 1998; 150(1):107–114.
91. Nase GP, Boegehold MA. Endothelium-derived nitric oxide limits sympathetic neurogenic constriction in intestinal microcirculation. Am J Physiol 1997; 273(1 Pt 2):H426–H433.
92. Lonart G, Johnson KM. Characterization of nitric oxide generator-induced hippocampal [^{3}H]norepinephrine release. I. The role of glutamate. J Pharmacol Exp Ther 1995; 275(1):7–13.
93. Calderone A, Thaik CM, Takahashi N, Chang DLF, Colucci WS. Nitric oxide, atrial natriuretic peptide, and cyclic GMP inhibit the growth-promoting effects of norepinephrine in cardiac myocytes and fibroblasts. J Clin Invest 1998; 101(4):812–818.
94. Colucci WS. Molecular and cellular mechanisms of myocardial failure. Am J Cardiol 1997; 80(11A):15L–25L.
95. Harper RA, Flaxman BA. Effect of pharmacological agents on human keratinocyte mitosis in vitro. II. Inhibition by catecholamines. J Cell Physiol 1975; 86(2 Pt 1):293–299.
96. Schallreuter KU. Epidermal adrenergic signal transduction as part of the neuronal network in the human epidermis. J Investig Dermatol Symp Proc 1997; 2(1):37–40.
97. Schallreuter KU, Lemke KR, Pittelkow MR, Wood JM, Korner C, Malik R. Catecholamines in human keratinocyte differentiation. J Invest Dermatol 1995; 104(6):953–957.

98. Flaxman BA, Harper RA. In vitro analysis of the control of keratinocyte proliferation in human epidermis by physiologic and pharmacologic agents. J Invest Dermatol 1975; 65(1):52–59.
99. Noiri E, Lee E, Testa J, Quigley J, Colflesh D, Keese CR, Giaever I, Goligorsky MS. Podokinesis in endothelial cell migration: role of nitric oxide. Am J Physiol. 1998; 274(1 Pt 1):C236–C244.
100. Ruschitzka FT, Noll G, Luscher TF. The endothelium in coronary artery disease. Cardiology 1997; 88 Suppl 3:3–19.
101. Ziche M, Morbidelli L, Masini E, Amerini S, Granger HJ, Maggi CA, Geppetti P, Ledda F. Nitric oxide mediates angiogenesis in vivo and endothelial cell growth and migration in vitro promoted by substance P. J Clin Invest 1994; 94(5):2036–2044.
102. Garcia-Cardena G, Folkman J. Is there a role for nitric oxide in tumor angiogenesis? J Natl Cancer Inst 1998; 90(8):560–561.
103. Gallo O, Masini E, Morbidelli L, Franchi A, Fini-Storchi I, Vergari WA, Ziche M. Role of nitric oxide in angiogenesis and tumor progression in head and neck cancer. J Natl Cancer Inst 1998; 90(8):587–596.
104. Thomsen LL, Miles DW. Role of nitric oxide in tumour progression: lessons from human tumours. Cancer Metastasis Rev 1998; 17(1):107–118.
105. Lala PK. Significance of nitric oxide in carcinogenesis, tumor progression and cancer therapy. Cancer Metastasis Rev 1998; 17(1):1–6.
106. Chinje EC, Stratford IJ. Role of nitric oxide in growth of solid tumours: a balancing act. Essays Biochem 1997; 32:61–72.
107. Kolb-Bachofen V, Fehsel K, Michel G, Ruzicka T. Epidermal keratinocyte expression of inducible nitric oxide synthase in skin lesions of psoriasis vulgaris. Lancet 1994; 344(8915):139.
108. Jun CD, Han MK, Kim UH, Chung HT. Nitric oxide induces ADP-ribosylation of actin in murine macrophages: association with the inhibition of pseudopodia formation, phagocytic activity, and adherence on a laminin substratum. Cell Immunol 1996; 174(1):25–34.
109. Noiri E, Peresleni T, Srivastava N, Weber P, Bahou WF, Peunova N, Goligorsky MS. Nitric oxide is necessary for a switch from stationary to locomoting phenotype in epithelial cells. Am J Physiol 1996; 270(3 Pt 1):C794–C802.
110. Grando SA, Horton RM, Pereira EF, Diethelm-Okita BM, George PM, Albuquerque EX, Conti-Fine BM. A nicotinic acetylcholine receptor regulating cell adhesion and motility is expressed in human keratinocytes. J Invest Dermatol 1995; 105(6):774–781.
111. Huang XZ, Wu JF, Cass D, Erle DJ, Corry D, Young SG, Farese RV Jr, Sheppard D. Inactivation of the integrin beta 6 subunit gene reveals a role of epithelial integrins in regulating inflammation in the lung and skin. Cell Biol 1996; 133(4):921–928.
112. Munzel T, Heitzer T, Harrison DG. The physiology and pathophysiology of the nitric oxide/superoxide system. Herz 1997; 22(3):158–172.
113. Dal Pra I, Tognana E, Guerriero C, Armato U. Investigations into mechanisms modulating proliferation, differentiation, and apoptosis in cultured liver, adrenal, skin, and bone cells. Ital J Anat Embryol 1997; 102(4):9–119.

114. Norsgaard H, Clark BF, Rattan SI. Distinction between differentiation and senescence and the absence of increased apoptosis in human keratinocytes undergoing cellular aging in vitro. Exp Gerontol 1996; 31(5):563–570.
115. Magal SS, Jackman A, Pei XF, Schlegel R, Sherman L. Induction of apoptosis in human keratinocytes containing mutated p53 alleles and its inhibition by both the E6 and E7 oncoproteins.,Int J Cancer 1998; 75(1):96–104.
116. Maruoka Y, Harada H, Mitsuyasu T, Seta Y, Kurokawa H, Kajiyama M, Toyoshima K. Keratinocytes become terminally differentiated in a process involving programmed cell death. Biochem Biophys Res Commun 1997; 238(3):886–890.
117. Hebestreit H, Dibbert B, Balatti I, Braun D, Schapowal A, Blaser K, Simon HU. Disruption of fas receptor signaling by nitric oxide in eosinophils. J Exp Med 1998; 187(3):415–425.
118. Kim YM, Talanian RV, Billiar TR. Nitric oxide inhibits apoptosis by preventing increases in caspase-3-like activity via two distinct mechanisms. J Biol Chem 1997; 272(49):31138–31148.
119. Mannick JB, Miao XQ, Stamler JS. Nitric oxide inhibits Fas-induced apoptosis. J Biol Chem 1997; 272(39):24125–24128.
120. Chlichlia K, Peter ME, Rocha M, Scaffidi C, Bucur M, Krammer PH, Schirrmacher V, Umansky V. Caspase Activation is required for nitric oxide-mediated, CD95(APO-1/Fas)-dependent and independent apoptosis in human neoplastic lymphoid cells. Blood 1998; 91(11):4311–4320.
121. Meldrum DR. Tumor necrosis factor in the heart. Am J Physiol 1998; 274(3 Pt 2):R577–R595.
122. Selleri C, Sato T, Raiola AM, Rotoli B, Young NS, Maciejewski JP. Induction of nitric oxide synthase is involved in the mechanism of Fas-mediated apoptosis in haemopoietic cells. Br J Haematol 1997; 99(3):481–489.
123. Stassi G, Maria RD, Trucco G, Rudert W, Testi R, Galluzzo A, Giordano C, Trucco M., Nitric oxide primes pancreatic beta cells for Fas-mediated destruction in insulin-dependent diabetes mellitus. J Exp Med 1997; 186(8):1193–1200.
124. Fukuo K, Hata S, Suhara T, Nakahashi T, Shinto Y, Tsujimoto Y, Morimoto S, Ogihara T. Nitric oxide induces upregulation of Fas and apoptosis in vascular smooth muscle. Hypertension 1996; 27(3 Pt 2):823–826.
125. Veves A, Akbari CM, Primavera J, Donaghue VM, Zacharoulis D, Chrzan JS, DeGirolami U, LoGerfo FW, Freeman R. Endothelial dysfunction and the expression of endothelial nitric oxide synthetase in diabetic neuropathy, vascular disease, and foot ulceration. Diabetes 1998; 47(3):457–463.
126. Bruch-Gerharz D, Fehsel K, Suschek C, Michel G, Ruzicka T, Kolb-Bachofen V. A proinflammatory activity of interleukin 8 in human skin: expression of the inducible nitric oxide synthase in psoriatic lesions and cultured keratinocytes. J Exp Med 1996 ; 184(5):2007–2012.
127. Qureshi AA, Lerner LH, Lerner EA. From bedside to the bench and back. Nitric oxide and the cutis. Arch Dermatol 1996; 132(8):889–893.

13

Taurine Chloramine, an Inhibitor of iNOS Expression and a Potential Modulator of Inflammation

Michael R. Quinn and Georgia B. Schuller-Levis
New York State Institute for Basic Research in Developmental Disabilities, Staten Island, New York

I. INTRODUCTION

Taurine (2-aminoethanesulfonic acid) has been demonstrated under diverse conditions to protect against tissue damage resulting from various insults, all of which have an inflammatory response as an integral pathogenic feature. The mechanism of the protective effects afforded by taurine has been attributed primarily to taurine acting as a detoxifying agent against hypochlorous acid by forming taurine chloramine (Tau-Cl). However, recent studies suggest that this is not the sole mechanism mediating the tissue protective effects of taurine. Indeed, Tau-Cl itself is not simply an end product of taurine interaction with hypochlorous acid; Tau-Cl is an effector molecule that inhibits the production of inflammatory mediators by activated immune responsive cells, e.g., macrophages, neutrophils, and glial cells.

The main objective of this chapter is to present recent findings supporting this view. It is not our intent to provide an inclusive review, but key contributions are cited as sources of background information.

II. TAURINE, A UNIQUE AMINO ACID WITH MANY ROLES

Taurine is the precursor amino acid for taurine chloramine formation and is a rather unique amino acid in its own right. Taurine is a β-amino acid that is present in high intracellular concentrations in the free amino acid pool of most mammalian cells. Since taurine exists as a sulfonic rather than as a carboxylic acid, it is not translationally incorporated into protein and is a zwitterion over the physiological range of pH (1). Taurine is considered an indispensable amino acid for felines (2) and a conditionally indispensable amino acid for human and nonhuman primates (3,4). The fundamental importance of taurine in the maintenance of normal retinal function and brain development (5) led to the inclusion of taurine as a supplement in virtually all normal human infant formulas that are currently marketed (4).

The purported physiological roles of taurine are numerous and range from such general functions as being important in development of the CNS (5) and immune system (6) to more mechanistic roles, e.g., acting as an agonist at inhibitory amino acid (GABA and glycine) gated chloride channels (7,8). Taurine is also thought to be an important cellular osmoregulator, to modulate intracellular free calcium levels, to influence protein phosphorylation, to stabilize membranes by interacting with constituent neutral phospholipids, and to act as an antioxidant (1).

III. TAURINE PROTECTS AGAINST TISSUE DAMAGE

It is of particular interest that taurine is reported to offer protection against tissue damage in a variety of conditions that involve potent inflammatory responses (Table 1). Exposure of humans and experimental animals to oxidant gases (NO_2, O_3, high O_2) results in pulmonary injury (29) that is morphologically characterized by destruction of endothelial cells, changes in bronchiolar epithelium, and edema accompanied by an influx of leukocytes into the alveolar spaces (30–32). Lung damage induced by inhalation exposure to NO_2 or O_3 is attenuated by administration of supplemental taurine prior to oxidant exposure (9,10). Taurine is also protective against tissue damage in other models of lung injury. For instance, intratracheal instillation of bleomycin or amiodarone causes lung damage through an L-arginine-dependent mechanism (33,34), and injury is attenuated by taurine administration (11,12). Taurine also protects against lung damage induced by paraquot (13).

The tissue-protective effects of taurine extend beyond models involving the lung (Table 1) and include such inflammatory disease states as psoriasis (35), ischemic-reperfusion injury (15,20), rheumatoid arthritis (22), atherosclerosis (17), and endotoxemia (14,24), where nitric oxide is an important mediator of tissue damage (36–39). The mechanisms most often invoked to account for the protective actions of taurine are membrane stabilization, regulation of Ca^{2+} fluxes, or antioxidant effects. Although metabolic precursors of taurine have antioxidant activity, they are normally present in relatively low amounts, and taurine inef-

Table 1 Taurine Protects Against Tissue Damage in a Variety of Paradigms that Involve an Inflammatory Reaction as a Major Etiological Component

Injury Model	Proposed mechanism(s) of taurine action
In vivo studies	
Lung damage, NO_2 inhalation in hamsters	Stabilizes lipid components of plasma membranes, antioxidant (9)
Lung damage, ozone inhalation in rats	Tau-Cl inhibits production of NO and other pro-inflammatory mediators (10)
Lung fibrosis, bleomycin or amiodarone instillation in hamsters	Antioxidant, membrane stabilization (11,12)
Kidney and lung damage, paraquot intoxication in beagles	Antioxidant, membrane stabilization (13)
Survival of mice after cecal ligation and puncture (sepsis)	Increased O_2^- and TNF-α, but decreased NO release by macrophages; increased Ca^{2+} flux and phagocytic activity (14)
Myocardial reperfusion injury in humans	HOCl detoxification via Tau-Cl formation, intracellular Ca^{2+} effects (15)
Congestive heart failure in humans	Inotropic effects (16)
Atherosclerotic lesions in cholesterol-fed rabbits	Antioxidant (17)
Carrageenan-induced paw oedema in rats	CNS inhibitory neurotransmitter agonist (18)
Hypoxia-induced learning impairment, convulsions and death in mice	CNS inhibitory activity, decreases free radicals, Ca^{2+} influx, HOCl (19)
In vitro studies	
Myocardial reperfusion injury in guinea pigs	Antioxidant, free radical scavenger, HOCl inactivator (20)
Rat alveolar macrophages exposed to ozone	Membrane stabilization, antioxidant decreasing lipid peroxidation (21)
Bovine collagen degradation by HOCl	Inert Tau-Cl formation to detoxify HOCl (22)
Cultured cat lung epithelial cell injury by HOCl exposure	Tau-Cl formation to detoxify HOCl (23)
Cell death in activated rat hepatocytes and oxygen radicals (24)	Decreases production of NO
Inhibition of respiratory burst in activated human PMN induced by activated PMN or arsenic + TNF-α	Tau-Cl formation stimulates MPO activity by removing HOCl feedback inhibition (25,26), decreases Ca^{2+} flux (27,28)

fectively reacts with hydroxyl radical, superoxide, and peroxide (40). However, taurine is very effective at attenuating the highly toxic effects of hypochlorous acid ($HOCl/OCl^-$) produced by activated polymorphonuclear leukocytes (pmn), eosinophils, and basophils (41–44). Tau-Cl formation may also account for the more general antioxidant characteristics attributed to taurine because $HOCl/OCl^-$ is involved in causing oxidative alterations of lipoproteins, a contributing factor to the development of atherosclerotic lesions (45–47). It is worth noting that other inflammatory mediators are also involved in the pathogenesis of atherosclerosis (48).

IV. TAURINE CHLORAMINE AND INFLAMMATION

Neutrophils and macrophages play a crucial role in all of the conditions listed in Table 1. Leukocytes maintain intracellular free taurine concentrations in the range of 20 to 50 mM (49–51), and pmn contain a halide-dependent myeloperoxidase (MPO) that catalyzes a reaction of chloride anion with hydrogen peroxide to form $HOCl/OCl^-$. Hypochlorous acid is a potent oxidant capable of causing indiscriminate damage to many essential cellular constituents (52). However, extensive tissue damage may be circumvented by the formation of less reactive oxidants such as chloramines, especially those that form relatively stable structures that are not lipophilic. The major chloramine formed in most biological systems is taurine monochloramine (Tau-Cl), a compound that is far less reactive than $HOCl/OCl^-$ (41–44,53).

Recent kinetic evidence suggests that halide-dependent MPO in pmn may form an enzyme-bound intermediate that directly chlorinates taurine to form Tau-Cl, thus bypassing synthesis of the highly reactive $HOCl/OCl^-$ (54). Although HOCl may not be synthesized under conditions that approximate the intracellular environment of neutrophils, MPO is released from activated pmn and catalyzes formation of HOCl under conditions of low extracellular free taurine concentrations. For taurine to act as an effective inactivator of HOCl during inflammation, extracellular taurine concentrations would need to be elevated. This was accomplished in most of the in vivo studies listed in Table 1 by providing supplemental taurine in the drinking water. Although extracellular taurine levels would be elevated systemically, Tau-Cl would be formed only at the site of inflammation by activated pmn.

Currently, the prevailing hypothesis explaining the protective effects of taurine is its detoxification of $HOCl/OCl^-$ through formation of Tau-Cl. However, recent evidence from our laboratories suggests that Tau-Cl, once formed, is more than an end product of $HOCl/OCl^-$ detoxification; Tau-Cl may be a significant biological effector molecule (55–59). Although it has been suggested that taurine stimulates MPO activity by forming Tau-Cl and removing HOCl product inhibition of MPO activity (25,26), some investigators report that Tau-Cl inactivates MPO (41,60). In either event, evidence from our laboratories suggests that Tau-Cl is transported into leukocytes and that if Tau-Cl is present either before or during activation, but not after, the production of inflammatory mediators is inhibited in a concentration-dependent manner (55–59). We have focused our studies on production of NO, TNF-α, and PGE_2 because they are important inflammatory mediators that are involved in mediating tissue damage during overt inflammatory reactions. We have used a murine macrophage-like cell line, RAW 264.7, for these studies but have obtained similar results using authentic macrophages (58) and glial cells (61), and have extended our findings to include Tau-Cl inhibition of O_2^- generation by stimulated pmn (58).

V. EFFECTS OF TAURINE CHLORAMINE ON INFLAMMATORY MEDIATORS

Tau-Cl formation requires the presence of the halide-dependent MPO system associated with pmn, and the quantity formed depends, in part, on the concentration of taurine that is accessible (23,43,62). Since macrophages and RAW 264.7 cells do not contain MPO activity, we synthesized Tau-Cl (63) and added known concentrations to cells in culture. RAW 264.7 cells (2×10^5 cells/well) were seeded into 96-well plates in phenol red-free DMEM containing 10% fetal calf serum, 2 mM glutamine, and antibiotics. After adherence overnight in a 5% CO_2 incubator (37°C), cells were activated with LPS (10 μg/mL) and recombinant murine IFN-γ (50 U/mL). Various concentrations of taurine or Tau-Cl were added separately at the time of activation and conditioned medium was assayed 24 hr later for NO_2^- using the Griess reagent. Standard curves were prepared using sodium nitrite.

Tau-Cl inhibited NO production by activated cells in a concentration-dependent way while taurine was without effect (Fig. 1). Half-maximal inhibition of NO production was achieved at approximately 0.5 mM Tau-Cl. Although extracellular taurine concentrations are usually in the 100 to 200 μM range, they may become elevated during inflammation; e.g., 1 to 20 mM extracellular taurine has been noted during chronic lung inflammatory conditions (23). Hence, physiologically relevant concentrations of Tau-Cl are likely to be attained within the microenvironment of cells that participate in the inflammatory response. In addition, macrophages also concentrate Tau-Cl by a transport system (55). RAW 264.7 cells have an uptake system for Tau-Cl (Fig. 2) that is sodium-dependent, requires energy, and is inhibited by low temperature (55). The transport system for Tau-Cl has a $K_m \approx 50$ μM, a $V_{max} \approx 85$ pmol/min/10^6 cells, and is relatively

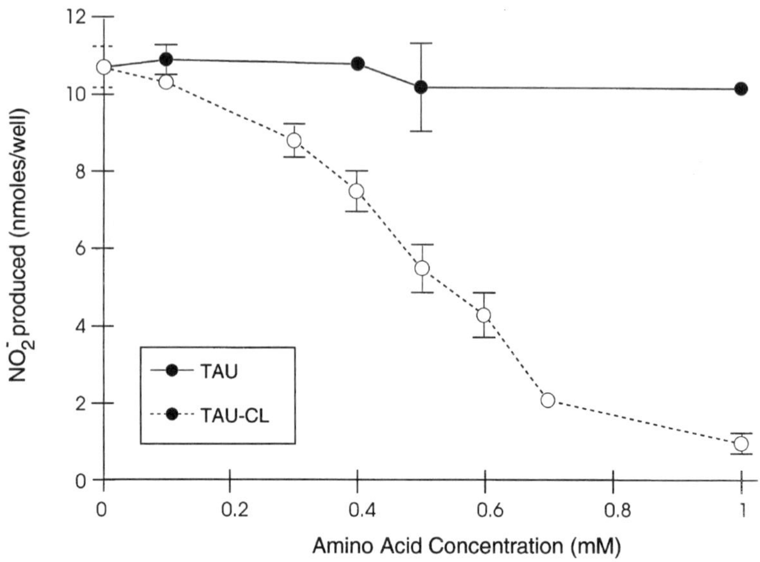

Figure 1 Tau-Cl inhibits NO_2^- production by RAW 264.7 cells activated with LPS (10 μg/mL) and rIFN-γ (50 U/mL). Cells were activated in the presence of various concentrations of Tau-Cl (-○-) or taurine (-●-). Conditioned medium was assayed for nitrite 24 hr after activation. Values represent the mean ± SD of triplicate samples with similar results obtained in 10 additional independent experiments. (From Ref. 55.)

unaffected by the presence of taurine. Likewise, transport of taurine has a $K_D \approx$ 17μM, a $V_{max} \approx$ 124 pmol/min 10^6 cells and is unaffected by the presence of Tau-Cl. These results suggest that the two uptake systems are distinct and provide a mechanism for Tau-Cl to be transported into RAW 264.7 cells, where it exerts its effects.

To gain insight into the mechanism utilized by Tau-Cl to inhibit NO production, Tau-Cl was added to cells at various times after activation with LPS and IFN-γ (Fig. 3). The most prominent inhibition of NO production occurred when Tau-Cl was added at the time of activation. Moderate inhibition was achieved if Tau-Cl was added within 3 hr of the activators, and relatively little effect was observed 6 to 24 hr after activation. This was examined further in Tau-Cl washout experiments (Fig. 4). Tau-Cl was added at the time of activation and after 6 hr was either removed or added back by replacing the culture medium with fresh medium containing activators ± Tau-Cl. For comparison, Tau-Cl was added after cells were activated for 6 hr in the absence of Tau-Cl. Nitric oxide production was measured 24 hr later. If Tau-Cl was removed after the first 6 hr of exposure in the presence of activators, then inhibition of NO production was intermediate between that obtained with the constant presence of Tau-Cl and the

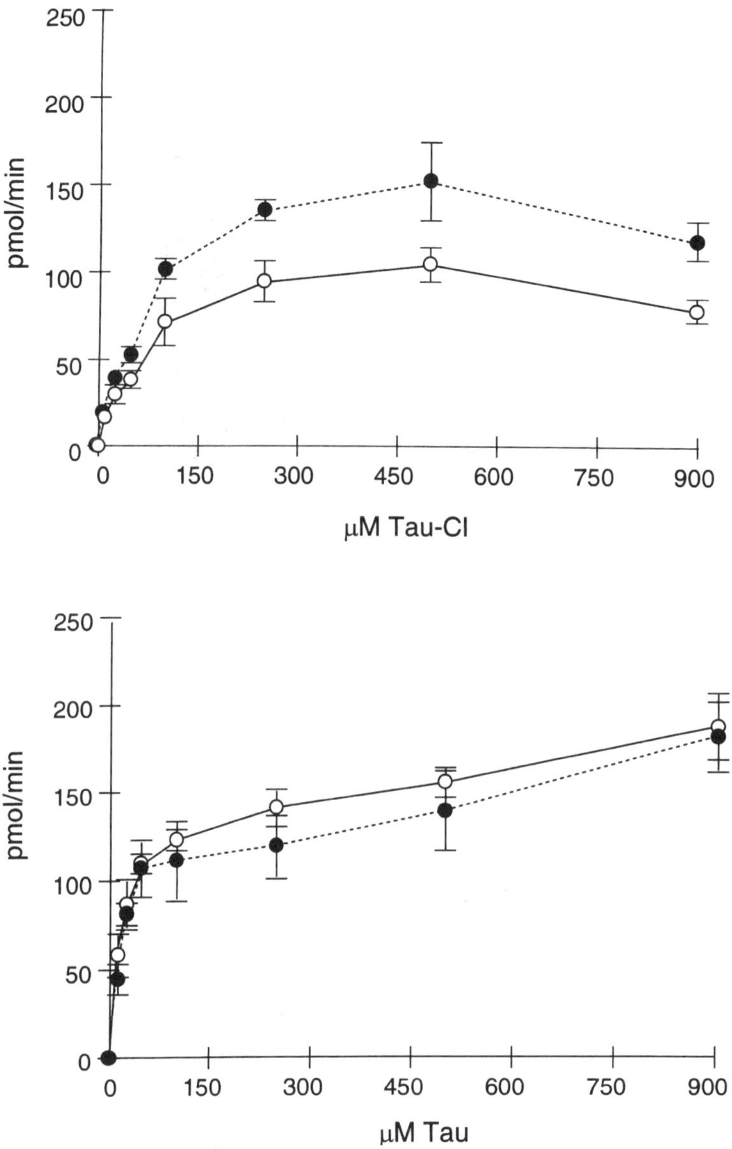

Figure 2 Saturation isotherms of (^{3}H)Tau-Cl and (^{3}H)taurine transport into RAW 264.7 cells. Top panel: (^{3}H)Tau-Cl uptake measured in the absence (-○-) and presence (-●-) of 500 μM unlabeled taurine. Bottom panel: (^{3}H)Taurine uptake measured in the absence (-○-) and presence (-●-) of 500 μM unlabeled Tau-Cl. Transport was measured after incubating at 37°C for 20 min. Values are corrected for nonspecific uptake measured at 0 to 4°C and represent the mean ± SD of four independent experiments. (From Ref. 55.)

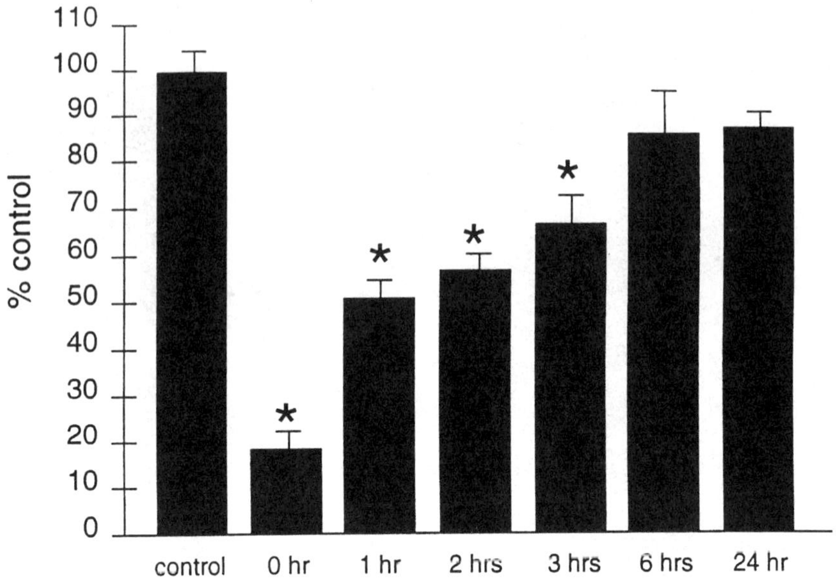

Figure 3 Inhibition of NO_2^- production by Tau-Cl added at various times after activation. RAW 264.7 cells were activated and 1 mM Tau-Cl was added coincidently (Ohr) or at various times thereafter. Control cells were not exposed to Tau-Cl. Nitrite was measured 24 hr after activation. Values are expressed as a percentage of the control (14.3 ± 0.6 nmol/well) and represent the mean ± SD of triplicate samples. Asterisk denotes a significant ($P < 0.05$) difference from control. (From Ref. 55.)

condition of Tau-Cl addition 6 hr after activation was initiated. These results suggested that Tau-Cl was exerting its effect at some early event in the induction of nitric oxide synthase type II (iNOS) expression, i.e., at the level of signal transduction, transcription of the iNOS gene, or translation of iNOS mRNA. Another explanation was that Tau-Cl inhibited activation in general because production of PGE_2 and TNF-α was inhibited with an effectiveness similar to that obtained for NO (Fig. 5). Taurine was without effect.

The effects of Tau-Cl on inflammatory cells is not restricted to macrophages; e.g., murine pmn production of superoxide anion is also inhibited by Tau-Cl (Fig. 6). However, unlike the results obtained with macrophages where preactivation exposure of cells to Tau-Cl inhibited subsequent production of NO and TNF-α (58), inhibition of O_2^- required Tau-Cl to be present during the entire period of stimulation; preactivation exposure to Tau-Cl was without effect. In this paradigm the effects of Tau-Cl were reversible because inhibition of O_2^- was not observed after Tau-Cl was removed from the media. In addition, the presence of taurine (1 mM) during pmn stimulation with phorbol myristate ac-

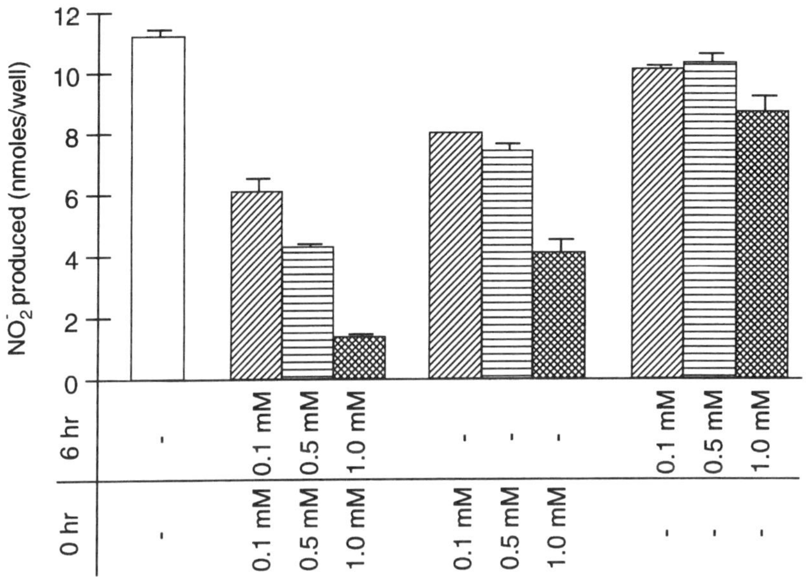

Figure 4 Tau-Cl exposure intervals exert different degrees of inhibition of nitrite production by activated RAW 264.7 cells. All cells were activated in the absence or presence of the indicated concentrations of Tau-Cl (0hr). After 6 hr the cells were washed and fresh medium containing the activators with or without the indicated concentrations of Tau-Cl was added to each well. Nitrite was measured in the conditioned media 24 hr later. Values represent the mean ± SD of triplicate samples. (From Ref. 55.)

etate resulted in significant inhibition (34%) of O_2^- production (58). This effect was attributed to Tau-Cl formation in vitro because halide-dependent MPO activity is associated with pmn (42,54,62). We also confirmed that under our assay conditions pmn, but not macrophages or RAW 264.7 cells, produce Tau-Cl when activated in the presence of taurine, and the amount of Tau-Cl formed is proportional to the extracellular concentration of taurine (Fig. 7). It seems likely that Tau-Cl inhibits generation of O_2^- by either inhibiting the assembly of the subunit components of NADPH oxidase haloenzyme, which happens only after pmn activation (64), or by interacting with the assembled multiprotein complex in the plasma membrane of activated pmn.

VI. MECHANISM(S) OF TAURINE CHLORAMINE ACTION

To address the mechanisms mediating the effects of Tau-Cl on the production of NO and PGE_2 we examined the expression of iNOS and cyclooxygenase-2

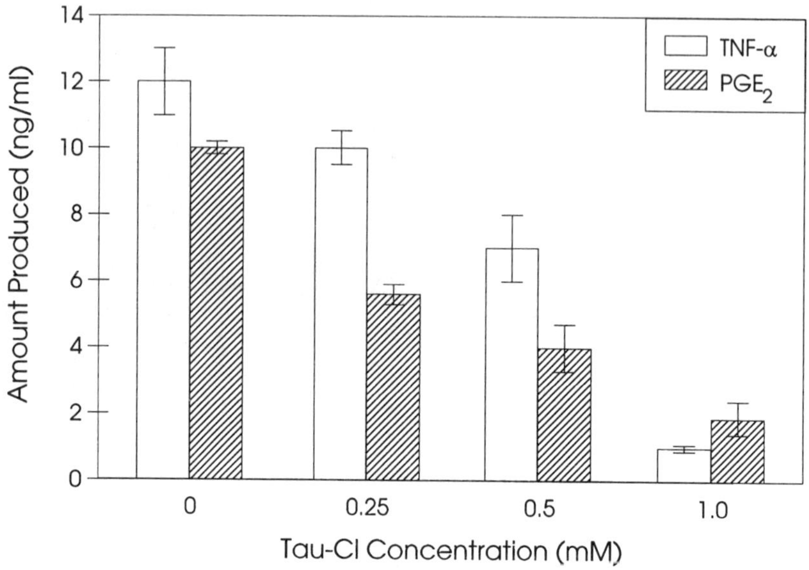

Figure 5 Tau-Cl inhibits accumulation of TNF-α and PGE_2 in the conditioned media of activated RAW 264.7 cells. Samples were analyzed for TNF-α by ELISA and for PGE_2 by EIA 24 hr after activation. Values are expressed as mean ± SD of triplicate samples with similar results obtained in three to five additional independent experiments.

(COX-2) by Western blot analyses (Fig. 8). COX-2 is an inducible enzyme, similar to iNOS, and is responsible for the enhanced production of prostaglandins during inflammation (66–69). Unlike COX-1, which is constitutively expressed in a variety of cell types, COX-2 expression is induced in such cells as macrophages, monocytes, and fibroblasts in response to endotoxin, various mitogens, and/or cytokines. Cell lysates were prepared 24 hr after cells were activated in the absence or presence of 0.8 mM taurine or Tau-Cl. Unactivated cells did not express iNOS or COX-2 (Fig. 8). The expression of both enzymes was diminished in cells that were activated in the presence of Tau-Cl, while taurine was without effect.

Pulse-chase experiments were conducted to determine the effects of Tau-Cl on the processing of TNF-α from the membrane associated, 27 kDa presecretory form in the cell lysate to the lower molecular weight forms that are secreted and accumulated in the conditioned media (56). The amount of radioactivity incorporated into protein in samples of cell lysates and the conditioned media was determined by precipitation with 10% TCA. Aliquots containing equal amounts of radiolabeled protein were immunoprecipitated with polyclonal rabbit antimurine TNF-α antibody and were analyzed by SDS-PAGE and autoradiography (Fig. 9). Synthesis of the membrane-bound, 27 kDa presecretory

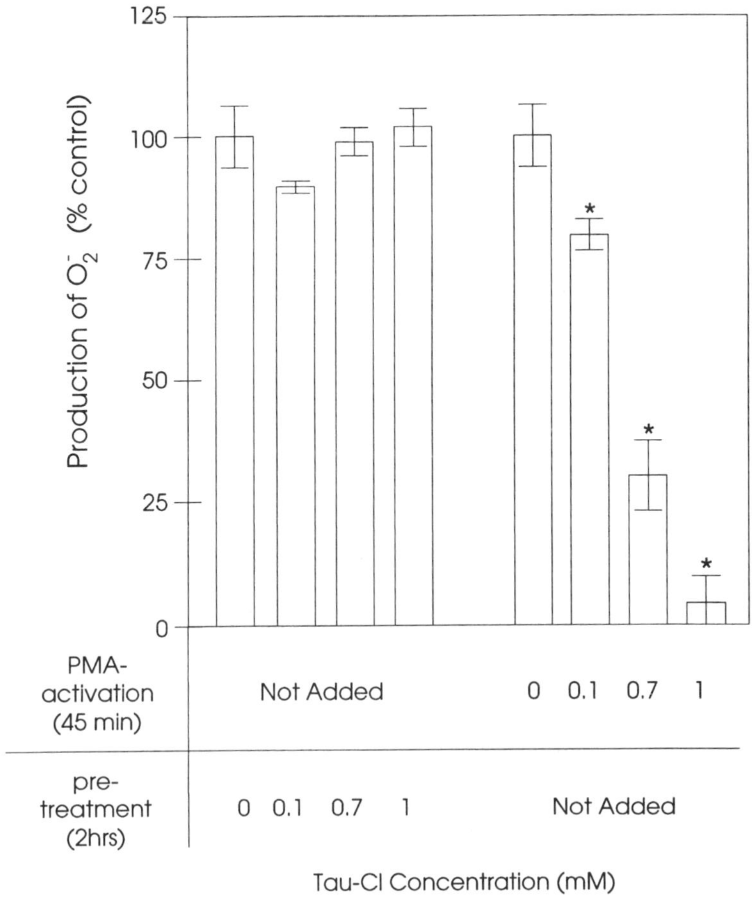

Figure 6 Exposure of PMN to Tau-Cl inhibits production of superoxide anion. Neutrophils were exposed to the indicated concentration of Tau-Cl for either 2 hr before or for 45 min during activation with phorbol myristate acetate (PMA). O_2^- production was measured by SOD inhibitable reduction of cytochrome c. Values represent the mean ± SD of triplicate samples and an asterisk denotes a significant ($P <$ .05) difference from control (2 ± 0.1 nmol/well). (From Ref. 58.)

form of TNF-α was inhibited by Tau-Cl (Fig. 9B; Ohr). The relative amount of the 17 kDa TNF-α recovered in the media immediately after pulse-labeling (Fig. 9B) was slightly higher than that of unactivated cells and remained depressed during the 16 hr postchase period. Protein synthesis in general was not influenced by the presence of Tau-Cl, because the amount of radioactivity recovered in the cell lysates before immunoprecipitation of TNF-α was not significantly affected by Tau-Cl (Table 2). These results demonstrate that synthesis of TNF-α in acti-

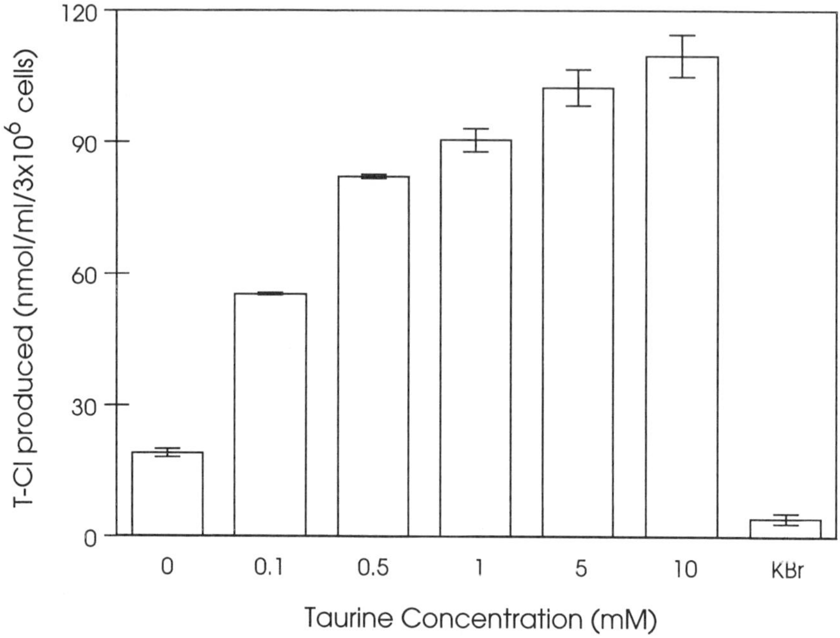

Figure 7 Human pmn produce Tau-Cl in proportion to the amount of taurine available. PMN were isolated from human peripheral blood and incubated in HBSS in the presence of the indicated concentrations of taurine and PMA. Tau-Cl was measured as described (53). KBr (10 μM) inhibits formation of Tau-Cl and was included as a control. Values represent the mean ± SD of triplicate samples with similar results obtained in four additional independent experiments.

vated cells is inhibited by Tau-Cl, similar to iNOS and COX-2 expression, and accounts for the decreased production of the 17 kDa and other multimeric forms of TNF-α in the conditioned media of these cells.

The influence of Tau-Cl on the kinetics of iNOS, COX-2, and TNF-α mRNA expression was examined by Northern blot analyses (Fig. 10). Cell cultures (1–5 × 10^7 cells/100 mm Petri dish) were analyzed using cDNA probes for iNOS, COX-2, and TNF-α after 4, 8, 16, and 24 hr of activation in the absence or presence of 0.8 mM Tau-Cl or taurine (56,57). α-Actin mRNA was used as an internal standard for normalizing the amount of mRNA loaded onto each gel lane. Unactivated cells did not express mRNA for iNOS, COX-2, or TNF-α at any of the times they were examined. Cells activated in the absence or presence of taurine expressed maximal or near-maximal levels of TNF-α and COX-2 transcript by 4 hr, which decreased thereafter. Transcripts for iNOS reached maximum levels of expression by 16 hr and remained elevated for up to 24 hr in the continuous presence of LPS and IFN-γ. Tau-Cl dramatically inhibited the expres-

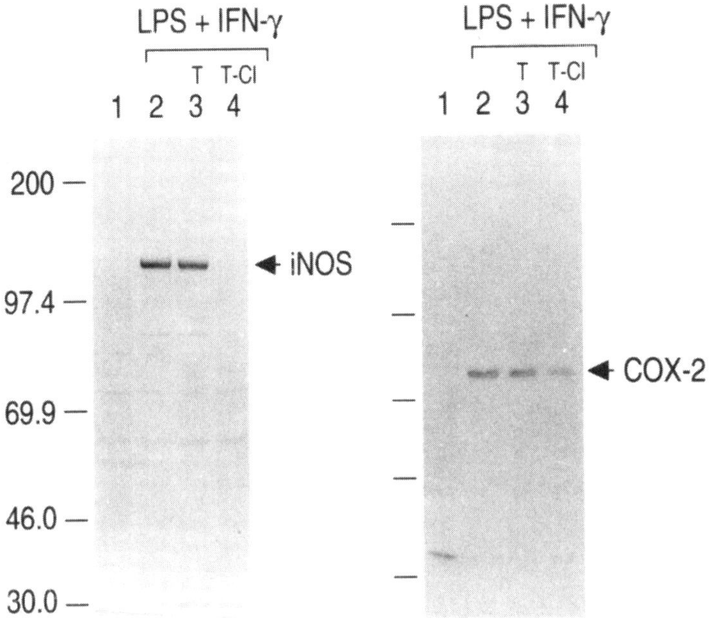

Figure 8 Western blot analyses of iNOS and COX-2 expression in lysates of RAW 264.7 cells. Lysates (25 μg protein/lane) from unactivated (lane 1), activated (LPS and rIFN-γ), (lane 2), and from cells activated in the presence of 0.8 mM taurine (lane 3) or Tau-Cl (lane 4) are shown. Cells were activated for 24 hr before lysates were prepared. (Adapted in part from Ref. 57.)

sion of iNOS mRNA at all times examined. While the kinetics of expression of COX-2 and TNF-α mRNAs were shifted to later time points by the presence of Tau-Cl, overall, the quantitative expression of these transcripts was not diminished. Unlike the activated cells that exhibited a gradual decline of TNF-α mRNA expression to undetectable levels (24 hr), as well as a decline in COX-2 mRNA, cells activated in the presence of Tau-Cl continued to express high levels of TNF-α and COX-2 transcripts after 24 hr of activation, suggesting either a decreased rate of message turnover or an increase in gene transcription.

Production of NO, PGE_2, and TNF-α is regulated by an integrated response that summarizes the molecular events of induction and suppression of synthesis and degradation of the respective mRNA transcripts and the posttranscriptional regulation of the resultant proteins (iNOS, COX-2, TNF-α) that mediate the inflammatory response of a cell. Tau-Cl inhibits production of NO, PGE_2, and TNF-α by activated cells through different mechanisms (Fig. 11). Tau-Cl inhibits transcription of the iNOS gene or some earlier event in the signal transduction pathway because iNOS mRNA is not detected in cells that are activated in the

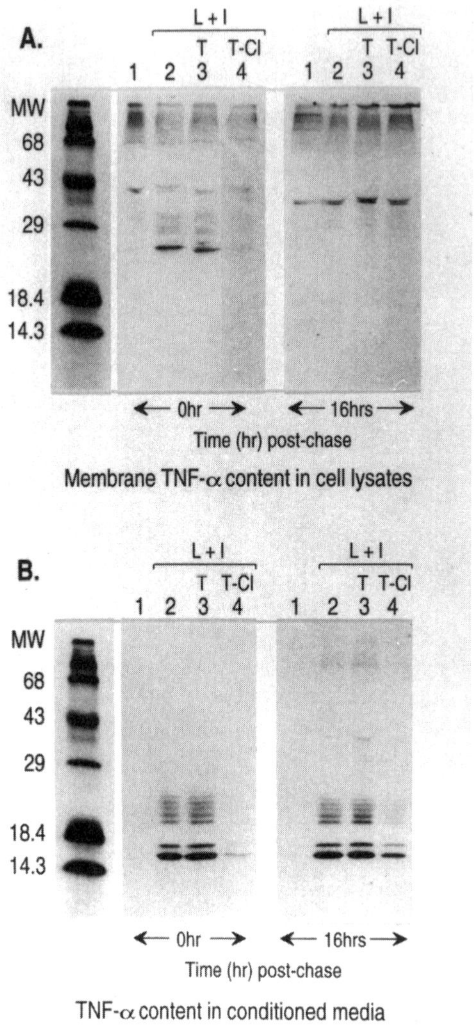

Figure 9 Tau-Cl inhibits metabolic labeling of TNF-α multimeric forms. Fluorograms of radiolabeled immunoprecipitated proteins were prepared from cell lysates (A) and conditioned media (B) obtained from RAW 264.7 cells. Cells were activated for 2.5 hr with LPS and IFN-γ (L + I) either alone (lane 2) or in the presence of 0.8 mM taurine (lane 3) or Tau-Cl (lane 4). Lane 1 is from unactivated cells. After the activation period cells were washed, incubated (30 min) in fresh media devoid of methionine and cysteine, and then pulse-labeled with (^{35}S)Tran for 1 hr. Label was removed by washing the cells. Cells were then either lysed for the 0 hr postchase analysis or allowed to continue in fresh media containing the indicated additives (activators, taurine, Tau-Cl) for an additional 16 hr. TNF-α was immunoprecipitated before separation on SDS-PAGE, and radiolabeled TNF-α was visualized by autoradiography. (From Ref. 56.)

Table 2 Radioactivity Recovered in TCA-Precipitates of Aliquots from Cell Lysates Analyzed in Figure 9

	Time postchase			
	0 hr		16 hr	
Treatment	cpm × 10^6	% Total cpm	cpm × 10^6	% Total cpm
Unactivated	2.2	90	2.0	96
LPS + IFN-γ	2.0	92	1.6	94
LPS + IFN-γ + taurine	1.9	91	1.5	94
LPS + IFN-γ + Tau-Cl	1.8	87	1.7	94

Cell lysate aliquots were taken after pulse-chase labeling with (^{35}S)labeled Tran (methionine + cysteine) before immunoprecipitation of TNF-α multimers. Values are expressed as total cpm recovered in the TCA precipitate and as the percent total cpm taken up into the cell that was incorporated into protein (TCA-ppt). Similar results were obtained in three additional independent experiments.

presence of Tau-Cl. The inhibition of iNOS mRNA expression by Tau-Cl is concentration-dependent (56), as is the inhibition of NO accumulation in the conditioned media and the expression of intracellular iNOS protein. However, the effects of Tau-Cl are not the result of a general effect on gene transcription because increases in COX-2 and TNF-α mRNA occur in the presence of Tau-Cl. In addition, the mode of RAW 264.7 cell activation is not an important consideration for Tau-Cl inhibition of iNOS mRNA expression because similar results are obtained using other activators in combination with IFN-γ (59). Inhibition of iNOS mRNA expression by Tau-Cl has been confirmed by others (70), but cells were analyzed after only 16 hr of activation.

Although the kinetics of expression of COX-2 and TNF-α mRNAs are altered in cells that are activated in the presence of Tau-Cl, once expressed, relatively high levels persist for longer periods of time than in cells not exposed to Tau-Cl. Whether this results from an increase in the stability of the transcripts or an increase in COX-2 and TNF-α gene transcription is uncertain. In any event, Tau-Cl appears to suppress the translation of COX-2 and TNF-α mRNAs because even though abundant amounts of COX-2 and TNF-α transcripts are induced and sustained in the presence of Tau-Cl, synthesis of COX-2 and TNF-α proteins is greatly inhibited. The suppressed translation of COX-2 and TNF-α mRNAs is not the result of inhibition of protein synthesis in general because the incorporation of (^{35}S)labeled amino acids into TCA-precipitated protein was not appreciably influenced by the presence of Tau-Cl (Table 2).

VII. GENERAL CONSIDERATIONS

Elevated extracellular taurine levels may protect host tissue from damage during inflammatory reactions by providing adequate precursor for taurine chloram-

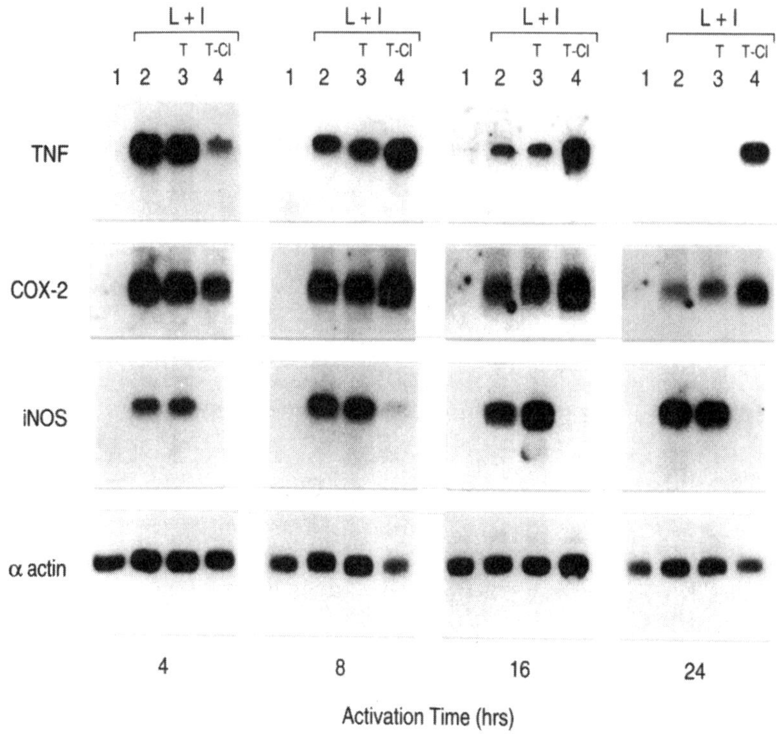

Figure 10 Kinetics of TNF-α, COX-2, iNOS, and α-actin mRNA expression in RAW 264.7 cells. Total RNA fractions were prepared from cells 4, 8, 16, and 24 hr after activation and size-fractionated on agarose gels before transfer to nylon membranes. Blots were probed with (^{32}P)dCTP random prime-labeled cDNA and visualized after autoradiography using Kodak XAR-5 film at –70°C. In each blot results are shown for cells that were either unactivated (lane 1), activated (lane 2), or activated in the presence of 0.8 mM taurine (lane 3) or Tau-Cl (lane 4). (From Ref. 56.)

ine formation at the site of injury if pmn are involved in the response. Tau-Cl formation not only circumvents the indiscriminate oxidation of host tissue that would otherwise arise from HOCl/OCl$^-$, as postulated by several groups in a variety of models (Table 1), but Tau-Cl also has the propensity to act as a modulator of the inflammatory response. The idea that Tau-Cl may act as a physiological modulator of inflammation was proposed in earlier studies, where it was demonstrated that Tau-Cl oxidizes chemotactic factors, i.e., the methionyl residue in f-met-leu-phe, was oxidized to the sulfoxide derivative and the chemotactic activity was subsequently inhibited (53). In addition, Tau-Cl was shown to have pro-inflammatory activity in the sense that it inactivated α-protease inhibitor, a

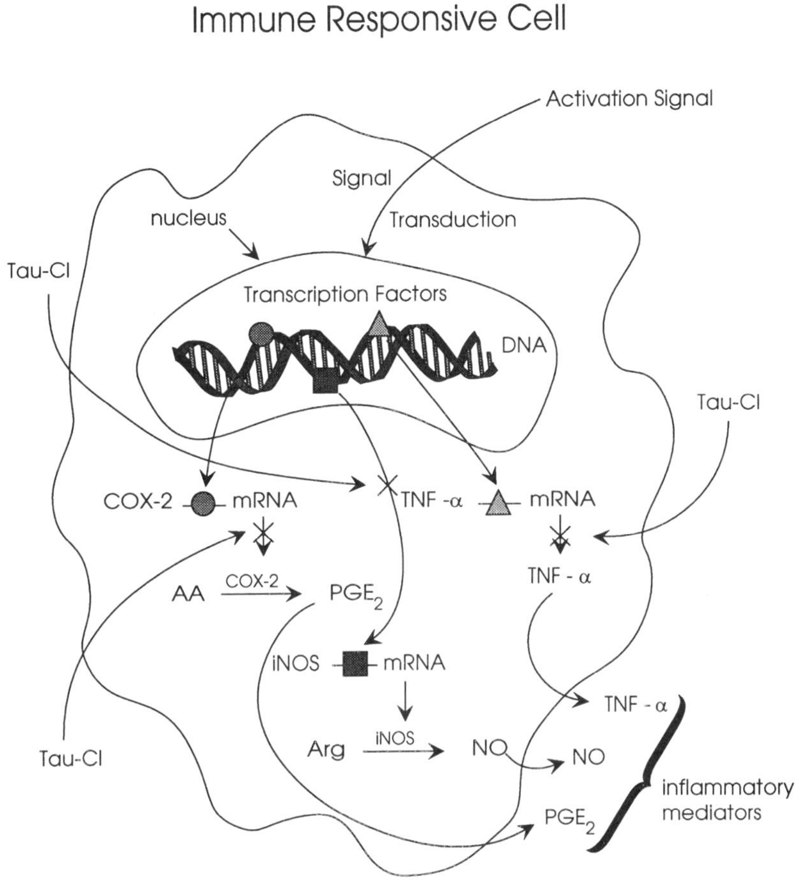

Figure 11 Schematic representation of mechanisms utilized by Tau-Cl to inhibit production of inflammatory mediators by immune responsive cells. The following abbreviations are used: X denotes the site of Tau-Cl action, COX-2 is cyclooxygenase-2, AA is arachidonic acid, and Arg is L-arginine.

serum protein that regulates leukocyte elastase activity (53). Others hypothesized that taurine stimulated MPO activity by removing HOCl feedback inhibition during the course of Tau-Cl formation and that this somchow led to increased production of HOCl and inactivation of neutrophil-derived leukotriene B_4, a chemotactic factor for recruiting leukocytes (25).

Studies from our laboratories demonstrated that Tau-Cl inhibited production of NO and TNF-α from activated macrophages (55) and that this inhibitory effect was mediated by transcriptional and posttranscriptional mechanisms (56).

Tau-Cl inhibition of COX-2 expression in activated cells was demonstrated to result from posttranscriptional effects as well (57). Tau-Cl also inhibited accumulation of IL-6, but not IL-1α, in the conditioned media of activated cells (70). The inhibitory effects of Tau-Cl on production of NO, TNF-α, and PGE_2 has been confirmed by others (70) and extended to other cells activated by various means (24,58,59,61,70). Tau-Cl also may attenuate inflammatory responses by inhibiting production of superoxide anion (58), and this may decrease the generation of $HOCl/OCl^-$ because superoxide apparently optimizes pmn production of $HOCl/OCl^-$ (71). The ability of taurine to attenuate the indiscriminate damage caused by $HOCl/OCl^-$ and for Tau-Cl, thus produced, to downregulate the production of inflammatory mediators, provides a unified hypothesis for explaining the tissue-protective effects of taurine listed in Table 1. The main criteria for administration of taurine to serve as a prophylactic agent in preventing tissue damage are that pmn, or other leukocytes that contain halide-dependent MPO, are involved at the site of injury, and that tissue damage results from production of inflammatory mediators. These criteria enable the formation of Tau-Cl to be localized to the site of injury, where it may attenuate the production of inflammatory mediators and protect against tissue damage.

Halide-dependent MPO is associated with neutrophils and catalyzes the formation of $HOCl/OCl^-$ from H_2O_2 and Cl^-. Neutrophils contain high intracellular taurine concentrations, and under in vitro conditions that approximate the intracellular environment of pmn, MPO may catalyze the formation of Tau-Cl directly, without formation of $HOCl/OCl^-$ (54). During inflammation, extracellular taurine concentrations may be of particular importance because activated neutrophils secrete up to 25% of their total MPO activity (72) and release it upon cell death. It is in the extracellular milieu that $HOCl/OCl^-$ formation by the MPO system may reach levels high enough to initiate indiscriminate oxidation of host tissue which leads to recruitment of macrophages to the site of injury, thereby propagating the inflammatory response leading to further tissue damage. Elevated concentrations of extracellular taurine would not only circumvent the tissue-damaging effects of $HOCl/OCl^-$ by forming Tau-Cl, a more stable and selective oxidant, but would also provide protection of host tissue by downregulating the production of inflammatory mediators.

ACKNOWLEDGMENTS

The authors appreciate the significant contributions of Dr. E. Park to the conduct of these studies and to Drs. C. Wright and C. Kim for their input into parts of the work reported herein. The technical assistance of Ms. J.-H. Jia and M. DeMasi and the secretarial help of Ms. J. Kay are gratefully appreciated. This work was supported by the New York State Office of Mental Retardation and

Developmental Disabilities, and by USPHS Research Grant No. HL-49942 from the NIH.

REFERENCES

1. Huxtable RJ. The physiological actions of taurine. Physiol Rev 1992; 72:101–163.
2. Hayes KC, Carey RE, Schmidt SY. Retinal degeneration associated with taurine deficiency in the cat. Science 1975; 188:949–951.
3. Gaull GE. Taurine in pediatric nutrition. Review and update. Pediatrics 1989; 83:433–442.
4. Sturman JA, Chesney RW. Taurine in pediatric nutrition. Pediatr Nutr 1995; 42:879–897.
5. Sturman JA. Taurine in development. Physiol Rev 1993; 73:119–148.
6. Schuller-Levis G, Mehta PD, Rudelli R, Sturman JA. Immunologic consequence of taurine deficiency in cats. J Leukoc Biol 1990; 47:321–331.
7. Quinn MR, Harris CL. Taurine allosterically inhibits binding of [^{35}S]-t-butylbicyclophosphorothionate (TBPS) to rat brain synaptic membranes. Neuropharmacology 1995; 35:1607–1613.
8. Schmieden V, Kuhse J, Betz H. Agonist pharmacology of neonatal and adult glycine receptor α subunits: identification of amino acid residues involved in taurine activation. EMBO J 1992; 11:2025–2032.
9. Gordon RE, Shaked AA, Solano DF. Taurine protects hamster bronchioles from acute NO_2-induced alterations. Am J Pathol 1986; 125:585–600.
10. Schuller-Levis G, Quinn MR, Wright C, Park E. Taurine protects against oxidant-induced lung injury: possible mechanism(s) of action. Adv Exp Med Biol 1994; 359:31–39.
11. Wang Q, Giri SN, Hyde DM, Nakashima JM. Effects of taurine on bleomycin-induced lung fibrosis in hamsters. Proc Soc Exp Biol Med 1989; 190:330–338.
12. Wang QJ, Hollinger MA, Giri SN. Attenuation of amiodarone induced lung fibrosis and phospholipidosis in hamsters by taurine and/or niacin treatment. J Pharmacol Exp Ther 1992; 262:127–132.
13. Izmi K, Nagata R, Motoya T, Yamashita J, Hirokane T, Naguta T, Satoh Y, Sawada Y, Ishibashi M, Yoshida H, Fukuda T. Preventive effect of taurine against acute paraquot intoxication in beagles. Jpn J Pharmacol 1989; 50:229–233.
14. Watson RWG, Redmond HP, McCarthy J, Bouchier-Hayes D. Taurolidine, an antilipopolysaccharide agent, has immunoregulatory properties that are mediated by the amino acid taurine. J Leukoc Biol 1995; 58:299–306.
15. Milei J, Ferreira R, Llesuy S, Forcada P, Covarrubias J, Boveris A. Reduction of reperfusion injury with preoperative rapid intravenous infusion of taurine during myocardial revascularization. Am Heart J 1992; 123:339–345.
16. Azuma J, Sawamura A, Awata N, Hasegawa H, Ogura K, Harada H, Ohta H, Yamauchi K, Kishimoto S. Double-blind randomized crossover trial of taurine in congestive heart failure. Curr Ther Res 1983; 34:543–557.

17. Petty MA, Kintz J, DiFrancesio GF. The effects of taurine on atherosclerosis development in cholesterol-fed rabbits. Eur J Pharmacol 1990; 180:119–127.
18. Bhattacharya SK, Sarkar MK. Effect of some centrally administered putative amino acid neurotransmitters on carrageenan-induced paw oedema in rats. J Pharm Pharmacol 1986; 38:144–146.
19. Malcangio M, Bartolin A, Ghelardini C, Bennardini F, Malmberg-Aiello P, Franconi F, Giotti A. Effect of IVC taurine on the impairment of learning, convulsions and death caused by hypoxia. Psychopharmacology 1989; 98:316–320.
20. Raschke P, Massoudy P, Becker BF. Taurine protects the heart from neutrophil-induced reperfusion injury. Free Radic Biol Med 1995; 19:461–471.
21. Banks MA, Porter DW, Martin WG, Castranova V. Ozone induced lipid peroxidation and membrane leakage in isolated rat alveolar macrophages: protective effects of taurine. J Nutr Biochem 1991; 2:308–313.
22. Davies JMS, Horwitz DA, Davies KJA. Potential roles of hypochlorous acid and N-chloramines in collagen breakdown by phagocytic cells in synovitis. Free Radic Biol Med 1993; 15:637–643.
23. Cantin AM. Taurine modulation of hypochlorous acid-induced lung epithelial cell injury in vitro, role of anion transport. J Clin Invest 1994; 93:606–614.
24. Redmond HP, Wang JH, Bouchier-Hayes D. Taurine attenuates nitric oxide- and reactive oxygen intermediate-dependent hepatocytic injury. Arch Surg 1996; 131:1280–1288.
25. McLoughlin DM, Stapleton PP, Bloomfield FL. Influence of taurine and a substituted taurine on the respiratory burst pathway in the inflammatory response. Biochem Soc Trans 1991; 19:73–78.
26. Stapleton PP, Bloomfield FJ. Effect of zwitterions on the respiratory burst. J Biomed Sci 1993; 3:79–84.
27. Wang JH, Redmond HP, Watson RWG, Condron C, Bouchier-Hayes D. The beneficial effect of taurine on the prevention of human endothelial cell death. Shock 1996; 6:331–338.
28. Wang JH, Redmond HP, Watson RWG, Condron C, Bouchier-Hayes D. Taurine protects against sodium arsenite-induced human endothelial cell apoptosis. Surg Forum 1995; 81:661–664.
29. Mustafa MG, Tierney DF. Biochemical and metabolic changes in the lung with oxygen, ozone, and nitrogen dioxide toxicity. Am Rev Respir Dis 1978; 118:1061–1090.
30. Rombout PJ, Dormans JA, Marra M, Van Esch GJ. Influence of exposure regimen on nitrogen dioxide-induced morphological changes in the rat lung. Environ Res 1986; 41:466–480.
31. Plopper CG, Chow CK, Dungworth DL, Brummer M, Nemeth TJ. Effect of low level of ozone on rat lungs. II. Morphological responses during recovery and re-exposure. Exp Mol Pathol 1978; 29:400–411.
32. Crapo JD, Barry BE, Chang LY, Mercer RR. Alterations in lung structure caused by inhalation of oxidants. J Toxicol Environ Health 1984; 13:301–321.
33. Huat A, Hacker M. Role of reactive nitrogen intermediate production in alveolar macrophage mediated cytostatic activity induced by bleomycin lung damage in rats. Cancer Res 1990; 50:7863–7866.

34. Berisha HI, Pakbaz H, Absood A, Said SI. Nitric oxide as a mediator of oxidant lung injury due to paraquat. Proc Natl Acad Sci USA 1994; 91:7445–7449.
35. Stapleton PP, Molloy AM, Rogers S, Bloomfield FJ. Neutrophil taurine in psoriasis. Irish J Med Sci 1996; 165:173–176.
36. Clancy RM, Abramson SB. Nitric oxide: a novel mediator of inflammation. Proc Soc Exp Biol Med 1995; 210:93–101.
37. McCartney-Francis N, Allen JB, Mizel DE, Albina JE, Xie Q-W, Nathan CF, Wahl SM. Suppression of arthritis by an inhibitor of nitric oxide synthase. J Exp Med 1993; 178:749–754.
38. McInnes IB, Leung BP, Field M, Wei XQ, Huang F-P, Sturrock RD, Kinninmonth A, Weidner J, Mumford R, Liew FY. Production of nitric oxide in the synovial membrane of rheumatoid and osteoarthritis patients. J Exp Med 1996; 184:1519–1524.
39. Bruch-Gerharz D, Fehsel K, Suschek C, Michel G, Ruzicka T, Kolb-Bachofen V. A proinflammatory activity of interleukin 8 in human skin: expression of the inducible nitric oxide synthase in psoriatic lesions and cultured keratinocytes. J Exp Med 1996; 184:2007–2012.
40. Aruoma OI, Halliwell B, Hoey BM, Butler J. The antioxidant action of taurine, hypotaurine and their metabolic precursors. Biochem J 1988; 256:251–255.
41. Zgliczynski JM, Stelmaszynska T, Domanski J, Ostrowski W. Chloramines as intermediates of oxidation reaction of amino acids by myeloperoxidase. Biochim Biophys Acta 1971; 235:419–424.
42. Weiss SJR, Klein A, Slivka A, Wei M. Chlorination of taurine by human neutrophils: evidence for hypochlorous acid generation. J Clin Invest 1982; 70:598–607.
43. Thomas EL, Grisham MB, Jefferson MM. Myeloperoxidase-dependent effect of amines on functions of isolated neutrophils. J Clin Invest 1983; 72:441–454.
44. Grisham MB, Jefferson MM, Melton DF, Thomas EL. Chlorination of endogenous amines by isolated neutrophils. J Biol Chem 1984; 259:10404–10413.
45. Panasenko OM, Evgina SS, Aidyraliev RK, Sergienko VI, Vladimirov YA. Peroxidation of human blood lipoproteins induced by exogenous hypochlorite or hypochlorite generated in the system of "myeloperoxidase + H_2O_2 + Cl^-". Free Radic Biol Med 1994; 16:143–148.
46. Hazell LJ, Arnold L, Flowers D, Waeg G, Malle E, Stocker R. Presence of hypochlorite-modified proteins in human atherosclerotic lesions. J Clin Invest 1996; 97:1535–1544.
47. Hazen SL, Hsu FF, Duffin K, Heincke JW. Molecular chlorine generated by the myeloperoxidase-hydrogen peroxide-chloride system of phagocytes converts low density lipoprotein cholesterol into a family of chlorinated sterols. J Biol Chem 1996; 271:23080–23088.
48. Ross R. The pathogenesis of atherosclerosis: a perspective for the 1990s. Nature 1993; 362:801–809.
49. Learn DB, Fried VA, Thomas EL. Taurine and hypotaurine content of human leukocytes. J Leukoc Biol 1990; 48:174–182.
50. Fukuda K, Hirai Y, Yoshida H, Hakajima T, Usii T. Free-amino acid content of lymphocytes and granulocytes compared. Clin Chem 1982; 28:1758–1761.
51. Green T, Fellman JH, Eicher AL, Pratt KJ. Anti-oxidant role and subcellular loca-

tion of hypotaurine and taurine in human neutrophils. Biochim Biophys Acta 1991; 1073:91–97.

52. Weiss SJ. Tissue destruction by neutrophils. N Engl J Med 1989; 320:365–376.
53. Weiss SJ, Lampert MB, Test ST. Long-lived oxidants generated by human neutrophils: characterization and bioactivity. Science 1983; 222:625–628.
54. Marquez LA, Dunford HB. Chlorination of taurine by myeloperoxidase: kinetic evidence for an enzyme-bound intermediate. J Biol Chem 1994; 269:7950–7956.
55. Park E, Quinn MR, Wright CE, Schuller-Levis G. Taurine chloramine inhibits the synthesis of nitric oxide and the release of tumor necrosis factor in activated RAW 264.7 cells. J Leukoc Biol 1993; 54:119–124.
56. Park E, Schuller-Levis G, Quinn MR. Taurine chloramine inhibits production of nitric oxide and TNF-α in activated RAW 264.7 cells by mechanisms that involve transcriptional and translational events. J Immunol 1995; 154:4778–4784.
57. Quinn MR, Park E, Schuller-Levis G. Taurine chloramine inhibits prostaglandin E_2 production in activated RAW 264.7 cells by post-transcriptional effects on inducible cyclooxygenase expression. Immunol Lett 1996; 50:185–188.
58. Kim C, Park E, Quinn MR, Schuller-Levis G. The production of superoxide anion and nitric oxide by cultured murine leukocytes and the accumulation of TNF-α in the conditioned media is inhibited by taurine chloramine. Immunopharmacology 1996; 34:89–95.
59. Park E, Schuller-Levis G, Jia J-H, Quinn MR. Preactivation exposure of RAW 264.7 cells to taurine chloramine attenuates subsequent production of nitric oxide and expression of iNOS mRNA. J Leukoc Biol 1997; 61:161–166.
60. Naskalski JW. Myeloperoxidase inactivation in the course of catalysis of chlorination of taurine. Biochim Biophys Acta 1977; 485:291–300.
61. Quinn MR, Schuller-Levis G, Park E, DeMasi M. Taurine chloramine inhibits prostaglandin E_2 production in activated macrophages and glial cells by inhibiting expression of inducible cyclooxygenase. Presented at Lipid Mediators: Recent Advances in Molecular Biology, Understanding of Regulation and Pharmacology. Keystone, CO, Jan 26–31, 1997.
62. Witko V, Nguyen AT, Descamps-Latscha B. Microtiter plate assay for phagocyte-derived taurine-chloramines. J Clin Lab Anal 1992; 6:47–53.
63. Thomas EL, Grisham MB, Jefferson MM. Preparation and characterization of chloramines. Methods Enzymol 1986; 132:569–571.
64. Chanock SJ, El Benna J, Smith RM. The respiratory burst oxidase. J Biol Chem 1994; 269:24519–24522.
65. Kujubu DA, Fletcher BS, Varnum BC, Lim RW, Herschman HR. TIS10, a phorbol ester tumor promoter-inducible mRNA from Swiss 3T3 cells, encodes a novel prostaglandin synthase/cyclooxygenase homologue. J Biol Chem 1991; 266:12866–12872.
66. Vane JR, Mitchell JA, Appleton I, Tomlinson A, Bishop-Bailey D, Croxtall J, Willoughby DA. Inducible isoforms of cyclooxygenase and nitric oxide synthase in inflammation. Proc Natl Acad Sci USA 1994; 91:2046–2050.
67. Crofford LJ, Wilder RL, Ristimaki AP, Sano H, Remmers EF, Epps HR, Hla T. Cyclooxygenase-1 and -2 expression in rheumatoid synovial tissues. J Clin Invest 1994; 93:1095–1101.

68. Riese J, Hoff T, Nordhoff A, DeWitt DL, Resch K, Kaever V. Transient expression of prostaglandin endoperoxide synthase-2 during macrophage activation. J Leukoc Biol 1994; 55:476–482.
69. Hempel SL, Monick MM, Hunninghake GM. Lipopolysaccharide induces prostaglandin H synthase-2 and mRNA in human alveolar macrophages and blood monocytes. J Clin Invest 1994; 93:391–396.
70. Marcinkiewicz J, Grabowska A, Bereta J, Stelmaszynska T. Taurine chloramine, a product of activated neutrophils, inhibits in vitro the generation of nitric oxide and other macrophage inflammatory mediators. J Leukoc Biol 1995; 58:667–674.
71. Kettle AJ, Winterbourn CC. Superoxide enhances hypochlorous acid production by stimulated human neutrophils. Biochim Biophys Acta 1990; 1052:379–385.
72. Smith RJ, Speziale SC, Bowman BJ. Properties of interleukin-1 as a complete secretogen for human neutrophils. Biochim Biophys Acta 1985; 130:1233–1240.

14

Modulation of the Nitric Oxide Pathway by Natural Products

Marion Man-Ying Chan
Temple University School of Medicine, Philadelphia, Pennsylvania

Dunne Fong
Rutgers University, Piscataway, New Jersey

I. INTRODUCTION

From neurotransmission to cardiovascular vasodilation to microbial killing, nitric oxide is involved in a diverse array of physiological and pathophysiological functions. In mammals including humans, NO is produced by three isoforms of nitric oxide synthases (NOS). The constitutive ones are known as NOS1 (nNOS or ncNOS, neuronal constitutive) and NOS3 (eNOS or ecNOS, endothelial constitutive); and the inducible one as NOS2 (iNOS, inducible). A variety of cell types produce NO. The level of NO generation is low for NOS1 and NOS3, but high for NOS2. High NO output is engaged during inflammation and infection (1–3). However, NO is a double-edged sword (4). Whereas NO is a regulator of a multitude of cellular functions, its overproduction in chronic inflammation may be detrimental. Excess NO can react with DNA, leading to mutations and eventually carcinogenesis (5,6). Hence, under specified pathophysiological conditions, selective modulation of the NO pathway is desirable.

Here, we attempt to review natural products that have been reported to modulate the NO pathway. A variety of natural products, ranging from crude extracts to purified compounds, have been shown to be effective. First, we will discuss natural products in studies on two commonly used in vitro systems, namely the inflammation-induced nitrite production by macrophages (for NOS2), and the endothelium-dependent vasorelaxation of aortic rings (for NOS3). Then we will summarize, for the better-characterized cases, natural product effects on

NOS gene expression, NOS enzyme activity, NOS protein level, and the NO product and related reactive nitrogen species (RNS). The mode of action of a few natural products on the NO pathway has already been documented. Many of them are dietary ingredients that may protect against cancer and heart diseases, such as soybean and red wine (7–10).

II. MODULATION OF MACROPHAGE NITRITE GENERATION BY NATURAL PRODUCTS

One commonly employed in vitro assay for NOS2 is the generation of NO by macrophages. Briefly, macrophages from various sources, or macrophagelike cell lines, are incubated in the presence of inducers—bacterial endotoxin lipopolysaccharide (LPS) and cytokine interferon-γ (IFN). The cells then produce NO that can be measured as nitrite, a stable NO product, using the Griess reagent.

By this approach we have found that the dietary phytochemicals epigallocatechin gallate (EGCG), carnosol, and curcumin inhibit nitrite production (11). Tea, the world's most widely consumed beverage, is manufactured from leaves of the plant *Camellia sinensis*. The major tea polyphenol, EGCG, has been implicated in the beneficial effects attributed to tea. Using peritoneal exudate macrophages from BALB/c mice and the inflammation-induced nitrite production assay, EGCG inhibits nitrite generation in a dose-dependent manner. At 1, 2.5, 5, and 10 μM, it inhibits by 6%, 36%, 65%, and 86%, respectively.

Rosemary is a commonly used herb from the plant *Rosmarinus officinalis*, and carnosol is a diterpene isolated from this plant. Similar to EGCG, carnosol inhibits murine macrophage nitrite production (at a somewhat higher dose—58% inhibition at 25 μM) (11).

Curcumin, the active principle from rhizome of the plant *Curcuma longa*, is commonly used in Indian cuisine (as curry) and as an anti-inflammatory ingredient in Indian traditional medicine (as turmeric) (12). We found that curcumin, at 1, 2, 5, 10 μM, inhibits nitrite production from murine peritoneal macrophages that have been stimulated with both LPS and IFN by 37%, 35%, 39%, and 46%, respectively. Joe and Lokesh (13) have also reported that curcumin, at 8 μM, inhibits nitrite production by 100% in a rat peritoneal macrophage system stimulated with LPS but not IFN. This difference in sensitivity is possibly due to the level of stimulation generated by LPS versus LPS and IFN dual stimulation. Controls have shown that these agents, EGCG, carnosol, and curcumin, have no effect on cell viability nor the nitrite detection assay at the doses used.

Capsaicin is the active principle of red pepper (cayenne or chili) from the plant *Capsicum frutescens*. In the same rat peritoneal macrophage system, it inhibits nitrite production by 100% at 10 μM (13).

Genistein and daidzein are the major isoflavones of soybean, from the plant *Glycine max*, that have been implicated in health benefits of soy products. Genistein is a protein tyrosine kinase inhibitor that also inhibits nitrite production in the murine system. Dong et al. (14), using LPS and IFN-stimulated peritoneal macrophages from both C3H/HeN and C3H/HeJ mice, showed that 25 μM genistein inhibits nitrite production by about 50%. In a system using concanavalin A and LPS-stimulated murine macrophages, Krol et al. (15) showed that 50 μM genistein inhibits nitrite production by 48%, and 50 μM daidzein inhibits by 94%.

Omega-3 polyunsaturated fatty acids (PUFA) contained in fish oil have been epidemiologically associated with reduced risk of colon cancer. The effect of some of these fatty acids on NO production has been investigated. In RAW 264 murine macrophagelike cells stimulated with LPS, docosahexaenoic acid (DHA), at 50 μM, reduces NO production to 15% of control. Eicosapentaenoic acid (EPA) and α-linolenic acid have similar inhibition, whereas linoleic acid (omega-6 PUFA) and oleic acid (omega-9 PUFA) have no effect (16).

In addition to natural products from the diet, compounds and crude extracts from herbal medicine have been shown to inhibit inflammation-induced nitrite production of macrophages. The Chinese herb ginseng, the most famous of all, has been widely used as a tonic remedy in China, Japan, and Korea for thousands of years. Ginsenosides are a mixture of saponins from the Korean red ginseng *Panax ginseng*. In a system of thioglycollate-elicited murine peritoneal macrophages stimulated with LPS and IFN, ginsenoside Rh_1 and Rh_2 decrease nitrite production in a dose-dependent manner. Ginsenoside Rh_2 is more effective than Rh_1: Rh2 produces 66.7% inhibition at 20 μM whereas Rh_1 produces 52.8% inhibition at 250 μM. Others, namely ginsenoside Rb_1, Rc, and Re, have no effect (17).

Sinomenine is an alkaloid from the plant *Sinomenium acutum* that has been traditionally used in the treatment of rheumatoid arthritis. It inhibits nitrite production by about 50% at 0.4 mM and 90% at 1 mM in a murine macrophagelike cell line RAW 264.7 induced with LPS and IFN (18). Costunolide is a sesquiterpene lactone from the plant *Magnolia sieboldii*. In another murine macrophagelike cell line J774.1 induced with LPS, it inhibits nitrite production by about 50% at 1 μg/mL and 96% at 100 μg/mL (19). Karasurin-A, a basic protein isolated from root tubers of the plant *Trichosanthes kirilowii,* is an abortifacient. It is a single-chain, ribosome-inactivating protein with similarity to the A chain of the ricin toxin. In the thioglycollate-elicited murine peritoneal macrophages induced with LPS, karasurin-A, at 100 to 10,000 ng/mL range, reduces nitrite production in a dose-dependent manner (20).

In traditional medicine, herbs may be used as a single plant species or in combination. Moreover, an herb may contain an array of compounds that work in concert. One such phytochemical extract is that from the plant *Ginkgo biloba* (maidenhair tree or Bai Guo). It is the best-selling herbal medicine in France and Germany (21). *Ginkgo biloba* extract EGb 761, used in Europe as a polyvalent therapeutic agent, has beneficial effects on the pathology of cardiovascular and cerebrovascular diseases (22). The extract EGb 761 contains several groups of major substances: flavonoid (quercetin, kampferol, and isorhamnetin), glycosides, and terpenoids (gingkolide and bilobalide). In the LPS and IFN-stimulated RAW 264.7 cells, it inhibits nitrite production by 50% at 100 μg/mL and by 80% at 200 μg/mL (23).

Another phytochemical extract is the traditional Chinese medicine Sho-Saiko-To (Xiao-Chai-Hu-Tang, or crude powder extract TJ-9). It is the most popular herbal medicine in Japan and is used to treat chronic liver diseases. Sho-Saiko-To is a combination of seven herbal components: Bupleuri Radix, Pinelliae Tuber, Scutellariae Radix, Zizyphi Fructus, Ginseng Radix, Glycyrrhizae Radix and Zingiberis Rhizoma. The characterized active principles include baicalin; baicalein; wogonin; glycyrrhizin; saikosaponin a, c, and d; and ginsenoside Rb_1 and Rg_1 (24). In murine J774A.1 macrophagelike cell line with LPS stimulation alone, 50 μg/mL of TJ-9 inhibits nitrite production by about 40%, a concentration that has no effect on cell viability and the Griess reagent assay (25). However, although inhibition is observed in stimulated macrophages, in unstimulated murine RAW 264.7 macrophagelike cells, TJ-9 stimulates nitrite production, 20.9 versus 8.5 μM nitrite, as compared to untreated control cells. Similar enhancement effect has been observed in another seven out of 18 traditional Chinese herbal medicines that have been studied (26). The significance of this phenomenom remains to be investigated.

Nitrite-enhancing effect is not only observed in plant extracts that contain multiple ingredients. Acemannan, an isolated component, has also been shown to have this effect. Acemannan is a polydispersed β-(1,4)-acetylated mannan, from the tropical cactus plant *Aloe vera*. Its hydrogel has been used to treat aphthous ulcers, wounds, and burns. In a chicken macrophagelike cell line HD11, acemannan stimulates nitrite production in a dose-dependent manner, and nitrite production from 1 mg/mL acemannan equals that of 1 μg/mL LPS (27). However, in the murine RAW 264.7 cells, it has no inducing effect, although at 10 to 50 μg/mL, acemannan enhances the effect of low dose of IFN (1 U/mL) (28). These results can probably be attributed to the fact that the NOS activating stimulants required by different cell lines and species are highly variable. Similar enhancement of nitrite production has also been seen in fungal β-glucan (29).

Thus, various natural products have modulatory effects on the inflammation-induced nitrite production by macrophages, so these substances may act on the inducible nitric oxide synthase (iNOS, or NOS2).

III. MODULATION OF ENDOTHELIUM-DEPENDENT VASODILATATION BY NATURAL PRODUCTS

Acetylcholine-induced blood vessel relaxation is dependent on the endothelial cells which are producers of endothelium-derived relaxing factor, a highly labile and diffusible vasodilator. It was the search of this factor that led to the discovery of the hypotensive physiological effect of NO. Endothelial cells produce NO at low levels through NOS3. The NO diffuses from endothelial cells to smooth muscle cells of the vessel wall, where it activates a soluble guanylate cyclase, the enzyme that produces cGMP. This increase in cGMP level ultimately leads to muscle relaxation (1).

A second assay for NO production is based on this in vitro vascular relaxation function. It measures the effect on NOS3 as opposed to LPS-stmulated macrophage nitrite production, which measures the effect on NOS2. Briefly, arteries such as aorta, normally from rats, are used. They are cut into rings, and pharmacological compounds such as acetylcholine or phenylephrine are added to induce contraction. Test substances are added to the medium and the endothelium-dependent relaxation or inhibition of contraction is measured, such as by recording the mechanical activity on a polygraph. Controls can be performed by stripping the endothelium from the aorta ring or by including compounds such as N-monomethyl arginine that are known to be NOS inhibitors.

Red wine and grape juice both enhance endothelium-dependent relaxation of rat aorta rings precontracted with phenylephrine (30). Further investigation with compounds purifed from grapes showed that tannin (1 μM), quercetin, and resveratrol (0.1 to 60 μM), also cause relaxation. This activity can be reversed by the NOS inhibitor N-monomethyl arginine or nitro-L-arginine (30,31). Quercetin, a flavonoid plant phenolic, is common in grapes and other fruits and vegetables (32). α-Tocopherol (vitamin E) is present in many fruits and vegetables. Relaxation of rat aortic rings precontracted with acetylcholine is improved, especially for those from old rats versus those from young rats, by pretreatment of tocopherol (100 μM) (33). Moderate alcohol consumption has been suggested to protect against coronary heart disease. The effect is more prominent in red wine, and compounds from wine and grapes (from the plant *Vitis vinifera*) may have beneficial effects.

Following the study performed with wine and grapes, a variety of other plant extracts have been tested in the rat aortic ring system (34). Aorta relaxation has been obtained with extracts from vegetables (including lima bean, corn, cranberry, eggplant, and blackeye pea), from fruits (including apple, guava, and plum), from nuts (including peanut, pecan, and walnut), from spices (such as cinnamon and garlic), and from green tea. Interestingly, a few extracts including ones from potato and banana have been shown to enhance contraction, although the active principles for this vasodilatation and contraction remain to be characterized (34,35).

In vitro vasodilatation effects have also been shown by compounds and plant extracts from traditional/herbal medicine. Ginsenosides, which inhibit NOS2-mediated NO production by macrophages (17), also relax NOS3-mediated NO production in rat aortic rings precontracted with phenylephrine (36).

The dry roots of the Chinese medicinal plant *Salvia miltiorrhiza* is the source of medication known as Dan Shen (red sage), used to treat coronary heart disease. Extract from Dan-Shen relaxes rat aortic rings precontracted with noradrenaline. The active principle has been characterized to be the polyphenolic compound lithospermic acid B (37).

The dry, unripe fruit of another Chinese medicinal plant, *Evodia rutaecarpa*, is the source of medication known as Wu-Chu-Yu, used to treat gastrointestinal disorders. Active principles include the alkaloids rutaecarpine, evodiamine, and dehydroevodiamine. All three alkaloids have capacities to relax rat aortic rings precontracted with phenylephrine, at a concentration of 100 μM for 92% to 100% relaxation (38).

In addition, the Chinese medicinal plant *Angelica pubescens* (Du Huo) is used to treat arthritis. Its active principle, osthole, is a coumarin. At 40 to 200 μM, it relaxes rat aortic rings precontracted with norepinephrine in a dose-dependent manner, by 43% to 76% (39). Many coumarins are contained in plants with potential therapeutic potentials (40).

Finally, the Brazilian herbal medicine Catuama, used to treat chronic diseases such as fatigue, is a concoction made from four medicinal plants: *Trichilia catigua*, *Paullinia cupana* (Guarana), *Ptychopetalum olacoides*, and *Zingiber officinale* (ginger). It induces relaxation of rat aortic rings precontracted with noradrenaline (41).

Various natural products have modulatory effects on endothelium-dependent vasodilatation, so their pharmacological activities may include action on the endothelial constitutive nitric oxide synthase (NOS3).

IV. MODULATION OF THE NO PATHWAY BY NATURAL PRODUCTS IN OTHER ASSAY SYSTEMS

The above-mentioned in vitro systems are the more common ones for assaying modulators of the NO pathway (for both iNOS and cNOS). However, in the literature there are also reports that use several other assays. These natural products include mustard oil, Zizyphi Fructus, and Nao Yi An.

Mustard oil induces inflammation in the rat hindpaw skin. There is edema, vasodilatation and a 2 to 3°C temperature increase. Its effect on NO is suggested by the parallel observation that the inflammation can be reduced by the NOS inhibitor N-nitro-arginine methy ester, suggesting involvement of NO (42).

The Chinese antiasthmatic herbal medicine Zizyphi Fructus is an extract of nutlets of Chinese date tree. In an in vitro system of cultured canine tracheal epithelium, addition of up to 100 μg/mL of the extract stimulates epithelial ciliary motility and increases NO production (as measured by an NO meter) (43). It is interesting to note that Zizyphi Fructus is one ingredient among seven of Sho-Saiko-To (TJ-9) (24).

The Chinese medicine Nao Yi An has multiple plant and animal components. The medicine inhibits NADPH-diaphorase activity, an accepted cytochemical marker of NOS1, which is induced in experimental intracerebral hemorrhage in the rat forebrain (44).

Thus, from crude extracts to purified compounds, natural products act on the NO pathway. In most instances, molecular mechanisms remain to be investigated. However, in several cases, the mode of action has been characterized. Modulation by natural products at specific levels of the NO pathway (NOS gene regulation and expression, NOS enzyme activity, NO and RNS products) will be discussed in the following sections.

V. MODULATION AT THE LEVEL OF NOS GENE EXPRESSION AND REGULATION

Expanding from the observation that nitrite production is modulated, the effect of natural products at the level of the NOS2 (iNOS) gene, which is inducible, has been studied. These studies include curcumin, *Gingko biloba* extract EGb 761, docosahexaenoic acid (DHA), acemannan, genistein, and labdane F2 (45). Among them, the best-characterized is probably curcumin. In our laboratory, we have characterized the effect of curcumin on the expression of iNOS gene and on tumor necrosis factor-α (TNF) and interleukin-1β (IL-1), two inflammation-induced cytokines that are required to enhance NOS2 gene expression. Curcumin, at 5 μM, inhibits LPS-induced production of TNF and IL-1 by human macrophagelike cell line MonoMac 6 (46). For gene transcription, the 5′ flanking sequence of both the human and murine iNOS genes contains consensus elements for the transcription factors nuclear factor kappa B (NFκB) and activating protein-1 (AP-1). Curcumin (5 μM) inhibits LPS-induced activation of NFκB in the same cellular system, as shown by standard gel mobility shift assay (46). Our finding has been confirmed in another cell line, the human myelomonoblastic leukemia cell line ML-1α, using other NFκB activators: TNF, phorbol ester, and hydrogen peroxide (47). In addition, AP-1, a fos-jun heterodimer, is also inhibited by curcumin. Curcumin inhibits c-jun mRNA expression in the NIH 3T3 murine fibroblast cell line (48).

As for NOS2 gene expression, curcumin reduces the amount of mRNA in ex vivo cultured BALB/c mouse peritoneal macrophages. Using competitive based RT-PCR, reductions of 42%, 53%, 60%, and 67% are obtained with 1, 2.5, 5, and 10 μM curcumin, respectively, from 0.045 attomole of NOS2 mRNA in the untreated sample to 0.026, 0.021, 0.018, and 0.015 attomole in the samples that have been treated (49). This finding concurs with the in vitro observation with RAW264.7, a cell line whose NOS2 expression is not tightly regulated. In the latter study, reduction in NOS2 mRNA and protein level was observed after treatment with 10 μM curcumin (50). Thus, regulation of NOS2 expression at the gene level is modulated by curcumin.

Similar studies on the effect of *Ginkgo biloba* extract EGb 761 have been performed on NFκB and AP-1. Ginkgo extract, at 10 μg/mL, inhibits the activation of AP-1 by inhibiting c-fos gene expression, but has no effect on NFκB activation in the human Jurkat T cell line or macrophages (23,51). In the LPS- and IFN-stimulated RAW 264.7 macrophagelike cells, RT-PCR has shown a dose-dependent (20 to 200 μg/mL) decrease in the induction of NOS2 gene expression, with no change in the control glyceraldehyde-3-phosphate dehydrogenase (GAPDH) gene.

For omega-3 polyunsaturated fatty acids, the ability of DHA to modulate NOS2 gene expression has been studied (16). In LPS-stimulated RAW 264 cells, DHA inhibits NOS2 gene expression in a dose-dependent manner at the mRNA level and correspondingly at the protein level.

Acemannan, which enhances NO production, also modulates the induction of NOS2 gene (28). In IFN (1 U/mL)-stimulated RAW 264.7 cells, acemannan enhances the mRNA level of NOS2 and expression of NOS2 protein. Preincubation with pyrrolidine dithiocarbamate (an NFkB inhibitor, at 50 μM) can inhibit this acemannan enhancement.

In addition, inhibition in NOS2 gene induction by genistein has been reported. In the rat C6-N glioma cell line stimulated with LPS and IFN, genistein (at 100 μM) inhibits NOS2 mRNA expression, as demonstrated by RT-PCR (52).

The diterpenoid labdane F2 is isolated from *Sideritis javalambrensis*, a medicinal plant with anti-inflammatory activities. At 3, 10, 30, and 100 μM, it causes a dose-dependent decrease in NOS2 protein expression in LPS-stimulated murine J774 cells, as detected by immunoblot analysis (45).

Thus, for most natural products, their mode of action at the gene level requires further study. Of the few with known activity toward NOS2, their effect on expression and regulation of NOS1 and NOS3 is unknown.

VI. MODULATION AT THE LEVEL OF NOS ENZYME ACTIVITY

Besides regulating gene expression, some natural products are found to be NOS enzyme inhibitors and activators. To measure enzyme activity, cell extracts exhibiting NOS activity are used to screen for modulators. Extract of endothelial cells can be used for NOS3, macrophages for NOS2, and brain for NOS1. Usually, enzyme activity is measured by detecting conversion of radiolabeled arginine, the substrate, to citrulline, the product that is produced at 1:1 ratio with NO. Inhibition of NOS enzymatic activity has been reported for tannin, quercetin, and *Ginkgo biloba* extract EGb 761, whereas a direct activation or enhancement has been seen in garlic and ginsenosides.

Tannin and quercetin are contained in grapes. These flavonoids directly inhibit NOS3 activity in cell extracts of bovine aortic endothelial cells. For NOS3, tannin at 10 μM produces 100% inhibition with an IC_{50} at 2 μM, whereas quercetin inhibits with an IC_{50} of 220 μM. They have a selectivity ratio of 20 to 30 for NOS3 versus NOS1 (from bovine brain extract) and NOS2 (from extract of interleukin 1β–stimulated rat aortic vascular smooth muscle cells). Inhibition of NOS3 by tannin shows noncompetitive kinetics. Tannin does not interfere with enzyme binding to arginine, NADPH, or tetrahydrobiopterin, and the activity is not affected by calmodulin or calcium level (53).

NOS2 activity of LPS- and IFN-induced RAW 264.7 macrophagelike cells is inhibited directly by the addition of *Ginkgo biloba* extract EGb 761 (23). Using the conversion of radiolabeled arginine to citrulline as assay, for 200 μg/mL *Ginkgo* extract, there is 39% inhibition (from a control value of 205 pmol/min/mg protein). Active principles for this action remain to be determined.

Activation of NOS activity by direct addition of garlic extract (from plant *Allium sativum*) is seen using homogenate of human placental villi (54). However, the active principle in the garlic extract for NOS activation is still undefined, although arginine itself, rich in garlic, has been ruled out.

Kim et al. (92) added labeled arginine and ginsenosides to cultured bovine aortic endothelial cells, lysed the cells after 15 min of incubation, and then measured the labeled product. Ginsenoside mixture and ginsenoside Rg_1 increased conversion, from 14.9% in the control to 20.3% and 22%, respectively. In addition, they also observed that Rg_1 increases the uptake of radiolabeled arginine to the endothelial cells. Thus, these data show that NOS3 activity is enhanced by the ginsenoside saponins (55).

Nonradiolabeled NOS enzyme activity assay has been used to investigate the action by genistein and ascorbic acid (vitamin C) on cells. In peritoneal exudate macrophages from C3H/HeN mice stimulated with LPS and IFN, genistein treatment (25 μM) reduces NOS2 activity (from 71.1 to 21.4 nmol [nitrite/nitrate]/mg protein) (14). In a human vascular endothelial cell line EA.hy 926 (a somatic cell hybrid) stimulated with LPS, IFN, and TNF, there is NOS2 activity that can be inhibited by ascorbic acid. At 50 μg/mL ascorbic acid, there is a decrease in enzyme activity, from 1.30 to 0.63 nmol/min/mg protein (56).

As seen from the data presented, natural products can modulate NOS enzymatic activity in cells and tissues. There is a strong interest in identifying compounds that may attenuate inflammation. Such an NOS isoform-specific compound remains to be identified.

VII. MODULATION AT THE LEVEL OF NO AND RNS PRODUCTS

NO and its related products are highly reactive and unstable. Soon after production NO reacts with other molecules to form RNS, such as with oxygen to form nitrogen dioxide and with superoxide to form peroxynitrite. It may also bind to free radical scavenging proteins such as hemoglobin. Many natural products, particularly those with antioxidative capacity, are NO scavengers. Many others react with RNS and derivatives from NO.

Direct scavengers of NO include the curcuminoids and the flavonoids. For curcuminoids, curcumin, demethoxycurcumin, bisdemethoxycurcumin, and diacetylcurcumin scavenge NO chemically produced from sodium nitroprusside (57). Curcumin shows a concentration-dependent NO scavenging, with 50% scavenged at 20.4 μM curcumin. For flavonoids, rutin, pelargonidin, cyanidin, and trihydroxyethyl quercetin react with NO gas (58). Pelargonidin is the most potent; at 5 μM, it decreases the half-life of NO by 50%.

For extract, *Ginkgo biloba* EGb761 has been shown to scavenge NO in two assay systems. It scavenges NO generated from sodium nitroprusside and inhibits NO conversion of oxy- to met-hemoglobin by 50%, at a concentration of 15 to 20 μg/mL (59). The *Ginkgo biloba* extract EGb 761, at 25 μg/mL, has also been shown to protect against peroxynitrite-induced cytotoxicity of bovine retinal pigmented epithelial cells in culture (60).

Compared to NO, peroxynitrite is an even more potent toxic species with a variety of activities including the nitration of tyrosine in proteins. Catechin polyphenols inhibit peroxynitrite-mediated tyrosine nitration of proteins (61). At 10 μM, the inhibition ranges from 20% to 38% for catechin, epicatechin, epigallocatechin, epicatechin gallate, EGCG, and gallic acid.

For nitrogen dioxide, curcumin scavenges nitrogen dioxide and protects hemoglobin from nitrogen dioxide-induced oxidation (62). Curcumin also inhibits nitrite-induced oxidation of hemoglobin (63). α–Tocopherol reacts with NO and NO products, although chemical details remain to be clarified (64,65). Besides α-tocopherol, γ-tocopherol also reacts with NO products (66,67).

Thus, in comparison to studies on natural product modulation of NOS gene expression and enzyme activity, many more natural products have been shown to possess activity to bind NO or RNS.

VIII. IN VIVO EFFECTS OF NATURAL PRODUCTS

Natural products that modulate the NO pathway at various levels, from gene expression and regulation, to NOS enzyme activity, to scavenging of NO and related products have been identified. Table 1 lists selected examples of these agents. All of these studies were performed in vitro. The action of these compounds in vivo would be needed to confirm the findings in animals.

For curcumin, we have recently found that it inhibits NOS2 induction in LPS-treated mice (49). In vivo inhibition of NOS2 expression in liver, spleen, and kidney is observed by the technique of RT-PCR. In rats fed with cod liver oil, which contains fatty acids of the omega series, there is decreased NO production by peritoneal macrophages (13). Oral administration of the Chinese medicine Sho-Saiko-To (TJ-9) to mice also leads to downregulation of nitrite production of peritoneal macrophages and protects mice from severe TNF-induced shock syndromes (68,69). In an aging study, TJ-960 is given to old mice by mixing it into the diet. NOS1 activities in the homogenate from different brain portions of the treated mice are compared to those from the young and untreated control animals using an NO analyzer. TJ-960 inhibits the increase of NOS1 activity normally seen in aging cerebral cortex (70). Intraperitoneal injection of glycyrrhizin, active principle from *Glycyrrhiza radix* (one component among 7 of Sho-Saiko-To), however, leads to enhanced nitrite production in murine peritoneal macrophages induced with LPS (71).

Intraperitoneal injection of β-glucan, also known as grifolan, isolated from the fungus *Grifola frondosa*, at 250 μg/mouse or zymocel, another purified insoluble glucan from a yeast, enhanced nitrite production of peritoneal macrophages. For grifolan, it enhances IFN production in liver and peritoneal macrophages of treated mice, as detected by RT-PCR, suggesting that IFN-mediated mechanism is involved in this enhanced NO synthesis (72,73). Thus, glucans and glycyrrhizin can activate macrophages in vivo. Based on the investigations cited above, it can be anticipated that more animal studies to demonstrate the effects of natural products on the NO pathway in vivo will soon appear in the literature.

Table 1 Modes of Action of Selected Natural Products on the Nitric Oxide Pathway

Natural product	Target enzyme	Induction or inhibition	Gene level: NOS-associated genes	Gene level: NOS gene	Enzyme level: Intact cells	Enzyme level: Cell extract	NO products: NO	NO products: NO-related products
Purified compounds								
Curcumin	iNOS	▼	+	+	+		+	+
Genistein	iNOS	▼		+	+			
Labdane F2	iNOS	▼		+				
Docosahexaenoic acid	iNOS	▼		+				
Acemannan	iNOS	▲	+	+				
Tannin	cNOS	▼				+		
Ginsenoside Rg_1	cNOS	▲			+			
Crude extracts								
Ginkgo	iNOS	▼	+	+		+	+	+
Garlic	cNOS	▲				+		

IX. CLINICAL SIGNIFICANCE, PERSPECTIVES AND FUTURE CHALLENGES

Another consideration is the relevance of the experimental data to human. Presently, many natural products inhibit LPS and IFN-induced NOS2 expression of murine macrophages or macrophagelike cell lines (as summarized in Sect. II). However, it has not been delineated that the regulation of NO production by human macrophages is identical to that of the murine counterpart (3,74). Thus, clinical significance of many of these natural products with known action on rodent cells awaits information on their activity in human cells.

Although further studies on the efficacy, safety, specificity, and active principles of NO-modulating effects of natural products on human cells need to be done, we must emphasize that the majority of natural products come from two sources—namely, dietary components and traditional herbal medications (see Table 1). Spices, drinks, food, and folk medicine have been consumed in different cultures for centuries. For most of these substances, the Food and Drug Administration has classified them as "GRAS" (generally regarded as safe). Furthermore, a majority of the natural products described above, such as curcumin, have been shown to possess cancer chemopreventive capacities in animal studies (mostly in mice and rats) (7,75). For this reason, clinical trials that include toxicology studies have already begun for some.

Increased NOS expression has been documented in some human and animal tumors (76–78). For human gastric cancer, one known risk factor is *Helicobacter pylori* infection. In a study involving 84 Colombian patients, using immunohistochemistry, there was increased NOS2 and nitrotyrosine staining and DNA fragmentation in gastric biopsy samples, indicating the presence of peroxynitrite production and apoptosis due to NO overproduction. Dietary supplements with β-carotene (30 mg/day for 4 to 12 months) leads to NOS2 decrease, and supplement with ascorbic acid (2 g/day for 4 to 12 months) leads to nitrotyrosine decrease. These results suggest that, besides antibiotics, natural products (β-carotene and ascorbic acid) may prevent the formation of potential carcinogens such as NO and peroxynitrite (79). Thus, inhibition of the NO pathway may be beneficial to some cancer therapy, although NO may mediate tumor cell killing in other cases.

Natural products that modulate the NO pathway come in a variety of chemical structures. Those with known chemical formulation, such as curcumin and genistein, may serve as lead compounds for the development of related molecules and synthetic analogs so as to obtain the synthetic versions that are more potent or more isoform-specific and thus more clinically useful and patentable. This approach is not unprecedented. For example, the synthetic flavone 3′-amino-4′-hydroxyflavone is more potent than the natural compound genistein in the inhibition of nitrite production by murine macrophages, 84% inhibition at 20 μM versus 48% at 50 μM (15). The natural coumarin osthole has endothelial-dependent vasodilatation activity, whereas the synthetic coumarin cloricromere can both relax rat aortic rings and inhibit NO production by LPS-stimulated J774 macrophagelike cells (80). Thus, it is feasible to obtain more desirable derivatives from natural products for modulating the NO pathway.

In conclusion, natural products provide a reservoir of different substances that may act as modulators of the NO pathway. A novel compound derived from natural products that is active in vivo and that specifically targets a single human NOS would be a clinically important achievement.

ACKNOWLEDGMENTS

The writing of this chapter was supported, in part, by grants from the American Institute for Cancer Research.

REFERENCES

1. Moncada S, Higgs A. The L-arginine-nitric oxide pathway. N Engl J Med1993; 329:2002–2012.
2. Billiar TR. Nitric oxide. Novel biology with clinical relevance. Ann Surg 1995; 221:339–349.
3. MacMicking J, Xie QW, Nathan C. Nitric oxide and macrophage function. Ann Rev Immunol 1997; 15:323–350.
4. Gross SS, Wolin MS. Nitric oxide: pathophysiological mechanisms. Ann Rev Physiol 1995; 57:737–769.
5. Ohshima H, Bartsch H. Chronic infections and inflammatory processes as cancer risk factors: possible role of nitric oxide in carcinogenesis. Mutat Res 1994; 305:253–264.
6. Tamir S, Tannenbaum SR. The role of nitric oxide (NO) in the carcinogenic process. Biochim Biophys Acta 1996; 1288:F31–F36.
7. Wattenberg L, Lipkin M, Boone CW, Kelloff GJ. Cancer Chemoprevention. Boca Raton, CRC Press, FL:1992.
8. Stavric B. Role of chemopreventers in human diet. Clin Biochem 1994; 27:319–332.
9. Messina M, Messina V, Setchell KDR. The Simple Soybean and Your Health. Garden City Park NY: Avery Publishing Group, 1994.
10. Jones F. The Save Your Heart Wine Guide. New York: St. Martin's Press, 1996.
11. Chan MM, Ho CT, Huang HI. Effects of three dietary phytochemicals from tea, rosemary and turmeric on inflammation-induced nitrite production. Cancer Lett 1995; 96:23–29.
12. Ammon HP, Wahl MA. Pharmacology of *Curcuma longa*. Planta Med 57:1–7, 1991.

13. Joe B, Lokesh BR. Role of capsaicin, curcumin and dietary n-3 fatty acids in lowering the generation of reactive oxygen species in rat peritoneal macrophages. Biochim Biophys Acta 1994; 1224:255–263.
14. Dong Z, Qi X, Xie K, Fidler IJ. Protein tyrosine kinase inhibitors decrease induction of nitric oxide synthase activity in lipopolysaccharide-responsive and lipopolysaccharide-nonresponsive murine macrophages. J Immunol 1993; 151:2717–2724.
15. Krol W, Czuba ZP. Threadgill MD. Cunningham BD. Pietsz G. Inhibition of nitric oxide (NO) production in murine macrophages by flavones. Biochem Pharmacol 1995; 50:1031–1035.
16. Ohata T, Fukuda K, Takahashi M, Sugimura T, Wakabayashi K. Suppression of nitric oxide production in lipopolysaccharide-stimulated macrophage cells by omega-3 polyunsaturated fatty acids. Jpn J Cancer Res 1997; 88:234–237.
17. Park YC, Lee CH, Kang HS, Kim KW, Chung HT, Kim HD. Ginsenoside-Rh_1 and Rh_2 inhibit the induction of nitric oxide synthesis in murine peritoneal macrophages. Biochem Mol Biol Int 1996; 40:751–757.
18. Liu L, Riese J, Resch K, Kaever V. Impairment of macrophage eicosanoid and nitric oxide production by an alkaloid from *Sinomenium acutum*. Arzneimittel-Forschung 1994; 44:1223–1226.
19. Park HJ, Jung WT, Basnet P, Kadota S, Namba T. Syringin 4-O-beta-glucoside, a new phenylpropanoid glycoside, and costunolide, a nitric oxide synthase inhibitor, from the stem bark of *Magnolia sieboldii*. J Nat Prod 1996; 59:1128–1130.
20. Terawaki K, Nose M, Kondo T, Kojima K, Mizukami H, Ogihara Y. Effect of karasurin A on nitric oxide production by murine macrophages and mitogenic response of murine splenocytes in vitro. Biol Pharm Bull 1997; 20:435–437.
21. Chevallier A. The Encyclopedia of Medicinal Plants. New York: DK Publishing, Inc., 1996.
22. Marcocci L, Packer L, Droy-Lefaix MT, Sekaki A, Gardes-Albert M. Antioxidant action of *Ginkgo biloba* EGb 761. Methods Enzymol 1994; 234:462–475.
23. Kobuchi H,. Droy-Lefaix MT, Christen Y, Packer L. *Ginkgo biloba* extract (EGb761): inhibitory effect on nitric oxide production in the macrophage cell line RAW 264.7. Biochem Pharm 1997; 53:897–903.
24. Yano H, Mizoguchi A, Fukuda K, Haramaki M, Ogasawara S, Momosaki S, Kojiro M. The herbal medicine sho-saiko-to inhibits proliferation of cancer cell lines by inducing apoptosis and arrest at the G0/G1 phase. Cancer Res 1994; 54:448–454.
25. Sakaguchi S, Furusawa S, Yokota K, Sasaki K, Takayanagi Y. Depressive effect of a traditional Chinese medicine (sho-saiko-to) on endotoxin-induced nitric oxide formation in activated murine macrophage J774A.1 cells. Biol Pharm Bull 1995; 18:621–623.
26. Izumi S, Ohno N, Miura NN, Takada K, Yadomae T. Stimulation of nitric oxide synthesis in macrophage cell line RAW 264.7 cells by several traditional Chinese herbal medicines. Drug Dev Res 1995; 36:112–118.
27. Karaca K, Sharma JM, Nordgren R. Nitric oxide production by chicken macrophages activated by Acemannan, a complex carbohydrate extracted from *Aloe Vera*. Int J Immunopharm 1995; 17:183–188.

28. Ramamoorthy L, Kemp MC, Tizard IR. Acemannan, a beta-(1,4)-acetylated mannan, induces nitric oxide production in macrophage cell line RAW 264.7. Mol Pharm 1996; 50:878–884.
29. Ohno N, Hashimoto T, Adachi Y, Yadomae T. Conformation dependency of nitric oxide synthesis of murine peritoneal macrophage by beta-glucans in vitro. Immunol Lett 1996; 52:1–7.
30. Fitzpatrick DF, Hirschfield SL, Coffey RG. Endothelium-dependent vasorelaxing activity of wine and other grape products. Am J Physiol 1993; 265(2 Pt 2):H774–H778.
31. Chen CK, Pace-Asciak CR. Vasorelaxing activity of resveratrol and quercetin in isolated rat aorta. Gen Pharmacol 1996; 27:363–366.
32. Formica JV, Regelson W. Review of the biology of quercetin and related bioflavonoids. Food Chem Toxicol 1995; 33:1061-1080.
33. Guarnieri C, Giordano E,. Muscari C, Grossi L, Caldarera CM. Alpha-tocopherol pretreatment improves endothelium-dependent vasodilation in aortic strips of young and aging rats exposed to oxidative stress. Mol Cell Biochem 1996; 157:223–228.
34. Fitzpatrick DF, Hirschfield SL, Ricci T, Jantzen P, Coffey RG. Endothelium-dependent vasorelaxation caused by various plant extracts. J Cardiovasc Pharmacol 1995; 26:90–95.
35. Ozturk Y, Aydin S, Kosar M, Baser KH. Endothelium-dependent and independent effects of garlic on rat aorta. J Ethnopharmacol 1994; 44:109–116.
36. Kim ND, Kang SY, Schini VB. Ginsenosides evoke endothelium-dependent vascular relaxation in rat aorta. Gen Pharmacol 1994; 25:1071–1077.
37. Kamata K, Iizuka T, Nagai M, Kasuya Y. Endothelium-dependent vasodilator effects of the extract from *Salviae miltiorrhizae* radix. A study on the identification of lithospermic acid B in the extracts. Gen Pharmacol 1993; 24:977–981.
38. Chiou WF, Liao JF, Chen CF. Comparative study of the vasodilatory effects of three quinazoline alkaloids isolated from *Evodia rutaecarpa*. J Nat Prod 1996; 59:374–378.
39. Ko FN, Wu TS, Liou MJ, Huang TF, Teng CM. Vasorelaxation of rat thoracic aorta caused by osthole isolated from *Angelica pubescens*. Eur J Pharmacol 1992; 219:29–34.
40. Hoult JR, Paya M. Pharmacological and biochemical actions of simple coumarins: natural products with therapeutic potential. Gen Pharmacol 1996; 27:713–722.
41. Calixto JB, Cabrini DA. Herbal medicine catuama induces endothelium-dependent and independent vasorelaxant action on isolated vessels from rats, guinea pigs and rabbits. Phytother Rese 1997; 11:32–38.
42. Lippe IT, Stabentheiner A. Holzer P. Participation of nitric oxide in the mustard oil-induced neurogenic inflammation of the rat paw skin. Eur J Pharmacol 232:113–120, 1993.
43. Tamaoki J, Kondo M, Tagaya E,. Takemura K, Konno K. Zizyphi fructus, a constituent of antiasthmatic herbal medicine, stimulates airway epithelial ciliary motility through nitric oxide generation. Exp Lung Res 1996; 22:255–266.
44. Peng ZC, Li XQ, Liang QH, Zhu CX, Luo TL, Yan XX, Hu SY. Induction of NADPH-diaphorase activity in the forebrain in a model of intracerebral hemorrhage

and its inhibition by the traditional Chinese medicine complex Nao Yi An. Brain Res Bull 1997; 42:119–128.

45. Pang L, de las Heras B, Hoult JR. A novel diterpenoid labdane from *Sideritis javalambrensis* inhibits eicosanoid generation from stimulated macrophages but enhances arachidonate release. Biochemical Pharmacology. 51:863–868, 1996.
46. Chan MM. Inhibition of tumor necrosis factor by curcumin, a phytochemical. Biochem Pharmacol 1995; 49:1551–1556.
47. Singh S, Aggarwal BB. Activation of transcription factor NF-kappa B is suppressed by curcumin (diferuloylmethane). J Biol Chem 1995; 270:24995–5000.
48. Huang TS, Lee SC, Lin JK. Suppression of c-Jun/AP-1 activation by an inhibitor of tumor promotion in mouse fibroblast cells. Proc Nat Acad Sci USA 1991; 88:5292–5296.
49. Chan MM, Huang HI, Fenton MR, Fong D. In vivo inhibition of nitric oxide synthase gene expression by curcumin, a cancer preventive natural product with anti-inflammatory properties. Biochem Pharmacol 1998; 55:1955–1962.
50. Brouet I, Ohshima H. Curcumin, an anti-tumour promoter and anti-inflammatory agent, inhibits induction of nitric oxide synthase in activated macrophages. Biochem Biophyl Res Commun 1995; 206:533–540.
51. Mizuno M, Droyl-Lefaix MT, Packer L. *Gingko biloba* extract EGb761 is a suppressor of AP-1 transcription factor stimulated by phorbol 12-myristate 13-acetate. Biochem Mol Biol Int 1996; 39:395–401.
52. Feinstein DL, Galea E, Cermak J, Chugh P, Lyandvert L, Reis DJ. Nitric oxide synthase expression in glial cells: suppression by tyrosine kinase inhibitors. J Neurochem 1994; 62:811–814.
53. Chiesi M, Schwaller R. Inhibition of constitutive endothelial NO-synthase activity by tannin and quercetin. Biochem Pharmacol 1995; 49: 495–5015.
54. Das I,. Hirani J, Sooranna S. Arginine is not responsible for the activation of nitric oxide synthase by garlic. J Ethnopharmacol 1996; 53:5–9.
55. Kim H, Chen X, Gillis CN. Ginsenosides protect pulmonary vascular endothelium against free radical-induced injury. Biochem Biophys Res Commun 1992; 189:670–676.
56. Galley HF, Walker BE, Howdle PD, Webster NR. Regulation of nitric oxide synthase activity in cultured human endothelial cells: effect of antioxidants. Free Rad Biol Med 1996; 21:97–101.
57. Sreejayan, Rao. Nitric oxide scavenging by curcuminoids. J Pharm Pharmacol 1997; 49:105–107.
58. van Acker SA, Tromp MN, Haenen GR, van der Vijgh WJ. Bast A. Flavonoids as scavengers of nitric oxide radical. Biochem Biophys Res Commun 1995; 214:755–759.
59. Marcocci L, Maguire JJ, Droy-Lefaix MT, Packer L. The nitric oxide-scavenging properties of *Ginkgo biloba* extract EGb 761. Biochem Biophys Res Commun 1994; 201:748–755.
60. Behar-Cohen FF, Heydolph S, Faure V, Droy-Lefaix MT, Courtois Y. Goureau O.Peroxynitrite cytotoxicity on bovine retinal pigmented epithelial cells in culture. Biochem Biophys Res Commun 1996; 226:842–9.

61. Pannala AS, Rice-Evans CA, Halliwell B, Singh S. Inhibition of peroxynitrite-mediated tyrosine nitration by catechin polyphenols. Biochem Biophys Res Comm 1997; 232:164–168.
62. Unnikrishnan MK, Rao MN. Curcumin inhibits nitrogen dioxide induced oxidation of hemoglobin. Mol Cell Biochem 1995; 146:35–37.
63. Unnikrishnan MK. Rao MN. Inhibition of nitrite induced oxidation of hemoglobin by curcuminoids. Pharmazie 1995; 50:490–492.
64. Gorbunov NV, Osipov AN, Sweetland MA, Day BW, Elsayed NM, Kagan VE. NO-redox paradox: direct oxidation of alpha-tocopherol and alpha-tocopherol-mediated oxidation of ascorbate. Biochem Biophys Res Commun 1996; 219:835–841.
65. Hogg N, Singh RJ, Goss SP, Kalyanaraman B. The reaction between nitric oxide and alpha-tocopherol: a reappraisal. Biochem Biophys Res Commun 1996; 224:696–702.
66. Cooney RV, Harwood PJ, Franke AA, Narala K, Sundstrom AK, Berggren PO. Mordan LJ. Products of gamma-tocopherol reaction with NO_2 and their formation in rat insulinoma (RINm5F) cells. Free Radic Biol Med 1995; 19:259–269.
67. Christen S, Woodall AA, Shigenaga MK, Southwellkeely PT, Duncan MW. Ames BN. Gamma-tocopherol traps mutagenic electrophiles such as NOx and complements alpha-tocopherol-physiological implications. Proc Nat Acad Sci USA. 1997; 94:3217–3222.
68. Izumi S,. Ohno N, Yadomae T. Down-regulation of LPS-induced nitric oxide synthesis of murine macrophages by oral administration of Sho-Saiko-To. Drug Dev Res 1997; 40:48–55.
69. Sakaguchi S, Furusawa S, Yokota K, Sasaki K, Takayanagi M, Takayanagi Y. Effects of antitumor activity and protection of shock symptoms by a traditional Chinese medicine (sho-saiko-to) in recombinant human tumor necrosis factor administered mice. Biol Pharm Bull 1996; 19:1474–1478.
70. Inada K, Yokoi I, Kabuto H, Habu H, Mori A, Ogawa N. Age-related increase in nitric oxide synthase activity in senescence accelerated mouse brain and the effect of long-term administration of superoxide radical scavenger. Mech Ageing Dev 1996; 89:95–102.
71. Kondo Y, Takano F. Nitric oxide production in mouse peritoneal macrophages enhanced with glycyrrhizin. Biol Pharm Bull 1994; 17:759–761.
72. Suzuki T, Ohno N, Chiba N, Miura NN, Adachi Y, Yadomae T. Immunopharmacological activity of the purified insoluble glucan, zymocel, in mice. J Pharm Pharmacol 1996; 48:1243–1248.
73. Ohno N, Egawa Y, Hashimoto T, Adachi Y, Yadomae T. Effect of beta-glucans on the nitric oxide synthesis by peritoneal macrophage in mice. Biol Pharm Bull 1996; 19:608–612.
74. Albina JE. On the expression of nitric oxide synthase by human macrophages. Why no NO? J Leukoc Biol 1995; 58:643–649.
75. Ho CT, Osawa T, Huang MT, Rosen RT. Food Phytochemicals for Cancer Research II. Teas, Spices, and Herbs. Washington DC, American Chemical Society, 1994.
76. Cobbs CE, Brenman JE, Aldape KD, Bredt DS, Israel MA. Expression of nitric oxide synthase in human central nervous system tumors. Cancer Res 1995; 55:727–730.

77. Thomsen LL, Miles DW, Happerfield L, Bobrow LG, Knowles RG, Moncada S. Nitric oxide synthase activity in human breast cancer. Br J Cancer 1995; 72:41–44.
78. Takahashi M, Fukuda K, Ohata T, Sugimura T, Wakabayashi K. Increased expression of inducible and endothelial constitutive nitric oxide synthases in rat colon tumors induced by azoxymethane. Can Res 1997; 57:1233–1237.
79. Mannick EE, Bravo LE, Zarama G, Realpe JL, Zhang XJ, Ruiz B,. Fontham ET, Mera R, Miller MJ, Correa P. Inducible nitric oxide synthase, nitrotyrosine, and apoptosis in *Helicobacter pylori* gastritis: effect of antibiotics and antioxidants. Can Res 1996; 56:3238–3243.
80. Zingarelli B, Carnuccio R, Di Rosa M. Cloricromene inhibits the induction of nitric oxide synthase. Eur J Pharmacol 1993; 243:107–111.

15

The Role of DNA Synthesis and Cytokine-Generated Oxidants in Hepatocyte Necrosis and Apoptosis

Joan P. Blonder
RxKinetix, Inc., Louisville, Colorado

David J. Sieg
The Scripps Research Institute, La Jolla, California

Danette R. Miller and Ruth E. Billings
Colorado State University, Fort Collins, Colorado

I. INTRODUCTION

A. Mechanisms of Liver Injury

Injury to the liver is a multifaceted process involving complex interactions among several different cell types and endogenous mediators. Cells involved in liver injury include liver cells: hepatocytes, Kupffer cells, and endothelial cells, as well as circulating leukocytes such as monocytes and neutrophils. Liver injury begins with a chemical, bacterial, or pathophysiological insult which initiates damage to liver cells, particularly hepatocytes. Initial damage to hepatocytes leads to release of cytoplasmic mediators resulting in Kupffer cell activation and leukocyte recruitment. Activated Kupffer cells and infiltrating leukocytes respond by releasing even more endogenous mediators, including reactive oxygen and nitrogen radicals (1,2,) proteases (3), leukotrienes (4), prostaglandins (5), and cytokines (6,7). Release of these mediators ultimately results in potentiation of the initial liver damage. In particular, cytokines released by activated Kupffer cells and recruited leukocytes play a pivotal role in exacerbation of a nontoxic insult to the liver.

B. Pro-inflammatory Cytokines

Cytokines are extremely potent, hormonelike proteins that act in concert to mediate the development and maintenance of an inflammatory or immune response whether beneficial or harmful to the organism (8). Tumor necrosis factor-α (TNFα) is an important "pro-inflammatory" cytokine released by Kupffer cells that is transiently induced and forms a complex network of mutual induction and overlapping activities with other cytokines (9). TNFα appears to be the initiator of inflammation and the primary mediator of liver damage. For example, during the onset of inflammation, TNFα is the first cytokine to appear in high concentrations in the plasma (10). TNFα then induces expression of other cytokines and numerous genes (9). Likewise, TNFα is responsible for inducing the significant liver injury caused by endotoxemia (11) and ischemia-reperfusion (12).

C. Mechanisms of TNFα-Induced Toxicity

TNFα exerts its actions through binding to two different cell receptors (TNF-R1 and TNF-R2), resulting in activation of a variety of signal transduction pathways, including activation of GTP-binding proteins, protein kinase C, phospholipase A_2 , and adenylate cyclase (13–15). TNFα-induced toxicity to hepatocytes and other cells is likely due to production of reactive oxygen intermediates (ROI) including superoxide anions (O_2^-), hydrogen peroxide (H_2O_2), and hydroxyl radicals (·OH), thus inducing oxidative stress within cells (16). Accordingly, alterations in cellular antioxidant defenses affect the actions of TNFα both in vivo and in vitro. For example, TNFα-induced toxicity increases when glutathione levels are decreased (16), whereas supplementing intracellular antioxidant levels using N-acetylcysteine prevents TNFα toxicity in vivo (17). In addition, TNFα-resistant cells are frequently sensitized to this cytokine by treatment with the protein synthesis inhibitor cycloheximide or the transcription inhibitor actinomycin D (13,18). This sensitization suggests that specific proteins are synthesized to protect cells against the toxic effects of TNFα.

D. Biochemical Properties of Nitric Oxide (NO)

TNFα initiates production of reactive nitrogen intermediates via induction of nitric oxide synthase (NOS2) protein and activity in a variety of cell types (19–21), including hepatocytes (22,23). NOS2 catalyzes production of NO from the oxidative conversion of arginine to citrulline. NO plays a significant role in mediating many physiological processes including vasodilatation (24), neurotransmission (25), macrophage- and neutrophil-mediated killing of pathogens (26,27), and regeneration of liver (28).

The biological role of NO in cytokine-mediated cellular injury is not clear. NO may exert both antioxidant and pro-oxidant properties. As an antioxidant, NO is cytoprotective to cells by reacting with O_2^-, thus preventing O_2^- conversion to the more reactive ·OH. In this case, the reaction of NO and O_2^- forms peroxynitrite ($ONOO^-$), which can rapidly and spontaneously rearrange to form nontoxic nitrite/nitrate (29). Subsequently, the cellular damage and cytotoxicity associated with O_2^- and O_2^--derived ROI is eliminated (30). Billiar *et al.* (31) demonstrate that inhibiting NO production dramatically increases hepatic damage induced by *Corynebacterium parvum* and lipopolysaccharide, presumably due to an increase in O_2^-.

As a pro-oxidant, NO may contribute to increased oxidative damage through formation of $ONOO^-$. $ONOO^-$ has a longer half-life than either NO or O_2^- (32). The rate constant for formation of $ONOO^-$ from NO and O_2^-- is three times greater than the rate constant for the removal of O_2^- by superoxide dismutase (SOD) and 100 times greater than the reaction between NO and heme compounds (29). Furthermore, $ONOO^-$ formation follows second-order kinetics and thus depends on the concentrations of both NO and O_2^-. Therefore, marginal changes in either radical concentration results in a substantial increase in $ONOO^-$. These characteristics are physiologically important since pro-inflammatory cytokines dramatically increase production of both O_2^- and NO, especially in inducible NOS cell systems, such as hepatocytes (16,23). Normally, NO concentrations are not detectable, but can increase over 50-fold upon cytokine stimulation (23). Thus, the formation of $ONOO^-$ in vivo is preferential.

Once formed, $ONOO^-$ can potentially oxidize a variety of biomolecules such as polyunsaturated fatty acids (33), sulfhydryl groups (34), and cellular DNA (35,36). Ischiropoulos *et al.* (37) show that $ONOO^-$ contributes to the oxidative injury of isolated rat lung from ischemia-reperfusion. Similarly, Ma *et al.* (38) demonstrate that $ONOO^-$ directly causes hepatocellular injury during reperfusion. In both studies, inhibiting NO production using NOS2 inhibitors protects against the oxidative injury to these organs.

E. Effects of ONOOG on Cellular DNA

$ONOO^-$ can interact with isolated or cellular DNA to cause base modifications. Inoue and Kawanishi (36) demonstrate that incubation of calf thymus DNA with $ONOO^-$, generated by the simultaneous production of NO and O_2^- via SIN-1 administration, results in oxidative DNA base modifications as indexed by 8-hydroxydeoxyguanosine (8-OH-dG) formation. Salgo *et al.* (39) show that $ONOO^-$ causes single-strand breaks in supercoiled plasmid DNA as determined by agarose gel electrophoresis. Furthermore, $ONOO^-$-mediated DNA strand breakage and base modifications occur in several different cell types including

cytes. These mitogens include epidermal growth factor (EGF) (59) and hepatocyte growth factor (HGF) (60).

Hepatocyte proliferation is also produced in vivo without loss of liver tissue. For example, lead (Pb^{2+}) induces hepatocyte proliferation without causing hepatocellular necrosis (61). The mechanism by which Pb^{2+} induces liver cell proliferation and initiates hepatocyte DNA synthesis is not known, but does not appear to involve induction of growth factors (62).

Additionally, hepatocyte culture conditions can induce DNA synthesis. For example, culturing hepatocytes at low cell densities (63,64) or on different substrata (59,65) causes hepatocytes to lose their liver-specific functions and begin proliferating. This change from differentiation to growth is evident by induction of DNA synthesis and inhibition of the synthesis of liver-specific gene products such as albumin, α-antitrypsin, and α-inhibitor III (60,63). Several of the dedifferentiating effects of cell density and extracellular matrix on hepatocellular physiology are attributed to cell-cell and cell-matrix interactions (59,60,65). However, the precise molecular mechanisms involved in induction of DNA synthesis by culturing conditions are not known.

J. Effects of DNA Synthesis on TNFα-Induced Toxicity

Active DNA replication affects cytokine-induced toxicity. For example, the sensitivity of proliferating ME-180 human cervical carcinoma cells to TNFα toxicity proportionately decreases with decreasing DNA synthesis (49). Likewise, TNFα-induced toxicity to Balb/c 3T3 cells depends on whether these cells are quiescent or dividing (66). The involvement of DNA synthesis on TNFα toxicity is not clearly understood, but may involve the physical state of DNA during replication. For example, actively replicating DNA is in an unwound state and therefore may be more susceptible to damage than compacted, nonreplicating DNA.

II. EXPERIMENTAL FINDINGS

This chapter explores the role of DNA synthesis and cytokine-induced oxidants in hepatocyte toxicity by TNFα and IFNγ. A cultured mouse hepatocyte model is used to clarify the precise mechanisms, as well as the mode, of cell death.

A. Induction of DNA Synthesis

Cultured mouse hepatocytes are stimulated to undergo DNA synthesis by either cell density or chemical treatment. Culturing at low cell density without chemical treatment or culturing at high cell density in the presence of either Pb^{2+}, EGF, or HGF induces DNA synthesis in hepatocytes (Fig. 1) as determined by significant increases in both ^{3}H-thymidine incorporation and DNA content.

B. Cytokine-Induced Hepatocyte Necrosis

Hepatocytes undergoing DNA synthesis differ in their sensitivity to cytokine-induced necrosis. Low cell density culture or Pb^{2+} administration sensitizes hepatocytes to the toxic actions of cytokines as determined by significant increases in lactate dehydrogenase (LDH) release into the media (Fig. 2). In contrast, EGF or HGF does not sensitize hepatocytes to cytokine-induced necrosis (Fig. 2). Since hepatocyte sensitivity to cytokine-induced necrosis depends on the manner in which DNA synthesis is induced, we examined differences occurring

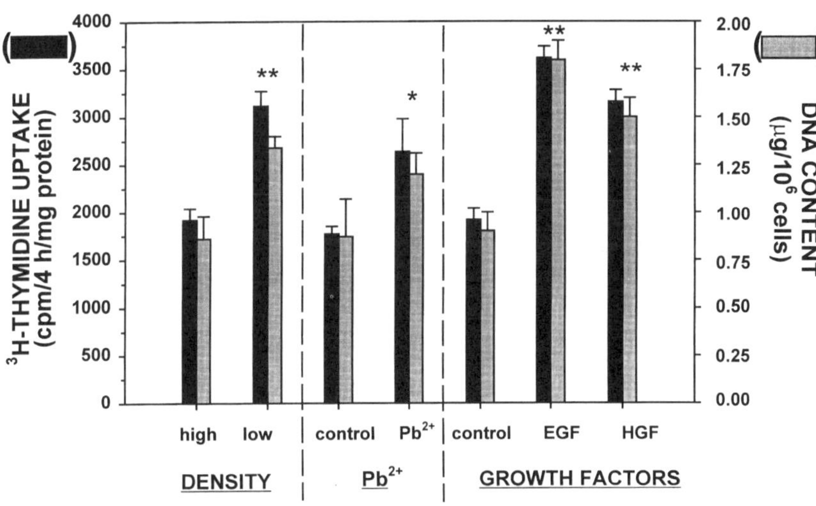

Figure 1 Conditions that induce DNA synthesis in cultured mouse hepatocytes. Hepatocytes were isolated from male Swiss-Webster mice by collagenase digestion and differential centrifugation. Hepatocytes were cultured in monolayer in RPMI 1640 media in six-well tissue culture plates at a density of either 0.6×10^6 cells/well for high density, Pb^{2+}, or growth factor studies or 0.2×10^6 cells/well for low-density studies. Where indicated, hepatocytes were incubated with lead acetate (Pb^{2+}; 50 µM), epidermal growth factor (EGF; 15 ng/mL), or hepatocyte growth factor (HGF; 15 ng/mL) and maintained in culture for 24 h. DNA synthesis was determined by quantitating ^{3}H-thymidine (2.0 mCi/mL) incorporation into hepatocytes. DNA content was determined by quantitating uptake of Hoescht dye (H33258) by hepatocytes. $*P < .05$. Pb^{2+}-induced DNA synthesis and content vs. control. $**P < .01$. Low density- or growth factor-induced DNA synthesis and content vs. control.

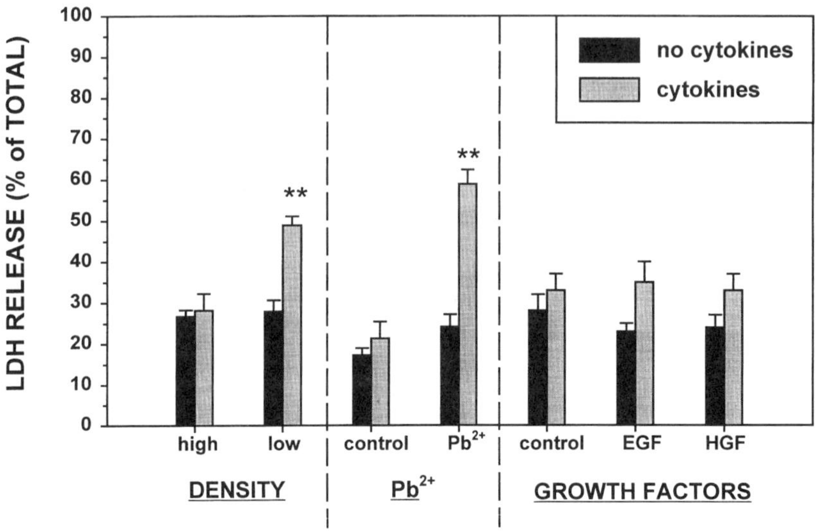

Figure 2 DNA synthesis by low-density culture or Pb^{2+} sensitizes hepatocytes to **cytokine-induced necrosis.** Hepatocytes were isolated and maintained in culture as described in Figure 1. Where indicated, IFNg (100 U/mL), lead acetate (Pb^{2+}, 50 μM), EGF (15 ng/mL), or HGF (15 ng/mL) was added 4 h prior to TNFα (0.1 μg/mL) treatment. Hepatocyte viability was determined by measuring LDH leakage into media at 24 h after TNFα addition. Total LDH in cells and media was determined by sonicating the culture well. $**P < .01$. Cytokines (IFNγ + TNFα) + Pb^{2+} or cytokines + low density vs. no cytokines.

among the various treatments used to induce DNA synthesis. Specifically, since cytokines induce both NO and O_2^- expression while growth factors inhibit cytokine-induced NO production, we compared levels of these cytokine-generated oxidants in the absence and presence of hepatocyte DNA synthesis.

C. Cytokine-Induced NO and O_2G Production

Cytokine-induced expression of NO in hepatocytes, determined by media nitrite levels, is altered by DNA synthesis (Fig. 3). IFNγ/TNFα-induced NO expression decreases in hepatocytes undergoing DNA synthesis initiated by Pb^{2+} or growth factor administration, but not by low-density culture. In fact, NO is completely inhibited by EGF or HGF (Fig. 3). This inhibition corresponds to complete inhibition of NOS2 protein (not shown) by these growth factors. Although cytokine-stimulated NO (Fig. 3) and NOS2 protein levels (not shown) are reduced by approximately 50% following Pb^{2+} administration, NO remains significantly elevated above levels observed in the absence of cytokines.

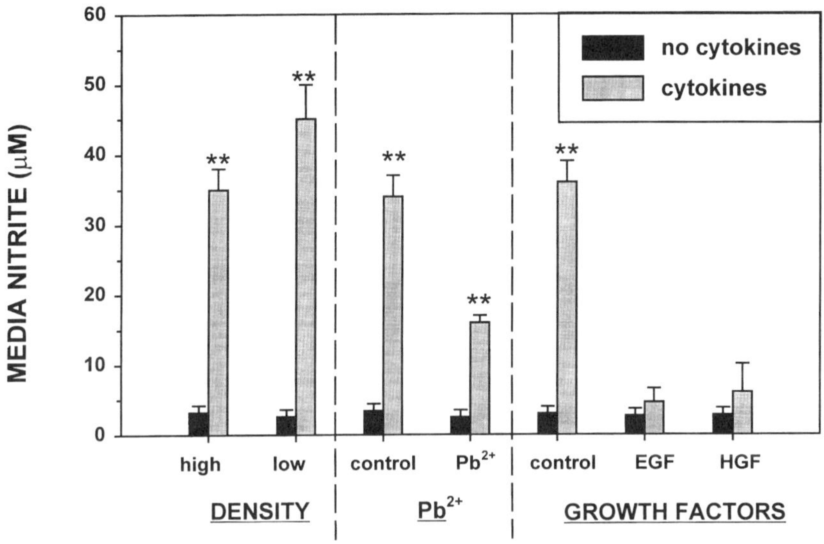

Figure 3 Nitric oxide formation by cytokine treatment in the absence or presence of DNA synthesis. Hepatocytes were isolated, cultured, and treated with cytokines and/or hepatomitogens as described in Figures 1 and 2. Nitric oxide levels were measured at 24 h by assessing media nitrite concentration using Greiss reagent. $^{**}P < .01$. Cytokines (IFNγ + TNFα) vs. no cytokines.

Cytokine-induced O_2^- production is the same in the absence or presence of hepatocyte DNA synthesis (Fig. 4). Intracellular production of this cytokine-generated oxidant was determined by lucigenin-derived chemiluminescence which specifically measures O_2^- produced via mitochondria (67).

D. Proposed Mechanism of Oxidant-Mediated DNA Damage and Apoptosis

Proposed interactions among cytokines, cytokine-induced oxidants, and DNA synthesis during induction of hepatocellular death are illustrated in Figure 5. Cytokine administration induces NO, via activation of NOS2, and O_2^-. The simultaneous production of both NO and O_2 leads to $ONOO^-$ formation which, in the presence of DNA synthesis by low-density culture or Pb^{2+} administration, causes oxidative DNA damage leading ultimately to hepatocyte death by necrosis. In the presence of DNA synthesis by growth factors, cytokine induction of NOS2 is blocked and thus, NO is not formed. In this situation, cytokines still cause O_2^- formation which can lead to hepatocyte death by apoptosis. These proposed interactions and conclusions were tested by examining the process by

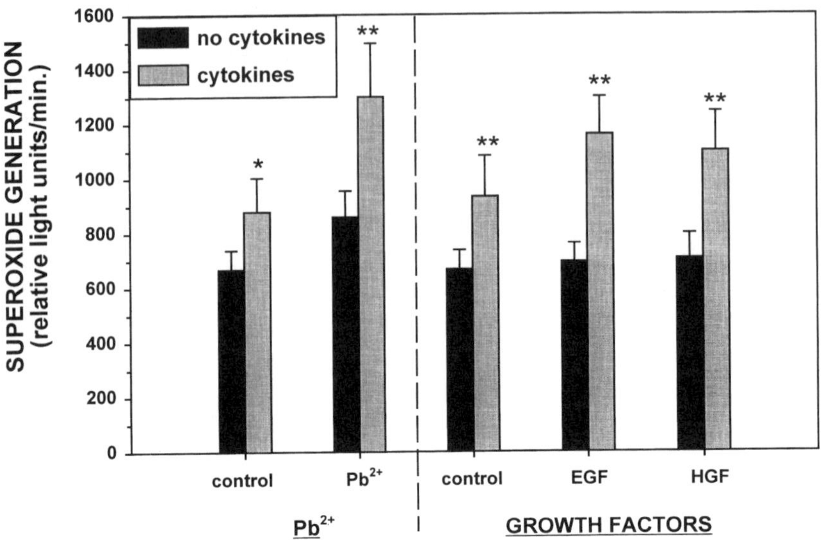

Figure 4 Superoxide generation by cytokine treatment in the absence or presence of DNA synthesis. Hepatocytes were isolated, cultured, and treated with cytokines and/or hepatomitogens as described in Figures 1 and 2. Intracellular superoxide generation was determined at 12 h by measuring lucigenin-derived chemiluminescence. $*P < .05$ or $**P < .01$. Cytokines (IFNγ + TNFα) vs. no cytokines.

which cytokines and the various DNA-synthesizing treatments initiated DNA damage. More specifically, DNA fragmentation characteristic of apoptosis was determined by field inversion gel electrophoresis (FIGE) (68) whereas oxidative DNA damage characteristic of necrosis was assessed by determining 8-OH-deoxyguanosine formation via HPLC (69,70).

E. Cytokine-Induced Apoptotic or Necrotic DNA Fragmentation

The type of DNA damage initiated by cytokines depends on the treatment used to induce DNA synthesis in hepatocytes. IFNγ/TNFα initiate some apoptotic DNA fragmentation in nonproliferating hepatocytes as indicated by distinct processing of the DNA which includes 50-kbp fragments (Fig. 6). However, cytokine administration in combination with Pb^{2+} treatment causes randomly sized DNA strand breakage as indicated by DNA smearing throughout the gel. This random DNA fragmenting is characteristic of oxidative damage rather than apoptosis. In separate studies in which EGF is added to stimulate DNA synthesis, cytokine ad-

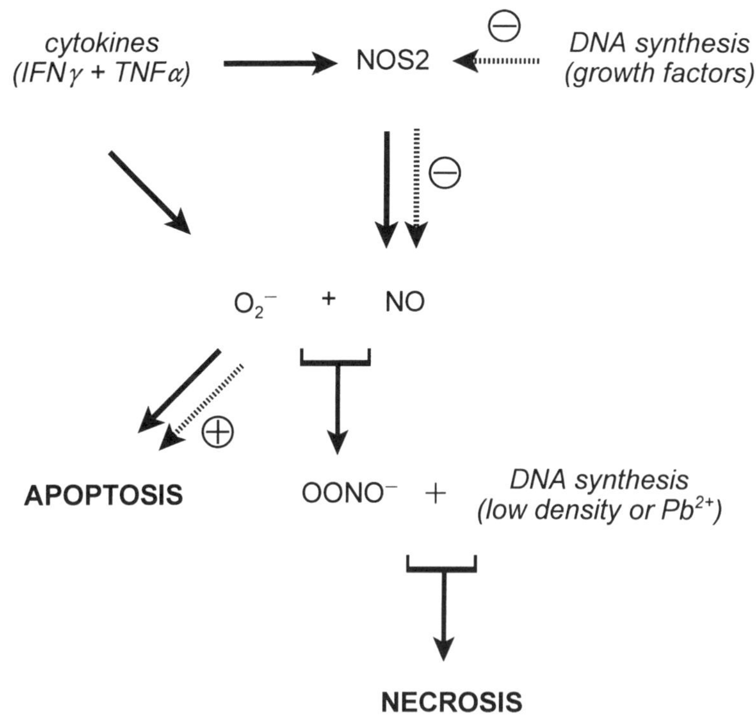

Figure 5 Proposed mechanism of oxidant-mediated DNA damage and apoptosis. Proposed interactions among the cytokine-induced oxidants, superoxide (O_2^-), nitric oxide (NO), and peroxynitrite ($ONOO^-$) that mediate either necrosis or apoptosis in hepatocytes are illustrated. TNFa initiates O_2^- and NO production, effects that are potentiated by IFNg. Simultaneous generation of O_2^- and NO leads to $ONOO^-$ formation. During mitogen-stimulated DNA synthesis, $ONOO^-$ causes oxidative DNA damage, leading to hepatocyte necrosis. Growth factors inhibit NOS2 such that NO is not produced and thus $ONOO^-$ is not formed. This pathway results in induction of apoptosis mediated via O_2^- or perhaps other ROI. In the absence of DNA synthesis, ONOOG formation protects against induction of apoptosis via removal of O_2^-.

ministration induces DNA fragmentation into 50-kbp fragments indicative of apoptosis. In fact, administration of interleukin-converting enzyme-1, an inhibitor of apoptotic DNA processing, to cytokine- and EGF-treated hepatocytes (EGF/cyt + inhib) eliminates the appearance of the 50-kbp band (Fig. 6). In these treatments, DNA is randomly fragmented. DNA from control cells exhibits a small amount of the 50-kbp fragments, demonstrating that some apoptosis may occur during isolation and culture of hepatocytes.

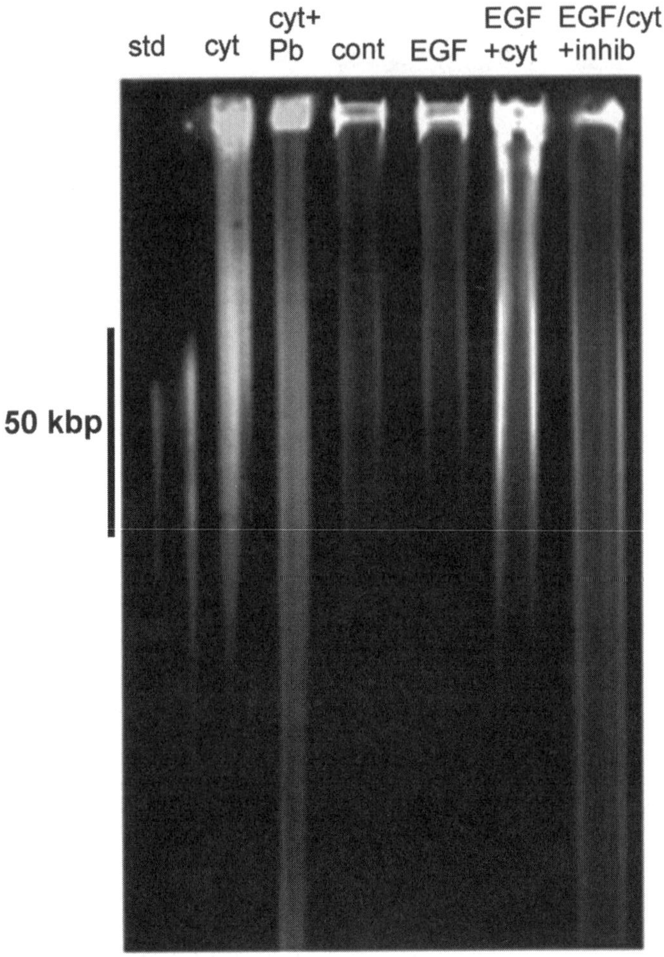

Figure 6 FIGE analysis of necrotic or apoptotic DNA fragmentation in hepatocytes. Hepatocytes were isolated, cultured, and treated with cytokines and/or hepatomitogens as described in Figures 1 and 2. At 12 h after TNFα addition, hepatocyte DNA fragmentation was determined using field inversion gel electrophoresis (FIGE) . A 50-kbp band identified by running a 50-kbp standard (std) indicates apoptotic DNA fragmentation whereas a smear throughout the gel indicates necrotic DNA fragmentation. A 50-kbp band is evident after treatment with cytokines (cyt) or EGF + cytokines (EGF + cyt). DNA smearing throughout the gel is evident after treatment with cytokines + Pb^{2+} (cyt + Pb) or treatment with the apoptosis inhibitor (+inhib) interleukin converting enzyme-1 in the EGF/cyt incubation.

Administration of oxidants confirm that $ONOO^-$ is the oxidant responsible for causing oxidative DNA damage in the presence of DNA synthesis (not shown). For example, administration of SIN-1 to generate $ONOO^-$ in Pb^{2+}-treated hepatocytes leads to DNA smearing during FIGE analysis. Furthermore, administration of DETA/NO to generate NO in hepatocytes treated with cytokines and EGF, where normally only O_2^- is produced, alters DNA fragmentation to resemble oxidative DNA smearing rather than apoptotic DNA fragmentation into distinct bands.

F. Cytokine-Induced Oxidative DNA Damage

Cytokine-induced necrosis in hepatocytes is due to oxidative DNA damage, indicated by 8-OH-deoxyguanosine formation, by $ONOO^-$. For example, IFNγ/TNFα addition to Pb^{2+}-treated hepatocytes causes significant increases in the level of 8-OH-deoxyguanosine expressed relative to deoxyguanosine (Fig. 7). These

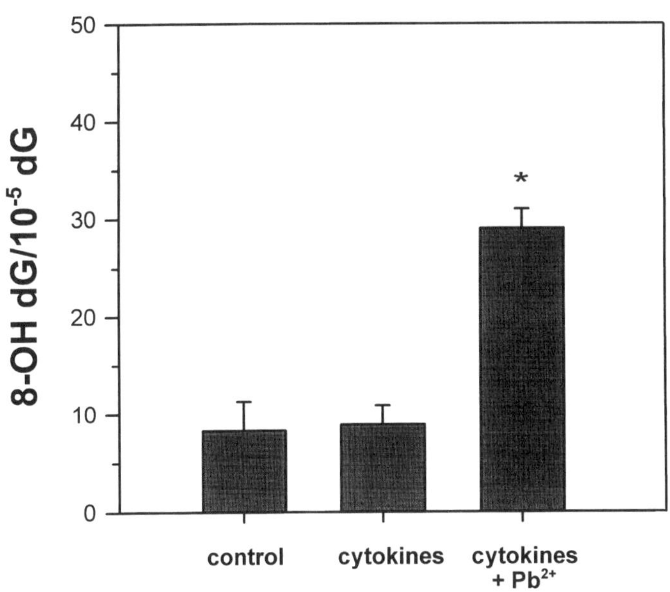

Figure 7 Oxidative DNA damage in hepatocytes treated with cytokines and Pb^{2+}. Hepatocytes were isolated, cultured, and treated with cytokines (IFNγ and TNFα) and Pb^{2+} as described in Figures 1 and 2. At 12 h after TNFα addition, oxidative DNA damage in hepatocytes was determined by measuring 8-hydroxy-deoxyguanosine (8-OH-dG) formation relative to deoxyguanosine (dG) via HPLC. $*P < .05$. Cytokines (IFNγ + TNFα) + Pb^{2+} vs. cytokines or control.

data correlate with FIGE analysis of DNA isolated from cytokine- and Pb^{2+}-treated hepatocytes, and therefore reinforce the finding that the mode of cell death is via oxidative DNA damage leading ultimately to necrosis. Administration of either oxidants or antioxidants further establish involvement of $ONOO^-$ in causing 8-OH-deoxyguanosine formation (not shown) and, thus, oxidative DNA damage. For example, administration of SIN-1 to generate $ONOO^-$ in Pb^{2+}-treated hepatocytes in the absence of cytokines causes even greater production of this DNA base modification.

III. DISCUSSION

This chapter details the role of cytokine (IFNγ/TNFα)-induced oxidants in both oxidative hepatocyte necrosis and in cytokine-induced apoptosis in a cultured mouse hepatocyte model. Specifically, the role of NO (generated from NOS2 induction), O_2^- (generated via mitochondria), and $ONOO^-$ (formed from the simultaneous generation of both NO and O_2^-) are examined. The studies show that the combination of cytokines (IFNγ/TNFα) causes either necrosis due to $ONOO^-$-mediated oxidative DNA damage or apoptosis via a O_2^-/ROI dependent mechanism. The mode of cell death depends on the specific culturing conditions of the cells, namely the addition of cytokines and the treatment used to induce DNA synthesis. Table 1 summarizes the culturing conditions under which apoptosis or oxidative DNA damage leading to hepatocyte death are observed. Included in this data summary are the actions of various inhibitors or scavengers of cytokine-derived oxidants as well as specific oxidant donors which are instrumental in clarifying the individual function of a particular oxidant.

Hepatocyte necrosis initiated by IFNγ/TNFα requires both induction of DNA synthesis and intracellular formation of $ONOO^-$. Under these conditions, $ONOO^-$ causes extensive DNA strand breaks that ultimately lead to hepatocyte necrosis. Induction of DNA synthesis appears to be the primary sensitizer of mouse hepatocytes to the DNA damaging and toxic effects of cytokine-induced $ONOO^-$, as nonproliferating hepatocytes are resistant to $ONOO^-$-mediated toxicity. Interestingly, induction of DNA synthesis could also affect $ONOO^-$ formation and thus, hepatocyte sensitivity. Low-density culture or Pb^{2+} treatment either does not affect or does not completely block $ONOO^-$ formation, and therefore these cells are sensitive to cytokines and die by necrosis. In contrast, growth factors cause DNA synthesis, but block cytokine induction of NOS2 (71), such that hepatocytes are resistant to cytokine-induced necrosis through the prevention of $ONOO^-$ formation. Additional experiments substantiate the role for $ONOO^-$ in cytokine-induced hepatocyte toxicity. Direct addition of NO to growth factor-treated cells results in extensive DNA strand breakage and cytokine sensitivity, presumably due to the intracellular formation of $ONOO^-$ (Table 1). Like-

Table 1 Summary of Treatments and Oxidants Mediating Apoptosis or Necrosis in Hepatocytes

Treatment	O_2^- Generation	NO Generation	$ONOO^-$ Formation	DNA Synthesis	Mode of Death
Cytokines	yes	yes	yes	no	apoptosis
Cytokines + O_2^- inhibitor	no	yes	no	no	none
Cytokines + PB^{2+}	yes	yes	yes	yes	necrosis[a]
Cytokines + PB^{2+} + NO inhibitor	yes	no	no	yes	apoptosis
Cytokines + EGF or HGF	yes	no	no	yes	apoptosis
Cytokines + EGF or HGF + NO	yes	yes	yes	yes	necrosis
Cytokines + EGF or HGF + O_2^- inhibitor	no	no	no	yes	none
Cytokines + low density	yes	yes	yes	yes	necrosis

Treatments and the oxidants formed which initiate hepatocyte death via apoptosis or necrosis are depicted. Mouse hepatocytes were cultured in 6-well culture plates at a density of 0.2×10^6 cells/well for low-density treatment or at 0.6×10^6 cells/well for all other treatments. Cytokines represent addition of TNFα and IFNγ. Superoxide (O_2^-) inhibitor is addition of Mn(III)tetrakis(1-methyl-4-pyrdyl)porphyrin) (MnTMPyP). Nitric oxide (NO) inhibitor is addition of either N-nitro-L-arginine methyl ester (L-NAME) or 2-phenyl-4,4,5,5-tetramethylimidazoline-1-oxy-3-oxide (PTIO). NO is addition of (Z)-1-[2-(aminoethyl)-N-(2-ammonioethyl)amino]diazen-1-ium-1,2-diolate (DETA/NO).

[a]The primary form of death in these cells is necrosis caused by oxidative DNA damage, although some apoptosis is observed.

wise, addition of $ONOO^-$ to hepatocytes stimulated to undergo DNA synthesis by any of the described conditions causes DNA damage and cytotoxicity (not shown).

DNA damage occurs prior to hepatocyte toxicity and therefore is a cause rather than a consequence of hepatocytes undergoing necrosis. The mechanism by which DNA synthesis confers sensitivity to cytokine-induced or SIN-1 generated $ONOO^-$ is not known, but may involve physical properties of replicating DNA or signaling pathways of hepatomitogens which render the DNA more susceptible to oxidant damage. It is unlikely that the observed $ONOO^-$-mediated DNA strand breakage directly causes hepatocyte toxicity through the disruption of genomic DNA (40,72). Rather, $ONOO^-$-mediated DNA strand breaks may result in subsequent activation of the nuclear enzyme poly ADP ribosyl synthetase (PARS) (73). The activation of PARS results in cleavage of NAD^+ into ADP-ribose and nicotinamide. PARS then covalently attaches multiple ADP-ribose groups (poly-ADP) to various proteins, including nuclear proteins, histones, and PARS itself (73). The physiological function of PARS is still unclear, but there are indications that PARS functions as a DNA repair enzyme (73), although cells from PARS knockout mice appear to have normal DNA repair characteristics (74). PARS may also play an important role in the regulation of gene expression and cell differentiation (75). Whatever the exact function, pronounced activation of PARS (due to extensive DNA strand breaks) can rapidly deplete the intracellular stores of NAD^+, slowing the rate of glycolysis, electron transport, and, therefore, ATP formation. Such a circumstance (known as the "PARS suicide hypothesis") leads to an energy-consuming, futile repair cycle that eventually leads to cellular energy depletion and cell death (42,72). Therefore, hepatocytes sensitized to cytokine toxicity via DNA synthesis in the mouse model may die via energy depletion by this pathway. Accordingly, ATP depletion in these hepatocytes occurs after DNA strand breakage (76).

Cytokines also induce apoptosis in cultured mouse hepatocytes. Cytokine-induced apoptosis occurs to some extent in nonproliferating hepatocytes and to a greater extent in growth factor-treated hepatocytes. Under these conditions, induction of apoptosis correlates with increased O_2^- generation and a state of cellular oxidative stress prior to the onset of apoptosis, suggesting an O_2^-/ROI-dependent mechanism. This hypothesis is supported by experiments in which O_2^- is scavenged using a superoxide dismutase mimetic compound, which completely inhibits cytokine apoptosis (Table 1). Similarly, TGFβ-induced apoptosis in fetal hepatocytes is preceded by ROI production and oxidative stress (77).

It is interesting that cytokine-induced NO production prevents the occurrence of apoptosis in mouse hepatocytes. That is, NO addition (via DETA/NO) completely inhibits cytokine-induced O_2^- generation and apoptosis whereas NOS2 inhibitors or NO scavengers potentiate cytokine-induced apoptosis (Table 1). These actions of NO may be due to scavenging of O_2^- to form $ONOO^-$ such that O_2^-/ROI are prevented from causing apoptosis. This protective role for NO against apoptosis contradicts other studies which associate NO production with apoptosis induction in many different cell types (78–80).

Similar to NO, the specific role of O_2^- in apoptosis is controversial. Several studies demonstrate that O_2^-/ROI are not involved in apoptosis. For example, apoptosis is induced in several different cell lines under extreme hypoxic conditions, such that O_2^-/ROI production, especially by mitochondria, is inhibited (81). Likewise, TNFα induces apoptosis in L929 cells independent of mitochondrial O_2^-/ROI production (51) and does not elicit O_2^- production in the mitochondrial matrices of mammalian cells (82). In contrast, studies also demonstrate that ROI are critical to induction of apoptosis. As previously mentioned, ROI production mediates TGFβ-induced apoptosis in fetal hepatocytes (77). Additionally, mitochondrial ROI generation constitutes an early feature of apoptosis and is an indicator of apoptosis in lymphocytes (83). The present findings clearly demonstrate that O_2^-/ROI production is critical to TNFα/IFNγ-induced apoptosis in cultured mouse hepatocytes.

O_2^- generation in mouse hepatocytes is presumably due to the actions of cytokines on the mitochondria, especially since the assay/compound (i.e., lucigenin) used to measure O_2^- production has been extensively characterized to accurately monitor mitochondrial O_2^- production (67). Furthermore, treatment with the mitochondrial complex I inhibitor, rotenone, completely inhibits cytokine-induced O_2^- generation in mouse hepatocytes (not shown). Additional studies support the hypothesis that O_2^- generation prior to the onset of apoptosis is derived from mitochondria. For example, TNFα increases mitochondrial O_2^- production in rat hepatocytes (84). Furthermore, mitochondria and mitochondrially derived factors/ROI appear to have central roles in apoptosis in other studies (83,85).

The exact pathway of cytokine induction of O_2^- via the mitochodria is not completely defined, but may involve specific apoptotic machinery (the ICE-like proteases/caspases) that functions both upstream and downstream of the mitochondria to induce apoptosis (55). In fact, activation of specific proteases/caspases is essential for induction of apoptosis (86). More specifically, cytokine-induced apoptosis involves activation of ICE and ICE-like proteases (87). For example, ICE protease inhibitors block murine liver injury and apoptosis by TNFα both in vivo and in vitro (88). Interestingly, in the mouse hepatocyte model, inhibiting these proteases/caspases with a nonspecific cleavage site-directed ICE inhibitor results in increased cytokine-induced necrosis. This action appears to result from switching from apoptosis to oxidative DNA damage and cell death, although no alterations in O_2^-/ROI generation are observed.

Interestingly, apoptosis occurs under conditions of mitogen-stimulated DNA synthesis. This characteristic is contradictory to observations in other cell types in which the withdrawal of growth factors, and thus decreased proliferation, leads to induction of apoptosis (46,89). These differences may be related to the ability of growth factors to downregulation NO production and thus prevent $ONOO^-$ formation, in order to protect hepatocytes from exacerbation of the initial liver injury during hepatocyte repair and regeneration. For example, induction of apoptosis by growth factors and cytokines may be necessary for removal of genetically damaged hepatocytes during liver regeneration and restoration of liver mass following a chemical insult.

The model presented in this chapter demonstrates that modifying culturing conditions in mouse hepatocytes enables investigation of the mechanisms of either cytokine-induced oxidative DNA damage, resulting in hepatocyte necrosis, or cytokine-induced apoptosis. Since both modes of cell death are evident in clinical situations (90), knowledge obtained via further dissection of the cytokine-induced signaling pathway(s) involved in either necrosis or apoptosis may enable clinically applicable interventions to be employed.

REFERENCES

1. Jaeschke H, Farhood A. Neutrophil and Kupffer cell-induced oxidant stress and ischemia-reperfusion injury in rat liver. Am J Physiol 1991; 260:452–479.
2. Laskin DL, Heck DE, Gardner CR, Feder LS, Laskin JD. Distinct patterns of nitric oxide production in hepatic macrophages and endothelial cells following acute exposure of rats to endotoxin. J Leukoc Biol 1994; 56:751–758.
3. Tanner A, Keyhani A, Reiner R, Goldstock G, Wright R. Proteolytic enzymes released by liver macrophages may promote hepatic injury in a rat model of hepatic damage. Gastroenterology 1981; 80:647–654.
4. Krawisz J, Sharon P, Stenson W. Quantitative assay for acute intestinal inflammation based on myeloperoxidase activity. Gastroenterology 1984; 87:1344–1350.
5. Dieter P, Schulze-Specking A, Decker K. Differential inhibition of prostaglandin and superoxide production by dexamethasone in primary cultures of rat Kupffer cells. Eur J Biochem 1986; 159:451–457.
6. Chensue S, Terbuh P, Remick DG, Scales W, Kunkel SL. In vivo biologic and immunohistochemical analysis of interleukin-1 alpha, beta and tumor necrosis factor during experimental endotoxemia. Am J Pathol 1991; 138:395–402.
7. Laskin DL. Role of macrophages and endothelial cells in hepatotoxicity. In: Hepatocyte and Kupffer Cell Interactions. Boca Raton, FL: CRC Press, 1992, pp. 147–168.
8. Beutler B, Cerami A. The biology of cachectin/TNF—a primary mediator of the host response. Annu Rev Immunol 1989; 7:625–655.
9. Akira S, Hirano T, Taga T, Kishimoto T. Biology of multifunctional cytokines: IL-1 and related molecules (IL-1 and TNFα). FASEB J 1990; 4:2860–2867.
10. Jaattela M. Biology of disease: biological activities and mechanisms of action of tumor necrosis factor-α/cachectin. Lab Invest 1991; 64:724–742.
11. Billiau AB, Vanderkerckhove F. Cytokines and their interactions with other inflammatory mediators in the pathogenesis of sepsis and septic shock. Eur J Clin Invest 1991; 21:559–573.
12. Colletti LM, Remick DG, Burtch GD, Kunkel SL, Strieter RM, Campbell DAJ. Role of tumor necrosis factor-α in the pathophysiologic alterations after hepatic ischemia/reperfusion injury in the rat. J Clin Invest 1990; 85:1936–1943.
13. Larrick JW, Wright SC. Cytotoxic mechanism of tumor necrosis factor-a. FASEB J 1990; 4:3215–3223.
14. Adamson GA, Billings RE. Tumor necrosis factor: receptor binding and expression of receptors in cultured mouse hepatocytes. J Pharmacol Exp Ther 1994; 269:367–373.
15. Shu H-B, Takeuchi M, Goeddel DV. The tumor necrosis factor receptor 2 signal

transducers TRAF2 and c-IAP1 are components of the tumor necrosis factor receptor 1 signaling complex. Proc Natl Acad Sci USA 1996; 93:13973–13978.

16. Adamson GA, Billings RE. Tumor necrosis factor induced oxidative stress in isolated mouse hepatocytes. Arch Biochem Biophys 1992; 294:223–229.
17. Zimmerman RJ, Marafino BJ, Chan A, Landre P, Winkelhake JL. The role of oxidant injury in tumor cell sensitivity to recombinant human TNF in vivo. J Immunol 1989; 142:1405–1409.
18. Leist M, Ganter F, Bohlinger I, Germann PG, Tiegs G, Wendel A. Murine hepatocyte apoptosis induced in vitro and in vivo by TNF-a requires transcriptional arrest. J Immunol 1994; 153:1778–1788.
19. Punjabi CJ, Laskin DL, Heck DE, Laskin JD. Production of nitric oxide by murine bone marrow cells. Inverse correlation with cellular proliferation. J Immunol 1992; 149:2179–2184.
20. Ignarro L, Buga G, Wood D, Byrns R, Chaudhuri G. Endothelium-derived relaxing factor produced and released from artery and vein is nitric oxide. Proc Natl Acad Sci USA 1987; 84:9265–9269.
21. McCall T, Boughton-Smith N, Palmer R, Whittle B, Moncada S. Synthesis of nitric oxide from L-arginine by neutrophils: release and interaction with superoxide anion. Biochem J 1989; 261:293–296.
22. Adamson GA, Billings RE. Cytokine toxicity and induction of NO synthase activity in cultured mouse hepatocytes. Toxicol Appl Pharmacol 1993; 119:100–107.
23. Duval DL, Sieg DJ, Billings RE. Regulation of hepatic nitric oxide synthase by reactive oxygen intermediates and glutathione. Arch Biochem Biophys 1995; 316:699–706.
24. Ignarro L. Biological actions and properties of endothelium-derived nitric oxide formed and released from artery and vein. Circ Res 1988; 65:1–21.
25. Lowenstein C, Kinerman J, Snyder S. Nitric oxide: a physiological messenger. Ann Intern Med 1994; 120:227–237.
26. Ialenti A, Ianaro A, Moncada S, DiRosa M. Modulation of acute inflammation by endogenous nitric oxide. Eur J Pharm 1992; 211:177–182.
27. Moncada S, Higgs A. The L-arginine-nitric oxide pathway. N Engl J Med 1993; 329:2002–2012.
28. Hortelano S, Dewez B, Genaro A, Diaz-Guerra M, Bosca L. Nitric oxide is released in regenerating liver after partial hepatectomy. Hepatology 1995; 21:776–786.
29. Pryor W, Squadrito G. The chemistry of peroxynitrite: a product from the reaction of nitric oxide with superoxide. Am J Physiol 1995; 268:L699–L722.
30. Wink D, Hanbauer I, Krishna M, DeGraff W, Gamson J, Mitchell J. Nitric oxide protects against cellular damage and cytotoxicity from reactive oxygen species. Proc Natl Acad Sci USA 1993; 90:9813–9817.
31. Billiar T, Curran R, Harbrecht B, Stuehr D, Demetris A, Simmons R. Modulation of nitrogen oxide synthesis in vivo: N^G-monomethyl-L-arginine inhibits endotoxin-induced nitrite/nitrate biosynthesis while promoting hepatic damage. J Leukoc Biol 1990; 48:565–569.
32. Beckman J, Beckman T, Chen J, Marshall P, Freeman B. Apparent hydroxyl radi-

cal production by peroxynitrite: implication for endothelial injury from nitric oxide and superoxide. Proc Natl Acad Sci USA 1990; 87:1620–1624.
33. Rubbo H, Radi R, Trujillo M, Telleri R, Kalyanaraman B, Barnes S, Kirk M, Freeman B. Nitric oxide regulation of superoxide and peroxynitrite-dependent lipid peroxidation. J Biol Chem 1994; 269:26066–26075.
34. Radi R, Beckman J, Bush K, Freeman B. Peroxynitrite oxidation of sulfhydryls: the cytotoxic potential of superoxide and nitric oxide. J Biol Chem 1991; 266:4244–4250.
35. Liu RH, Hotchkiss JH. Potential genotoxicity of chronically elevated nitric oxide: a review. Mutat Res 1995; 339:73–89.
36. Inoue S, Kawanishi S. Oxidative DNA damage induced by simultaneous generation of nitric oxide and superoxide. FEBS Lett 1995; 371:86–88.
37. Ischiropoulos H, Al-Mehdi A, Fisher A. Reactive species in ischemic rat lung injury: contribution of peroxynitrite. Am J Physiol 1995; 269:L158–L164.
38. Ma T, Ischiropoulos H, Brass C. Endotoxin-stimulated nitric oxide production increases injury and reduces rat liver chemiluminescence during reperfusion. Gastroenterology 1995; 108:463–469.
39. Salgo M, Stone K, Squadrito G, Battista J, Pryor W. Peroxynitrite causes DNA nicks in plasmid pBR322. Biochem Biophys Res Commun 1995; 210:1025–1030.
40. Szabo C, Zingarelli B, O'Connor M, Salzman AL. DNA strand breakage, activation of poly(ADP-ribose) synthetase, and cellular energy depletion are involved in cytotoxicity in macrophages and smooth muscle cells exposed to peroxynitrite. Proc Natl Acad Sci USA 1996; 93:1753–1758.
41. Delaney CA, Tyrberg B, Bouwens L, Vaghef H, Hellman B, Eizirik DL. Sensitivity of human pancreatic islets to peroxynitrite-induced cell dysfunction and death. FEBS Lett 1996; 394:300–306.
42. Szabo C. DNA strand breakage and activation of poly-ADP ribosyltransferase: a cytotoxic pathway triggered by peroxynitrite. Free Radic Biol Med 1996; 21:855–869.
43. Sheehan KFC, Schreiber RB. The synergy and antagonism of interferon-gamma and TNF. In: Tumor Necrosis Factors. New York: Raven Press, 1992, pp. 145–178.
44. Talmadge JE, Bowersox O, Tribble H, Lee SH, Shepard M, Liggett D. Toxicity of tumor necrosis factor is synergistic with gamma-interferon and can be reduced with cyclooxygenase inhibitors. Am J Pathol 1987; 128:410–425.
45. Hecker M, Preiss C, Klemm P, Busse R. Inhibition by anitoxidants of nitric oxide synthase expression in murine macrophages: role of nuclear factor kB and interferon regulatory factor 1. Br J Pharmacol 1996; 118:2178–2184.
46. Corcoran GB, Fix L, Jones DP, Moslen MT, Nicotera P, Oberhammer FA, Buttyan R. Apoptosis: molecular control point in toxicity. Toxicol Appl Pharmacol 1994; 128:169–181.
47. Kroemer G, Petit P, Zamzami N, Vayssiere J, Mignotte B. The biochemistry of programmed cell death. FASEB J 1995; 9:1277–1287.
48. Wright SC, Kumar P, Tam AW, Shen N, Varma M, Larrick JW. Apoptosis and DNA fragmentation precede TNF-induced cytolysis in U937 cells. J Cell Biochem 1992; 48:344–355.

49. Manchester KM, Heston WD, Donner DB. Tumor necrosis factor-induced cytotoxicity is accompanied by intracellular mitogenic signals in ME-180 human cervical carcinoma cells. Biochem J 1993; 290:185–190.
50. Bellomo G, Perotti M, Taddei F, Mirabelli F, Finardi G, Nicotera P, Orrenius S. Tumor necrosis factor a induces apoptosis in mammary adenocarcinoma cells by an increase in intranuclear free Ca^{2+} concentration and DNA fragmentation. Carcinogenesis 1992; 52:1342–1346.
51. O'Donnell VB, Spycher S, Azzi A. Involvement of oxidants and oxidant-generating enzyme(s) in tumor-necrosis-factor-α-mediated apoptosis: role of lipoxygenase pathway but not mitochondrial respiratory chain. Biochem J 1995; 310:133–141.
52. Leist M, Ganter F, Jilg S, Wendel A. Activation of the 55 kDa TNF receptor is necessary and sufficient for TNF-induced liver failure, hepatocyte apoptosis, and nitrite release. J Immunol 1995; 154:1307–1316.
53. Leist M, Gantner F, Kunstle G, Bohlinger I, Tiegs G, Bluethmann H, Wendel A. The 55-kD tumor necrosis factor receptor and CD independently signal murine hepatocyte apoptosis and subsequent liver failure. Mol Med 1996; 2:109–124.
54. Lui Z, Hus H, Goeddel DV, Karin M. Dissection of TNF receptor 1 effector functions: JNK activation is not linked to apoptosis while NF-κB activation prevents cell death. Cell 1996; 87:565–576.
55. Zamzami N, Susin SA, Marchetti P, Hirsch T, Gomez-Monterrey I, Castedo M, Kroemer G. Mitochondrial control of nuclear apoptosis. J Exp Med 1996; 183:1533–1544.
56. Bortner CD, Oldenburg NBE, Cidlowski JA. The role of DNA fragmentation in apoptosis. Trends Cell Biol 1995; 5:21–26.
57. Dive C, Gregory CD, Phipps DJ, Evans DL, Milner AE, Wyllie AH. Analysis and discrimination of necrosis and apoptosis (programmed cell death) by multiparameter flow cytometry. Biochim Biophys Acta 1992; 1133:275–285.
58. Michalopoulos GK. Liver regeneration: molecular mechanisms of growth control. FASEB J 1990; 4:176–187.
59. Tomomura A, Sawada N, Sattler GL, Kleinman HK, Pitot HC. The control of DNA synthesis in primary cultures of hepatocytes from adult and young rats: interactions of extracellular matrix components, epidermal growth factor, and the cell cycle. J Cell Physiol 1987; 130:221–227.
60. Takehara T, Matsumoto K, Nakamura T. Cell density-dependent regulation of albumin synthesis and DNA synthesis in rat hepatocytes by hepatocyte growth factor. J Biochem 1992; 112:330–334.
61. Columbano A, Ledda GM, Sirigu P, Perra T, Pani P. Liver cell proliferation induced by a single dose of lead nitrate. Am J Pathol 1983; 110:279–283.
62. Shinozuka H, Kubo Y, Katyl SL, Coni P, Ledda-Columbano GM, Columbano A, Nakamura T. Roles of growth factors and of tumor necrosis factor-α on liver cell proliferation induced in rats by lead nitrate. Lab Invest 1994; 71:35–41.
63. Ben-Ze'ev A, Robinson G, Bucher N, Farmer S. Cell-cell and cell-matrix interactions differentially regulate the expression of hepatic and cytoskeletal genes in primary cultures of rat hepatocytes. Proc Natl Acad Sci USA 1988; 85:2161–2165.
64. Enat R, Jefferson D, Ruiz-Opazo N, Gatmaitan Z, Leinwand L, Reid L. Hepatocyte

proliferation in vitro: its dependence on the use of serum-free hormonally defined medium and substrata of extracelluar matrix. Proc Natl Acad Sci USA 1984; 81:1411–1415.

65. Mooney D, Hansen L, Vacanti J, Langer R, Farmer S, Ingber D. Switching from differentiation to growth in hepatocytes: control by extracellular matrix. J Cell Physiol 1992; 151:497–505.
66. Palombella VJ, Vilcek J. Mitogenic and cytotoxic actions of tumor necrosis factor in BALB/c 3T3 cells. J Biol Chem 1989; 264:18128–18136.
67. Rembish SJ, Trush MA. Further evidence that lucigenin-derived chemiluminescence monitors mitochondrial superoxide generation in rat alveolar macrophages. Free Radic Biol Med 1994; 17:117–126.
68. Luo Y, Kessel D. Detection of early and late stage apoptosis with field inversion gel electrophoresis. BioTechniques 1996; 21:812–816.
69. Floyd R, Watson J, Wong P, Altmiller D, Rickard R. Hydroxyl free radical adduct of deoxyguanosine: sensitive detection and mechanisms of formation. Free Radic Res Commun 1986; 1:163–172.
70. Fischer-Nielsen A, Poulsen HE, Loft S. 8-Hydroxydeoxyguanosine in vitro: effects of glutathione, ascorbate, and 5-aminosalicylic acid. Free Radic Biol Med 1992; 13:121–126.
71. Heck DE, Laskin DL, Gardner CR, Laskin JD. Epidermal growth factor suppresses nitric oxide and hydrogen peroxide production by keratinocytes: potential role for nitric oxide in the regulation of wound healing. J Biol Chem 1992; 267:21277–21282.
72. Zingarelli B, O'Connor M, Wong H, Salzman AL, Szabo C. Peroxynitrite-mediated DNA strand breakage activates poly-adenosine diphosphate ribosyl synthetase and causes cellular energy depletion in macrophages stimulated with bacterial lipopolysaccharide. J Immunol 1996; 156:350–358.
73. Ueta K, Hayashi O. ADP-ribosylation. Annu Rev Biochem 1985; 54:73–100.
74. Wang Z-Q, Auer B, Stingl L, Berghammer H, Haidacher D, Schweiger M, Wagner EF. Mice lacking ADPRT and poly(ADP-ribosyl)ation develop normally but are susceptible to skin disease. Genes Dev 1995; 9:510–520.
75. Smulson ME, Kang VH, Ntambi JM, Rosenthal DS, Ding R, Simbulan CM. Requirement for the expression of poly(ADP-ribose) polymerase during the early stages of differentiation of 3T3-L1 preadipocytes, as studied by antisense RNA induction. J Biol Chem 1995; 270:119–127.
76. Sieg DJ, Billings RE. Lead/cytokine-mediated oxidative DNA damage in cultured mouse hepatocytes. Toxicol Appl Pharmacol 1997; 142:106–115.
77. Sanchez A, Alvarez AM, Benito M, Fabregat I. Apoptosis induced by transforming growth factor-b in fetal hepatocyte primary cultures: involvement of reactive oxygen intermediates. J Biol Chem 1996; 271:7416–7422.
78. Fehsel K, Kroncke K, Meyer KL, Huber H, Wahn V, Kolb-Bachofen V. Nitric oxide induces apoptosis in mouse thymocytes. J Immunol 1995; 155:2858–2865.
79. Blanco FJ, Ochs RL, Schwarz H, Lotz M. Chondrocyte apoptosis induced by nitric oxide. Am J Pathol 1995; 146:75–85.
80. Laskin DL, Rodriguez del Valle M, Heck DE, Hwang SM, Ohnishi ST, Durham SK,

Goller NL, Laskin JD. Hepatic nitric oxide production following acute endotoxemia in rats is mediated by increased inducible nitric oxide synthase gene expression. Hepatology 1995; 22:223–234.

81. Muschel RJ, Bernhard EJ, Garza L, McKenna WG, Koch CJ. Induction of apoptosis at different oxygen tensions: evidence that oxygen radicals do not mediate apoptotic signaling. Cancer Res 1995; 55:995–998.
82. Gardner PR, White CW. Failure of tumor necrosis factor and interleukin-1 to elicit superoxide production in the mitochondrial matrices of mammalian cells. Arch Biochem Biophys 1996; 334:158–162.
83. Zamzami N, Marchetti P, Castedo M, Decaudin D, Macho A, Hirsch T, Susin SA, Petit PX, Mignotte B, Kroemer G. Sequential reduction of mitochondrial transmembrane potential and generation of reactive oxygen species in early programmed cell death. J Exp Med 1995; 182:367–377.
84. Duval DL, Miller DR, Collier J, Billings RE. Characterization of hepatic nitric oxide synthase: identification as the cytokine-inducible form primarily regulated by oxidants. Mol Pharmacol 1996; 50:277–284.
85. Marchetti P, Castedo M, Susin SA, Zamzami N, Hirsch T, Macho A, Haeffner A, Hirsch F, Geuskens M, Kroemer G. Mitochondrial permeability transition is a central coordinating event of apoptosis. J Exp Med 1996; 184:1155–1160.
86. Zhivotovsky B, Burgess DH, Vanags DM, Orrenius S. Involvement of cellular proteolytic machinery in apoptosis. Biochem Biophys Res Commun 1997; 230:481–488.
87. Li P, Allen H, Banerjee S, Seshadri T. Characterization of mice deficient in interleukin-1β converting enzyme. J Cell Biochem 1997; 64:27–32.
88. Kunstle G, Leist M, Uhlig S, Revesz L, Feifel R, MacKenzie A, Wendel A. ICE-protease inhibitors block murine liver injury and apoptosis caused by CD95 or by TNF-a. Immunol Lett 1997; 55:5–10.
89. Roberts RA, Soames AR, Gill JH, James NH, Wheeldon EB. Non-genotoxic hepatocarcinogens stimulate DNA synthesis and their withdrawal induces apoptosis, but in different hepatocyte populations. Carcinogenesis 1995; 16:1693–1698.
90. Krams SM, Egawa H, Quinn MB, Garcia-Kennedy R, Martinez OM. Apoptosis as a mechanism of cell death in liver allograft rejection. Transplantation 1995; 59:621–625.

Index